GENETIC ELEMENTS
PROPERTIES AND FUNCTION

GENETIC ELEMENTS

PROPERTIES AND FUNCTION

SYMPOSIUM HELD ON OCCASION OF THE THIRD MEETING OF THE FEDERATION OF EUROPEAN BIOCHEMICAL SOCIETIES, ORGANIZED BY THE POLISH BIOCHEMICAL SOCIETY, WARSAW, APRIL 4th–7th, 1966

Edited by
D. SHUGAR

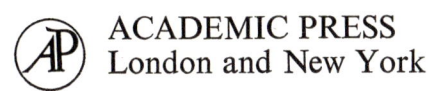

ACADEMIC PRESS
London and New York

PWN—POLISH SCIENTIFIC PUBLISHERS
Warszawa

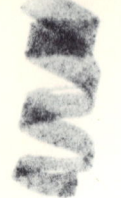

ACADEMIC PRESS INC. (LONDON) LTD
Berkeley Square House
Berkeley Square
London, W. 1

U.S. Edition published by
ACADEMIC PRESS INC.
111 Fifth Avenue
New York, New York 10003

Copyright © 1967 by the Polish Biochemical Society

Sale by PWN in Poland, by Academic Press in all other countries

Library of Congress Catalog Card Number: 66-30398

Printed in Poland (DRP)

FOREWORD

Felix qui potuit rerum cognoscere causas.
Virgilius

Rapid progress in elucidation of molecular mechanisms of heredity is the dominating feature of contemporary biochemistry. New facts are published at an ever increasing rate, new hypotheses are propounded, abolished or supported at decreasing time intervals.

In pursuit of specific knowledge, the notion of the basic unity of all biological sciences is apt to recede. To the speakers of this Symposium, however, the fundamental role of genetic elements in biological phenomena was paramount. The fascinating features of the genetic code, the mode of operation of molecules involved in expressing the gene action, the intriguing properties of extra-nuclear self-replicating particles, together with the physical chemistry of genetic elements, were subjects of contributions presented by outstanding scientists.

The Symposium took place during the Third Meeting of the Federation of European Biochemical Societies. Thus the latest achievements in the field of molecular genetics were disseminated not only amongst the specialists, but also amongst many other biochemists who pursue different avenues of research.

This book, as well as two others, namely "The Biochemistry of Mitochondria" and "The Biochemistry of Blood Platelets" emerged out of the proceedings of the Third Meeting of the Federation of European Biochemical Societies, which was organized by the Polish Biochemical Society. As the officers of the Federation, we are happy to present them to the scientific community. We hope that the Third FEBS Meeting may be considered as a step towards fulfilment of the purpose of the Federation which is "to promote the science of biochemistry, and in particular to encourage closer contacts between European biochemists".

K. Zakrzewski, Warsaw, *Chairman*
Warsaw, April, 1966   W. J. Whelan, London, *Secretary-General*

Editor's Preface

When initially adopted, the title "Properties and Function of Genetic Elements" seemed quite appropriate for a Symposium. On further reflection, however, it appears to be rather inadequate. We are witnesses today to an explosion in the field of molecular biology and its applications to the fundamental processes of cellular genetics, the impact of which shows no signs of abatement. Developments during the past ten years have been phenomenal, and have been accentuated by the veritable invasion of this exciting field by physicists, chemists, quantum chemists and theoretical physicists, so that interdisciplinary collaboration is becoming more and more both a requirement and standard practice.

This has of necessity led to an increasing degree of specialization, with the result that each of us is being gradually relegated into his own little groove. The day has now long since passed when one could browse at leisure through a scientific journal; it has even become something of a major effort merely to keep track of the titles of pertinent papers in one's own field of specialization. In these circumstances it is not at all surprising that we have come to rely to an increasing extent on review articles; but even these, in turn, are slowly losing some of their value because of the lengthy periods still involved in standard printing procedures.

The subject matter and speakers for this Symposium were selected with the foregoing in mind. Insofar as possible, an attempt was made to select topics of direct interest, not only to the specialist, but to as broad a number of participants as could be envisaged. Furthermore, each speaker was requested to include in his presentation, where possible, at least a short up-to-date background review of the subject matter dealt with, as well as some of the perspectives for further research. This will be seen to be reflected in the written texts, in some of which the authors have expanded the section devoted to a background review. It is perhaps worth adding that, during the course of the Symposium, conversations with members of the audience elicited a favourable response to this type of presentation at such large conferences.

If this volume, in addition to being of interest to specialists in such fields as the structure of nucleic acids, the genetic code, mechanisms of protein synthesis, etc., proves to be of value to a much broader audience, this is due entirely to the efforts of those who participated in the Symposium. I should like also to acknowledge the collaboration of the Foreign Language technical editorial staff of Państwowe Wydawnictwo Naukowe (Polish Scientific Publishers), and the unstinted help of my wife and of Dr A. Michałowski in the preparation of the author and subject indexes.

D. Shugar

Institute of Biochemistry and Biophysics, Academy of Sciences
Department of Biophysics, University of Warsaw

CONTENTS

FOREWORD. By Chairman and Executive Secretary of F.E.B.S.   v
EDITOR'S PREFACE   vii

GENETIC ELEMENTS

Enzymatic Replication of Polydeoxynucleotides. By F. J. Bollum   3
The Molecular Structures of the Nucleic Acids and their Biological Significance. By W. Fuller   17
The Structure of Nucleic Acids in Solution, as Determined by X-ray Scattering Techniques: DNA, RNA, Poly A. By V. Luzzati, J. Witz and A. Mathis   41
Active Sites of RNA's. By G. L. Brown, Sheila Lee and D. Metz   57
Mitochondrial DNA and other Forms of Cytoplasmic DNA. By P. Borst, A. M. Kroon and G.J.C.M. Ruttenberg   81
Mechanisms of Viral RNA Replication. By E. M. Martin   117
Studies on Temperature Conditional Mutants of *E. Coli* with a Modified Aminoacyl RNA Synthetase. By M. Yaniv and F. Gros   157
The Mechanism of Peptide Bond Formation in Protein Synthesis. By R. E. Monro, B. E. H. Maden and R. R. Traut   179

THE GENETIC CODE

Chairman's Introductory Remarks. By F. Sanger   207
Polynucleotide Synthesis and the Genetic Code—II. By H. Gobind Khorana   209
Analysis of the Genetic Code by Amino Acid Adapting. By H. Matthaei, G. Heller, H. P. Voigt, R. Neth, G. Schöch and H. Kübler   233
Studies on the Translation of the Genetic Message with Synthetic Polynucleotides. By M. A. Smith, M. Salas, M. B. Hille, W. M. Stanley, Jr., A. J. Wahba and S. Ochoa   251
Invited Comments. By P. Leder and B. F. C. Clark   263

NUCLEOTIDE SEQUENCES IN RNA

Chairman's Introductory Remarks. By S. Ochoa   269
On the Primary Structure of Transfer Ribonucleic Acids. By H. G. Zachau, D. Dütting and H. Feldmann   271
Primary Structure of the Valine Transfer RNA. Partial Reconstruction of the Molecule. By A. A. Bayev, T. V. Venkstern, A. D. Mirzabekov, A. I. Krutilina, V. A. Axelrod, L. Li and V. A. Engelhardt   287
Fractionation of Radioactive Nucleotides. By F. Sanger and G. G. Brownlee   303
Properties and Function of Methyl-deficient Phenylalanine Transfer RNA. By U. Z. Littauer and M. Revel.   315

AUTHOR INDEX   331
SUBJECT INDEX   349

GENETIC ELEMENTS

Enzymatic Replication of Polydeoxynucleotides

F. J. Bollum

Department of Biochemistry, University of Kentucky Medical School, Lexington, Kentucky 40506, U.S.A.

Current theory of genetic activity predicts that the hereditary material, DNA, contains information for the control and accomplishment of self replication. An analysis of episomal mutants in bacteria provides reasonable evidence that control mechanisms exist (1). The existence of genetic control over DNA synthesis limits enzyme studies to rapidly dividing tissues, e.g. *E. coli* (2), calf thymus (3) or systems where the control mechanisms can be overcome, as in regenerating (4) liver or bacteriophage infection (5). The accomplishment of the replication process depends finally on proteins, however, and studies on DNA polymerase enzymes are therefore an important source of information about the mechanism of replication. It should be made clear at the outset, earlier statements to the contrary (1), that known DNA polymerases show some elements of control, particularly on the limit of synthesis, and they also provide some information about the chemical initiation of DNA synthesis.

Two kinds of DNA polymerase have been isolated: an enzyme isolated from bacterial sources (2) will initiate replication on native DNA and continue until deoxynucleotide equivalent to many copies of original template has been polymerized. DNA polymerase isolated from calf thymus gland (3) and from phage infected bacteria (4) will not initiate synthesis on native DNA. When single stranded polydeoxynucleotide is supplied to the latter enzymes, replication proceeds only until deoxynucleotide equivalent to the original template has been polymerized (3, 6). The detailed nature of the products of the two kinds of DNA polymerase has been described (7, 8).

The cessation of replication in the initiation defective enzymes may be due to a built-in control element, or to the absence of suitable initiation conditions. The latter possibility seems most likely from the experiments to be described here. To use genetic terminology, DNA polymerases may be found to be capable of initiation (I^+) or incapable of initiation (I^-) on native DNA templates. To demonstrate I^- enzymes *in vitro* the initiation steps must be carried out non-enzymatically, and these procedures will be described later. A possible model for I^+ and I^- enzymes follows from the nature of the defect in I^-; that is, that the enzymes must be bifunctional, having a catalytic site K and an initiator site I. The complete enzyme is KI and in I^+ enzymes K and I are connected. In I^- enzymes K and I are dissociable and only K is isolated on purification. The letters I and K may correspond to different peptide chains.

The experiments presented in this discussion will demonstrate the mechanism of initiation on some simple model templates where the chemical reactions of initiation are readily susceptible to analysis. The overall result suggests that at least some of the chemical reactions of initiation are similar in I⁻ and I⁺ enzymes. The major difference lies in the inability of I⁻ enzymes to carry out the chemical reactions of initiation on double-stranded templates. The missing factor in I⁻ enzymes seems to be a component required for strand separation during initiation and replication. Continued analysis of DNA synthesis defective genes in phage T₄ (genes 39, 1, 32, 41–46 and 47) (9) may provide genetic evidence for this hypothetical second component, I, of DNA polymerase.

Experimental

The analysis of the chemical factors for initiation of DNA synthesis has been carried out with calf thymus DNA polymerase, an enzyme that is best described as I⁻. The general reaction for DNA polymerase is shown in Figure 1. I⁻ enzymes are similar to I⁺ enzymes in that all the deoxynucle-

$$\begin{matrix}\text{dATP}\\\text{dCTP}\\\text{dGTP}\\\text{dTTP}\end{matrix} + \text{Template DNA} \xrightarrow[E]{Mg^{++}} \text{DNA product} + \text{PPi}$$

Figure 1. DNA polymerase reaction.

oside triphosphate substrates complementary to the template are required for synthesis. I⁻ differs from I⁺ in requiring a single-stranded template, I⁺ works on both single-stranded and native templates. It has also been found that I⁻ does not form a strong complex with DNA (10), I⁺ does (11, 12).

DNA Templates. The requirement for single-stranded (denatured) template is demonstrated in Figure 2 (taken from Ref. 13). Analysis for acid insoluble radioactivity in this kind of experiment is carried out on filter paper discs as described earlier (13). Density gradient analysis of the product formed in I⁻ catalyzed reactions shows that the daughter strand is firmly bound to the parent chain and we assume that this is due to a "hairpin" structure (Figure 3a) of the hybrid molecule. A second possible structure not excluded by the analysis is shown in Figure 3b. Electron microscopy* of the calf thymus DNA polymerase product shows *no* branching. Nearest neighbor frequencies have been presented earlier (14).

Homopolymer Templates. The initiation reaction may be analyzed in more detail with certain homopolydeoxynucleotide templates synthesized by a terminal deoxynucleotidyl transferase (15). These synthetic polydeoxynucleotides have a completely defined structure and they demonstrate some new characteristics of the initiation reaction. First of all, when all four deoxynucleoside triphosphates are present, synthetic polydeoxynucleotides

---

* By Dr. Walther Stoeckenius, Rockefeller Institute.

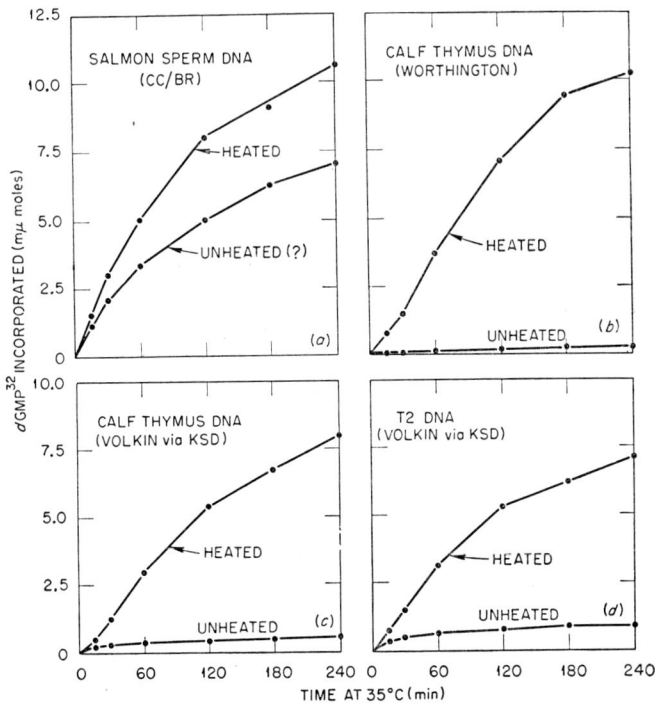

FIGURE 2. Action of calf thymus DNA polymerase on native and denatured templates.

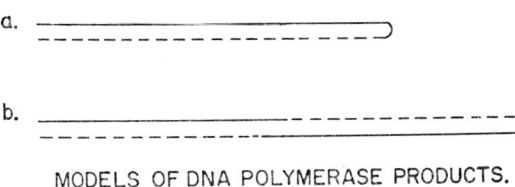

MODELS OF DNA POLYMERASE PRODUCTS.

FIGURE 3. Two models for DNA polymerase products. The solid line is the template chain and the dashed line newly synthesized material.

are essentially inactive as templates (Figure 4). Thus single-strandedness is a necessary but not sufficient condition for template activity.* Poly dA chains can be initiated by adding an external initiator (16), for example, hexathymidylate d(pT)$_6$ (Figure 4). Oligodeoxynucleotide initiators of this kind must be complementary (d(pA)$_6$ does not work with poly dA) and have a chain length greater than 5 (Figure 5). If dATP is the only substrate

* Optical rotatory dispersion studies by Dr. Paul Ts'o, Johns Hopkins University, have indicated a lack of secondary structure in poly dA at pH 7, in marked contrast to polyriboadenylate.

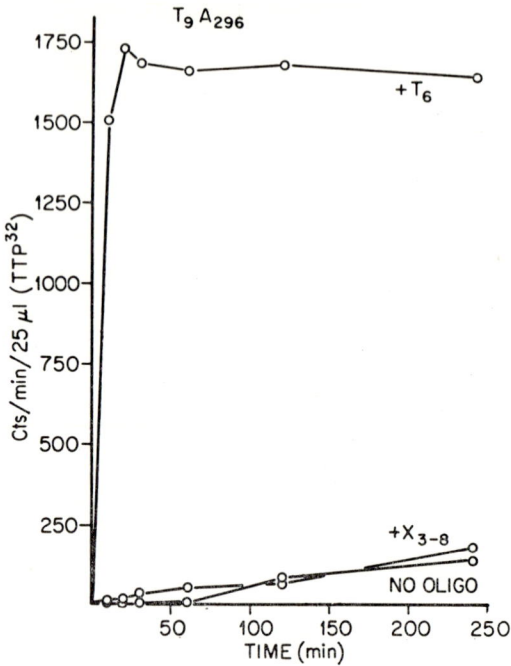

FIGURE 4. Polydeoxyadenylate template for DNA polymerase. Each reaction contained all four dXTP's (^{32}P–dTTP), polydeoxyadenylate and calf thymus DNA polymerase in addition to the oligodeoxynucleotides indicated. X refers to heterogeneous oligodeoxynucleotide. Hexadeoxyadenylate is also inactive as initiator for poly dA.

present in the polymerization mixture, self initiation takes place after a lag period of about 15 minutes (Figure 6). Similar results are obtained whether poly dA or poly dT is used as a template. Acetylation of the 3'-hydroxyl of the template completely blocks self initiation (Table I). Acetylation of

TABLE I. Acetylated Poly dT as a Template for Calf Thymus DNA Polymerase

	^{14}C-dATP Incorporated, mμ moles
Poly dT	4.92
Poly dT+d(pA)$_6$	8.20
Poly dT-3'-O-Acetyl	0.21
Poly dT-3'-O-Acetyl+d(pA)$_6$	3.30

Reaction mixtures contained 14 mμ moles of poly dT or acetylated poly dT in addition to 1 mM MgCl$_2$; 40 mM potassium phosphate, pH 7.0; 25 mμmoles of dATP and 80 μg DNA polymerase (calf thymus). Incubation was for 2 hours at 35° and samples were processed as described previously (13). Poly dT synthesized by a terminal deoxynucleotidyl transferase (10) was acetylated in aqueous solution (22), precipitated with ethanol, and non-acetylated chains were removed by treatment with E. coli exonuclease I (23). Exonuclease I was the generous gift of Dr. I. R. Lehman.

ENZYMATIC REPLICATION OF POLYDEOXYNUCLEOTIDES 7

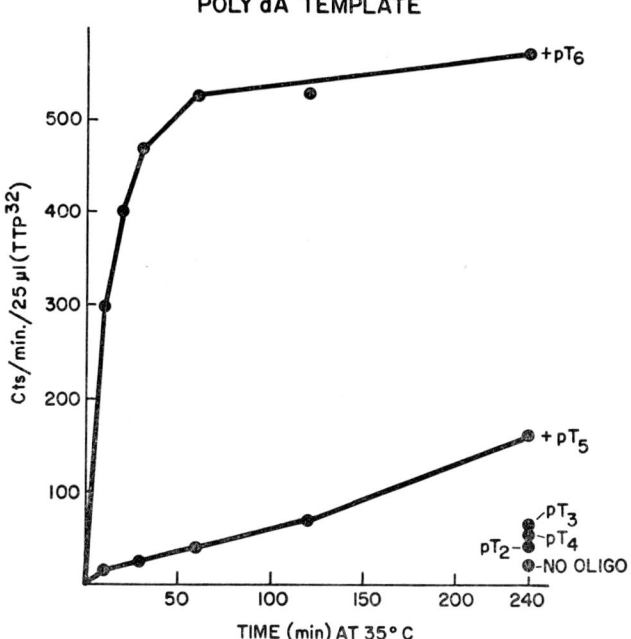

FIGURE 5. Chain length requirements for oligodeoxynucleotide initiation on a homopolymer template. Each reaction mixture contained all four dXTP's (^{32}P–dTTP), polydeoxyadenylate and calf thymus DNA polymerase in addition to oligodeoxythymidylate of the chain lengths indicated.

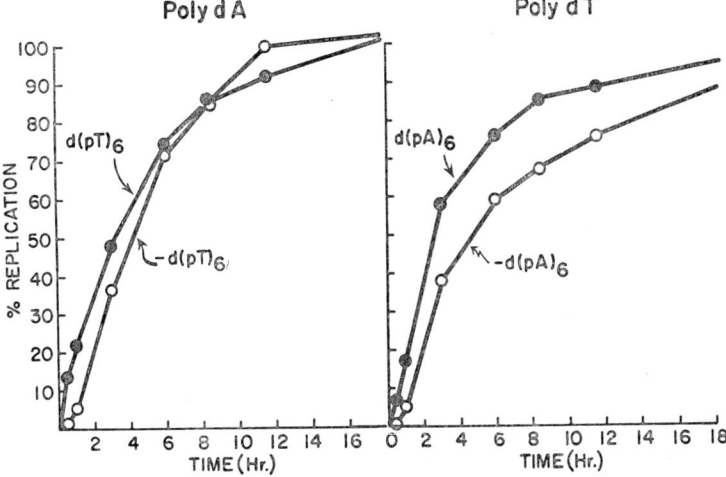

FIGURE 6. Self initiation of polydeoxyadenylate and polydeoxythymidylate. Each reaction contained poly dA or poly dT, a single, complementary ^{14}C–dXTP, and calf thymus DNA polymerase.

the 3′-hydroxyl of the template does not inhibit oligodeoxynucleotide initiation (17). All of the reactions stop when the dA:dT complex is complete.

The products of polydeoxyadenylate replication were studied in more detail using cesium chloride density gradients to detect the presence of intermediate polymer states during the replication. For this study polydeoxyadenylate was incubated with dTTP in the presence of calf thymus DNA polymerase, and then samples taken at various stages of replication were deproteinized, dialyzed and banded in cesium chloride with suitable markers. Reaction kinetics similar to Figure 6 (uninitiated) were observed. Figure 7 shows that poly dA:dT has a buoyant density less than that of

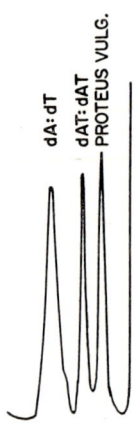

FIGURE 7. Density of dA:dT and dAT:dAT in neutral CsCl gradient. Microdensitometer tracing of photograph taken after 20 hr at 44,770 r.p.m., 25°, pH 8.1. *Proteus vulgaris* DNA was used as a density reference.

the alternating copolymer poly dAT:dAT in the cesium chloride gradient at neutral pH. The alternating copolymer was therefore used as a density marker in the gradients. Figure 8 shows that at 0% replication a broad band of poly dA exists in the reaction mixture and it is well separated from the dAT:dAT marker at density 1.698. As the replication proceeds to 25% the poly dA band disappears and a new component appears precisely at the reference density. At 50% replication, as measured by the incorporation of ^{14}C thymidylate, *all* of the polydeoxyadenylate has disappeared into the new band. It is now clear that our choice of a density marker was not a particularly good one, but it is nevertheless possible to interpret the experiments. At 90% replication in Figure 8 we see that material has now moved out of the $\varrho_{25} = 1.698$ band into a new band of intermediate density between the starting poly dA and the reference density of 1.698. As the replication proceeds to 100% all of the new material is banding between the starting density 1.640 and the reference density 1.698. The density of the final band is 1.647, identical with the density of poly dA:dT.

The density gradient diagrams of Figure 8 have been analyzed quantitatively by measuring the area under each component and Table II presents these data. The results confirm and augment the qualitative interpretation of the density gradient diagrams in that the percent replication of the template measured by ^{14}C-thymidylate incorporation is roughly one-half of the percent replication measured by poly dA disappearance. The polydeoxyadenylate disappearance is quantitatively accounted for by the appearance of new

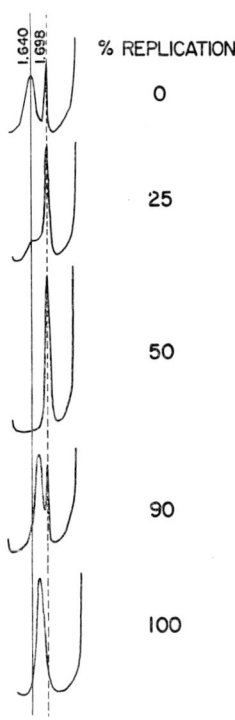

FIGURE 8. Density gradient analysis of polydeoxyadenylate replication. Microdensitometer tracings of photographs taken after 20 hr at 50,740 r.p.m., 25°, pH 8.1. Poly dAT:dAT was added to each cell, except 100% replication sample, as a density reference. Percent replication measured by incorporation of ^{14}C–dTTP.

material at density 1.698 in the early stages of the replication, and the appearance of the poly dA:dT band at density 1.647 is accounted for by the disappearance of material from the band at density 1.698.

There are three unusual findings in this replication experiment. The first is that polydeoxyadenylate is disappearing at twice the rate of replication. All of the polydeoxyadenylate and the dA:dT product are being transferred to a single density material, and this material first goes to a high density

TABLE II. Replication of Polydeoxyadenylate

t (Minutes)	% Replication† (^{14}C–dTMP)	% Poly dA ($\rho_{25} = 1.640$)	% Poly dA·dT ($\rho_{25} = 1.647$)	% Poly dA$_2$·dT ($\rho_{25} = 1.698$)
0	0	100	0	0
75	7.5	87.9	0	0
105	12.4	80.3	0	9.4
150	19.6	76.6	0	30.9
195	27.6	44.6	0	51.5
270	41.7	12.8	0	82.9
390	56.8	0	0	90.5
840	91.0	0	75	20.9
∞	100.0	0	100	0

† Percent replication measured by incorporation of ^{14}C–dTMP, individual polymer components estimated by area measurements of the separated components in CsCl density gradients.

(1.698) and then becomes lighter (1.647). Secondly, the highest density band has a narrower band width than either the poly dA starting material or the dA:dT product. Third, there does not appear to be appreciable material of intermediate densities formed during the replication. The first point we tentatively interpret as being due to the formation of a complex, heretofore unrecognized, of poly dA$_2$:dT. This could be the result of two simultaneous reactions:

(1) poly dA+dTTP → poly dA:dT

(2) poly dA+dA:dT → poly dA$_2$:dT

The postulated complex may not necessarily exist in the reaction mixture, but it appears under the conditions of cesium chloride centrifugation. It would disappear by reaction:

(3) poly dA$_2$:dT+dTTP → 2 poly dA:dT

Point two we interpret as being due to the reactions described above, that is, the intermediate complex of density 1.698 has a narrower band width than the final product because it is of higher molecular weight. Thus poly dA$_2$:dT, the postulated three-stranded complex* is of higher molecular weight than the final product poly dA:dT, a two-stranded (not four-stranded) product. Observation three we interpret as being due to nonsynchronous replication of the poly dA population. That is, DNA polymerase complexes with a polydeoxyadenylate molecule and completes the replication of that molecule before dissociating and replicating a second poly dA molecule. In the self-initiated replication there are only two bulk classes of polymer molecules—those that are not being replicated and those that are completely replicated. Partially replicated molecules are a minor category in the population. Aside from the demonstration of unexpected polymer complexes this experiment is interesting because it demonstrates that the DNA polymerase remains on the template until the replication is completed, then dissociates and goes on to the replication of another template.

The nature of the final dA:dT products synthesized in complementary oligodeoxynucleotide-initiated or self-initiated reactions are indistinguishable in neutral CsCl or by melting profile. They are readily distinguished in alkaline CsCl density gradients. The oligodeoxynucleotide-initiated products show a clear separation of the dA and dT chains of the product (Figure 9, upper), while self-initiated products show a band of intermediate density, indicating that dA chains are continuous with dT (Figure 9, lower). Homopoly dA · dT is distinguished from alternating copolymer dAT:dAT by melting behavior, T_m of dAT:dAT is 65° and dA:dT 5° higher in SSC, and by buoyant density in CsCl (Figure 7).

Modified DNA Templates. Single chain DNA can be modified by adding

---

* Note that this postulate implies that I$^-$ enzymes must replicate the third strand of three-stranded complexes if they actually occur in the polymerization mixture. This result can be demonstrated experimentally with the dA:dT$_2$ complex observed earlier, e.g., poly dA:dT$_2$+dATP → 2 dA:dT.

a stretch of homopoly dA at the 3'-hydroxyl end with terminal deoxynucleotidyl transferase. The modified DNA (DNA·pA$_n$) exhibits initiation properties similar to the synthetic polydeoxynucleotide (16). The replication of the heteropolymer part of the modified DNA is controlled by the homopolymer region, containing only dA residues. The properties of DNA polymerase products on modified DNA templates will probably be similar to the poly dA:dT result.

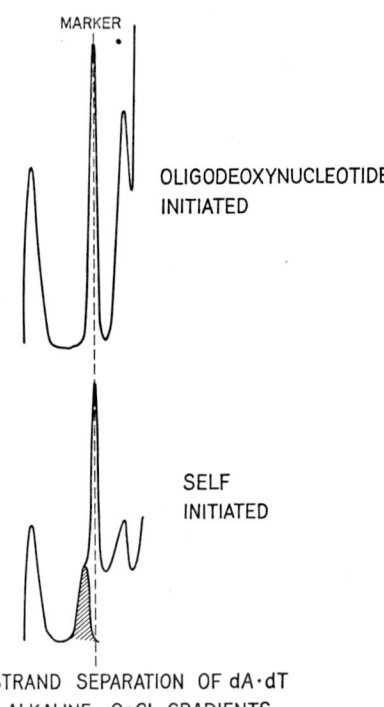

FIGURE 9. Strand separation of poly dA:dT. Microdensitometer tracings of photographs taken after 20 hr at 50,740 r.p.m., 25°, pH 12.4. Poly dAT:dAT was added to each cell as a density reference. The position of non-separable dA:dT was determined in a separate run.

Taken altogether the results suggest that the ability of single-stranded DNA to serve as a template is the result of a fold at the end of the chain forming a short, two-stranded initiator region and a free 3'-hydroxyl that is continuous with the main chain.

DISCUSSION

There are three experimental points to be emphasized.
 1. Initiation-defective enzymes are sensitive to secondary structure and cannot initiate or replicate through hydrogen-bonded templates.

2. When artificial initiation procedures are used the replication proceeds, but regardless of the initiation procedure replication continues only until the hydrogen-bonded product is completed. All reaction then stops.

3. The necessary *and* sufficient condition for initiation on a single chain template is a stretch of complementary base pairs that has a free 3′-hydroxyl group facing the single-stranded region to be replicated.* Models for such sites on the various templates used are depicted in Figure 10. These configurations are also sufficient for the limited (low temperature) reaction (18) of *E. coli* DNA polymerase (I$^+$ enzyme).

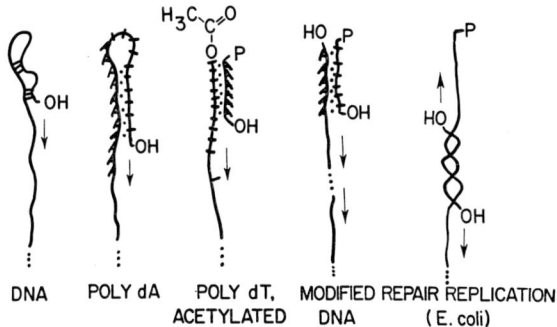

FIGURE 10. Models of sites on various templates which fulfil the conditions for initiation of replication.

The detailed nature of the initiation reaction in I$^-$ enzyme systems has some intrinsic interest, but it is not essential to attempt to make a detailed extrapolation to *in vivo* replication at this time. It should be evident that the chemical details of initiation in I$^-$ and I$^+$ DNA polymerases are quite similar.

The control function inherent in I$^-$ enzymes is worth consideration in extrapolating to biological systems. The properties of known DNA polymerases have elements of theoretical interest and these basic properties should not be dismissed in superficial ways since it is only by the analysis of such gene products that we get any information about genes and the control of gene activity. Secondly, the known DNA polymerases (I$^-$) are probably responsible for DNA synthesis *in vivo*, as has recently been demonstrated in Am and ts mutants of T$_4$ and T$_5$ bacteriophage (19, 20, 21; see also Addendum). To classify the known DNA polymerases as repair enzymes is simply redundant. DNA synthesis is an element of one (of a variety of hypothetical) repair processes.

Understanding of the complex overall mechanism of DNA replication is still fragmentary, both *in vivo* and *in vitro*. Aspects of the problem for future consideration are the chemical and biological properties of the enzyme systems already partially described, the replication defective mutants of F-factor and T-phages in *E. coli*, and the structure of the replicating units.

* The possibility of an analogous complementary 5′-triphosphate-ended initiator site remains to be clarified.

Solution of these problems will expand our view of the general problem of template polymerizations and provide a sound experimental basis for theoretical considerations.

ACKNOWLEDGMENT

The research described was started in the Biology Division of Oak Ridge National Laboratory, (Operated for the U.S. Atomic Energy Commission by Union Carbide Corporation) and is currently supported by a grant from the U.S. Public Health Service (CA-08487-01). The experiments described in this paper are collaborative efforts with Dr. G. R. Cassani and Mr. G. E. Houts. Suggestions of Dr. Rollin Hotchkiss and Mr. Clarence Colby, Jr. have contributed to the design and analysis of some experiments.

REFERENCES

1. Jacob, F., Brenner, S. and Cuzin, F., *Cold Spring Harbor Symp. Quant. Biol.* **28**, 329 (1963).
2. Kornberg, A., *Enzymatic Synthesis of DNA*, John Wiley and Sons, New York (1962).
3. Bollum, F. J., *Progress in Nucleic Acid Research* (J. N. Davidson & W. E. Cohn, eds.), Academic Press, N. Y. **1**, 1 (1963).
4. Bollum, F. J. and Potter, V. R., *J. Am. Chem. Soc.* **79**, 3603 (1957).
5. Aposhian, H. V. and Kornberg, A., *J. Biol. Chem.* **237**, 519 (1962).
6. Orr, C. W. M., Herriott, S. T. and Bessman M., *J. Biol. Chem.* **240**, 4652 (1965).
7. Bollum, F. J., *Cold Spring Harbor Symp. Quant. Biol.* **28**, 21 (1963).
8. Richardson, C. C., Schildkraut, C. L. and Kornberg, A., *Cold Spring Harbor Symp. Quant. Biol.* **28**, 9 (1963).
9. Epstein, R. H., Bolle, A., Steinberg, C. M., Kellenberger, E., Boy de la Tour, E., Chevalley, R., Edgar, R. S., Susman, M., Denhardt, G. H. and Lielausis, A., *Cold Spring Harbor Symp. Quant. Biol.* **28**, 375 (1963).
10. Yoneda, M. and Bollum, F. J., *J. Biol. Chem.* **240**, 3385 (1965).
11. Billen, D., *Biochem. et Biophys. Acta* **68**, 342 (1963).
12. Kadoya, M., Mitsiu, H., Takagi, Y., Otaka, E., Suzuki, H. and Osawa, S., *Biochem. et Biophys. Acta* **91**, 36 (1964).
13. Bollum, F. J., *J. Biol. Chem.* **234**, 2733 (1959).
14. Bollum, F. J., *J. Cell. Comp. Physiol.* **62**, Suppl. 1, 61 (1963).
15. Bollum, F. J., Groeniger, E. and Yoneda, M., *Proc. Nat. Acad. Sci. U.S.* **51**, 853 (1964)
16. Bollum, F. J., *Science* **144**, 560 (1964).
17. Cassani, G. R. and Bollum, F. J., *Federation Proc.* **25**, 708 (1966).
18. Richardson, C. C., Inman, R. B. and Kornberg, A., *J. Mol. Biol.* **9**, 46 (1964).
19. Dirksen, M. L., Hutson, J. C. and Buchanan, J. M., *Proc. Nat. Acad. Sci. U.S.* **50**, 507 (1963).
20. Wiberg, J. S., Dirksen, M. L., Epstein, R. H., Luria, S. E. and Buchanan, J. M. *Proc. Nat. Acad. Sci. U.S.* **48**, 293 (1962).
21. De Waard, A., Paul, A. V. and Lehman, I. R., *Proc. Nat. Acad. Sci. U. S.* **54**, 1241 (1965).
22. Stuart, A. and Khorana, H. G., *J. Am. Chem. Soc.* **85**, 2346 (1963).
23. Lehman, I. R. and Nussbaum, A. L., *J. Biol. Chem.* **239**, 2628 (1964).

DISCUSSION

(Chairman, D. SHUGAR)

D. FAN: Could you please tell us what happens when you try to synthesize DNA with the initiation positive polymerase from product of initiation negative polymerase action on poly-dA. For instance, do you get pieces which are twice as large as the original input primer?

F. J. BOLLUM: Yes, larger, but not necessarily twice as large.

ADDENDUM (added in proof): It should be pointed out that the genetic experiments are independent evidence for the participation of these enzymes in DNA replication. Physiological evidence for the appearance of DNA polymerase during DNA synthesis in liver regeneration (4) and bacteriophage infection (Kornberg, A., Zimmerman, S. B., Kornberg, S. R., and Josse, J., Proc. Nat. Acad. Sci., U.S., 45 : 772 (1959) has been available for some time.

The Molecular Structures of the Nucleic Acids and their Biological Significance

W. FULLER

Biophysics Department, University of London King's College, 26–29 Drury Lane, London, W.C.2., England

INTRODUCTION

An essential requirement for understanding fully the biological functioning of a molecule is the determination of its three dimensional structure. Information on the conformation of the molecule and its stability as a function of the molecular environment and on the way the molecule interacts with other molecules may lead to a hypothesis concerning its role in a particular biological phenomenon, e.g. the Crick–Watson model for DNA indicated how the genetic information might be stored and copied. A great many physical and chemical techniques have been developed for studying macromolecular structure, and this paper reviews recent X-ray diffraction and molecular model building studies of the structure of nucleic acid molecules in fibres. The X-ray technique is particularly powerful if the molecule can be purified and induced to form single crystals or crystalline fibres. Since much of the relevance of such an analysis rests on the assumption that the molecular structure is not significantly changed during specimen preparation, it is important that the material be studied at various stages of extraction. The construction of a model of a structure allows the results of the various physical and chemical techniques, which have been used to study it, to be correlated with each other and, what is particularly important, with the wealth of stereochemical data on covalent bond lengths and angles, hydrogen bond geometry, van der Waals radii and so on, which has come from structural studies on small molecules and model compounds. Recently, improved estimates have become available of the relative importance of the various forces (e.g. electrostatic interactions, hydrogen bonds, van der Waals dispersion forces, hydrophobic interactions) which are important in defining molecular conformation and molecular interaction. Calculations are now being attempted of the conformational energy of macromolecules as a function of their environment. Such calculations will allow the stability of particular molecular conformations and particular intermolecular interactions to be estimated and hence contribute to an understanding in dynamical, rather than purely statical terms, of the role of the molecule in biological processes.

Deoxyribonucleic Acid Structure

One of the most important features in the early X-ray diffraction patterns obtained by Wilkins and his collaborators (1, 2) from fibres of sodium deoxyribonucleic acid (DNA) was the sharpness of the diffraction maxima. This indicated a high degree of regularity in the molecular packing and hence a highly regular structure for the DNA molecule which did not differ significantly with the source of the DNA. This regularity, coupled with that in

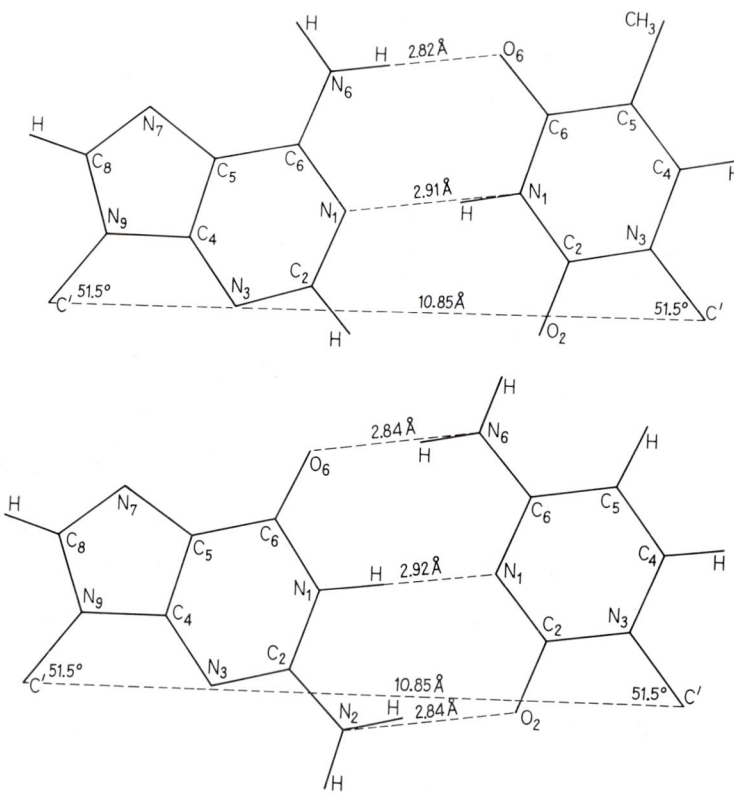

FIGURE 1. Watson–Crick base-pairing scheme for DNA. (Refined by Arnott.)

the base compositions observed by Chargaff (3), was reconciled with the structural irregularity essential for unrestricted information storage, by Watson and Crick (4). Because the adenine–thymine and guanine–cytosine base-pairs holding the two polynucleotide chains together in this model have such similar geometry (Figure 1), the sugar-phosphate chain is independent of the base-pair sequence. This step-ladder structure is coiled up into a helix so that the base-pairs lie on top of each other like steps in a spiral staircase with the sugar-phosphate chains as bannisters. The structure is sensitive to the water content of its environment, changing from a structure with 10 nucleotide-pairs per helix pitch (structure B) to one with 11 (structure A) as the relative humidity of the fibre environment is reduced from 92% to 75%.

Although the inclination of the base-pairs to the helix axis differs in the two structures (5, 6, 7) both have the hydrophobic surfaces of successive base-pairs lying on top of each other. As well as being favoured from hydrophobic considerations, such an arrangement allows strong interaction between the dipole moments of successive base-pairs. DeVoe and Tinoco (8) have estimated the energy from dipole–dipole interactions for the sixteen possible base-pair sequences (N.B. AT followed by TA is distinguished from TA followed by AT). They find that the weakest stacking interaction occurs for the sequence $\uparrow {}^{AT}_{TA} \downarrow$ and the strongest for $\uparrow {}^{CG}_{GC} \downarrow$. (AT denotes the base-pair adenine–thymine, GC the base-pair guanine–cytosine. The arrows designate the direction of the chain i.e. ${}^{T}_{A} \downarrow$ represents T-sugar 3'-phosphate-5' sugar-A). Such weak spots along the DNA helix might correspond to mutational "hot spots". The variation in the melting temperature of DNA as a function of its guanine–cytosine content was originally attributed to the extra hydrogen bond of the guanine–cytosine base-pair. However, DeVoe and Tinoco's calculations show that the dipole–dipole interaction within a base-pair gives rise to -3.9 kcal/base-pair for guanine–cytosine, but only $+0.2$ kcal/base-pair for adenine–thymine. Since, after strand separation, the base hydrogen bonding groups will be able to form hydrogen bonds with water molecules, interbase hydrogen bonding may make a relatively small contribution to the stabilization of the helical over the random coil structure, and be much less important than that due to dipole–dipole interaction within a base-pair.

The sensitivity of the DNA structure to the relative humidity of the fibre environment is easy to understand, qualitatively at least, in terms of hydrophobic interaction. What is much harder to account for is the molecule's conformational sensitivity to the ionic concentration of the fibre. Since the DNA phosphate groups ionize at about pH 2, the alkali metal salts are usually studied. The sodium, potassium, rubidium and cesium salts of DNA behave similarly but differ from the lithium salt. Anions such as chloride, citrate and acetate at concentrations of the order of 5% also influence both molecular conformation and molecular packing. The sodium salt of DNA gives the A to B structural transition when the relative humidity is increased from 75% to 92% if there is between 5–8% by weight of chloride in the fibre. If the chloride concentration is less than 5% the A conformation persists at 92% and if it is greater than 8% the B conformation persists at 75%. The existence of the A to B transition has been known for many years (2), but it is only recently that the importance of the anion concentration has been emphasized by Cooper and Hamilton (9). The lithium salt of DNA does not assume the A conformation at any humidity. At 92% the molecules are in the B conformation and packed in a similar semi-crystalline array to that in which sodium DNA molecules pack at this humidity. When the humidity is reduced to 66%, and if there is about 5% of chloride in the fibre, the molecules remain in the *B* conformation but pack in a fully crystalline arrangement. If the chloride concentration is less than this the molecules assume a third conformation called *C* which is non-integral with 28 nucleotide-pairs in 3 turns of the helix (10). It is not clear in what way these ions are affecting the conformational energy of the DNA. It may be due to purely electrostatic effects or to the different hydrations of the various ions modifying the hydrophobic contribution to the molecular stability. Whilst the primary purpose of the cations is clearly to neutralize the phosphate groups, the anions may perform what is essentially a geometrical role, filling spaces which are left when DNA molecules in a particular conformation pack in a regular array.

There is no evidence that any but the *B* conformation of DNA has any biological importance. In sperm heads and in material of chromosomal origin the DNA is bound to proteins, and X-ray evidence indicates that it is in the *B* conformation (11). However, it may well be that under some biological conditions the nucleic acid environment is such that a conformational change occurs. Molecular models of the *A* and *B* conformations of DNA are illustrated in Figures 2 and 3.

The relative width of the two grooves of the DNA molecule formed by the sugar-phosphate chains is, in the *B* conformation, about 2 to 1. X-ray diffraction studies of sperm heads and extracted nucleoprotamine show that, in sperm, the highly basic protamine is wound, as an extended polypeptide chain, around the small groove of the DNA with the positively charged guanidinium groups of the arginine side-chains interacting with DNA phosphate groups and replacing the sodium ions (12, 13). X-ray diffraction patterns from nucleohistone extracted from chromosomes are less well defined. Here the DNA is complexed with the basic protein, histone, but the conformation of the protein and its location on the DNA molecule have still not been determined with certainty. Recent studies by Pardon, Richards and Wilkins (14) have interpreted the medium and low angle X-ray diffraction pattern from nucleohistone in terms of a coiled coil in which each DNA molecule, along with its attached histone, is coiled into a super helix of radius 50 Å and pitch 120 Å.

Ribonucleic Acid Structure

The variety in the types of ribonucleic acid (RNA) occurring in the cell make extraction of a homogeneous preparation a difficult procedure, with the result that until quite recently (and probably to some extent even now) physical and chemical studies have been made on heterogeneous preparations. The first RNA diffraction patterns were obtained by Rich and Watson (15). The patterns were poorly defined, but did not differ significantly with the source of the RNA (plant viral, ribosomal). They indicated that the RNA probably had a two stranded helical structure of the DNA type. The lack of definition in the patterns was attributed to irregularities in the structure. Subsequent studies by a number of workers supported the view that RNA molecules from a wide range of sources have a rather similar structure. However, a detailed interpretation of this pattern in terms of a molecular

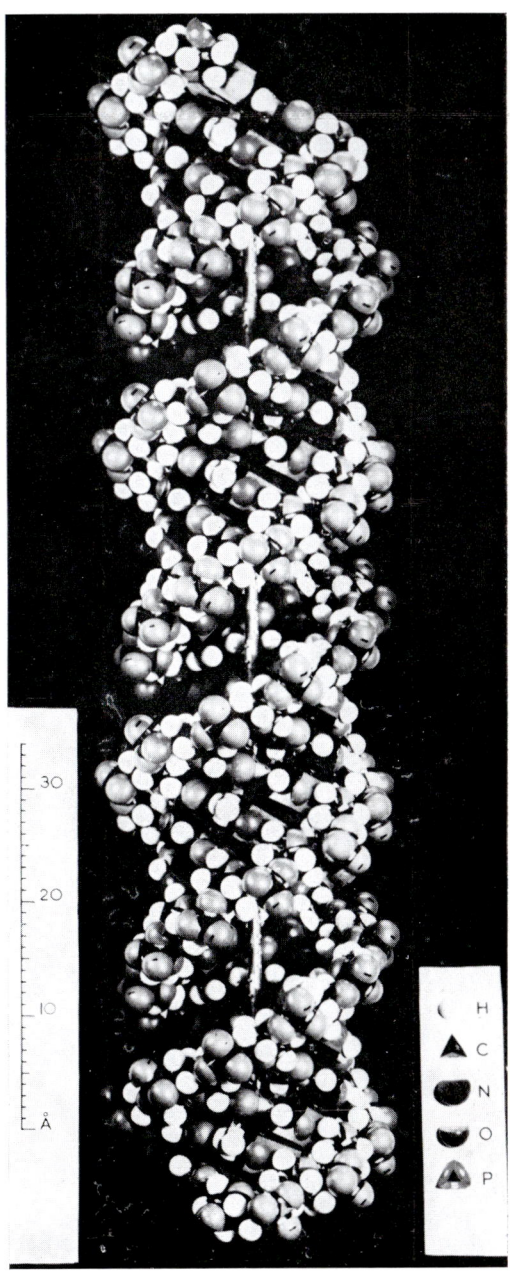

FIGURE 2. Molecular Model of the DNA *A* conformation.

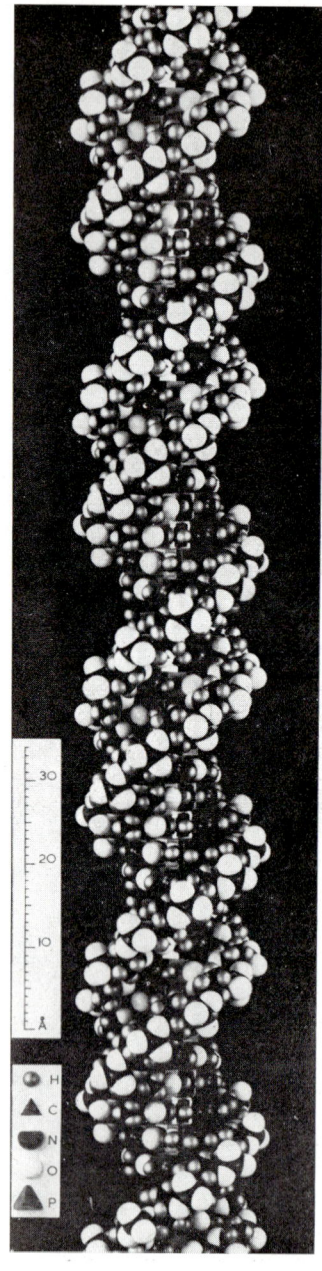

FIGURE 3. Molecular Model of the DNA *B* conformation.

model was not achieved until Spencer, Fuller, Wilkins and Brown (16) obtained much improved patterns from a yeast RNA which was first believed to be intact transfer RNA, but which was later shown to be fragmented RNA (17), probably of ribosomal origin (18). The definition in these patterns was adequate to show that these molecules had a structure with a regularity approaching that of DNA. The intensity distribution in the pattern indicated that the molecular structure was similar to that of A DNA. It is difficult to see how this similarity can occur unless the base-pairing in the RNA is of the same Watson–Crick type as in DNA. Molecular model building showed that the extra hydroxyl group of the RNA sugar could be accommodated without difficulty in the A DNA model.

Although the spots in the diffraction pattern from this material were much better defined than in those obtained previously, the overall intensity distribution was very similar. It was, therefore, reasonable to assume that the other RNA preparations also contained molecules with helical regions very similar to those of A DNA. The lack of definition in the patterns from these preparations could be due to either irregularity in the helical regions or to irregular packing of regions which were themselves regular. The two sugar-phosphate chains in the DNA molecule have sequences of atoms running in opposite directions. A helical region in an RNA molecule with this structural feature could be formed by hydrogen bonding between different polynucleotide strands or by a single polynucleotide strand folding back on itself. Even if the base sequence on one side of the loop so formed is not the Watson and Crick complementary sequence to that on the other side, hydrogen bonds might still be formed between opposite bases, e.g. adenine forms two perfectly good hydrogen bonds with guanine, and similarly cytosine and uracil. Such base-pairs would vary in size and the resulting sugar-phosphate chain would be much less regular than that in DNA. However, satisfactory base-pair stacking could be achieved and the average conformation along the chain would be expected to be similar to that of the regular RNA. Even if the base sequence in the loops is such that a regular conformation can be assumed, poorly defined patterns might still result because of the intrusion of the irregular pieces of chain connecting the regular helical loops.

Whilst the yeast RNA diffraction patterns were much better defined than any patterns previously obtained, they were still semi-crystalline. In addition the shortness of the helical regions resulted in layer lines 6–11 being rather broad. This led to uncertainty in determining the number of nucleotide-pairs per helix pitch. Fully crystalline patterns have since been obtained by Langridge and Gomatos (19) from reovirus RNA, Tomita and Rich (20) from wound tumour viral RNA, Langridge et al. (21) from MS2 virus replicative intermediate RNA, and Sato et al. (22) from rice dwarf virus RNA. In all these RNAs the base composition is compatible with Watson–Crick complementary base-pairing throughout the molecule (23, 24). Langridge and Gomatos (19) suggested that the RNA helix might be 10-fold with the bases tilted at between 10 and 15° to the horizontal (this might be compared with the 11-fold A DNA model which has a base tilt of 20°). Arnott, Hutchinson,

Spencer, Wilkins, Fuller and Langridge (25) have built both 10- and 11-fold models for RNA. The 10-fold model (base tilt 11°) was built using in the main data from the yeast RNA patterns. However, a subsequent more extensive analysis showed that an 11-fold helical model (base tilt 14°)

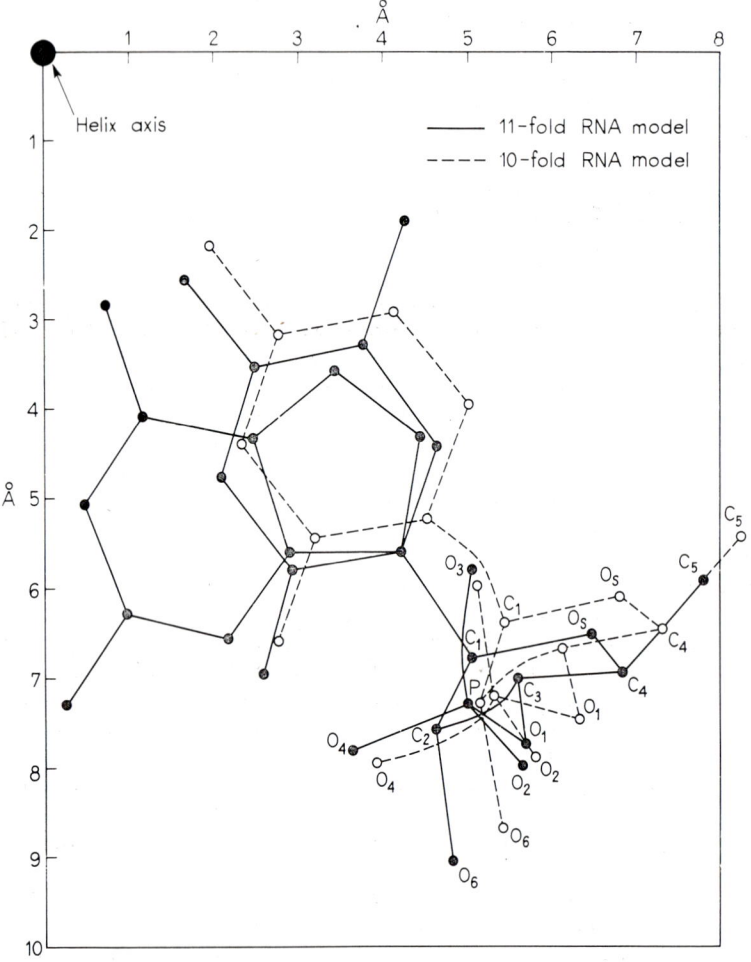

FIGURE 4. Similarity of the ten- and eleven-fold RNA models as illustrated by the projection of a single nucleotide onto a plane perpendicular to the helix axis.

was in as good an agreement with both the reovirus and yeast RNA patterns as the 10-fold one. This is not too surprising when it is realized that the difference in the rotation per nucleotide-pair for 10- and 11-fold models is only 2.3°. The sugar-phosphate chain linking successive bases has therefore only to stretch over approximately an extra 1/3 Å if the model is changed from 11- to 10-fold (Figure 4). This need not involve any significant change in the conformation of the sugar-phosphate chain.

One interesting feature of both the 10- and 11-fold models is that, although helical models can be constructed with the ribose OH hydrogen bonded to another part of the same molecule, it has not yet been found possible to build such a model which is both stereochemically satisfactory and compatible with the X-ray data (unpublished studies with S.D. Dover). It appears likely, therefore, that the OH group is free to form intermolecular hydrogen bonds. If the ribose OH were free to form a hydrogen bond within the RNA helix this would provide an explanation which is otherwise so far lacking of why RNA double helices are stable at higher temperatures than DNA double helices of equivalent base composition (26). This study refers to RNA work in the solid state and it is therefore not possible to exclude the possibility that the RNA helix in dilute solution, when its stability is assessed, contains intramolecular hydrogen bonds and that when the RNA solution is concentrated the helical conformation alters reversibly and intermolecular hydrogen bonds form. It may, in fact, be rather naive to assume that extra stability implies a hydrogen bond within the molecule rather than to a solvent molecule. The gain in energy in forming one rather than the other is probably not large, and it is not unlikely that the increased stability of RNA over DNA is due to some other stereochemical consideration so far undefined. Unlike DNA, RNA does not show a marked structural change with the relative humidity of the fibre. So far no observations have been published which show that the molecular conformation of RNA varies with the ionic content of the fibre.

Regular two-stranded viral RNA can crystallize in two distinct forms called α and β. Both are hexagonal (for α, $a = 44$ Å, $c = 30$ Å; for β, $a = 40$ Å, $c = 30$ Å) with three molecules per unit cell. The α-RNA diffraction pattern was first indexed by Rich and Tomita (20) and a preliminary analysis of the β pattern from rice dwarf virus RNA has been published by Sato et al. (22). They suggest that the RNA molecule is helical with ten residues per helix pitch of 30.5 Å and that each of the three molecules in the unit cell is displaced along the c axis by $c/3$ with respect to its neighbours. Although they did not describe a detailed molecular model, they deduced from the X-ray data that each RNA molecule contained two polynucleotide strands separated along the helix axis by 13 Å and with the phosphorus atoms at a radius of 9 Å or 12 Å. These parameters were obtained from a cylindrical Patterson synthesis. Such syntheses are often difficult to interpret for structures of this kind and this difficulty is reflected in the ambiguity in the determination of the phosphorus radius. Infrared dichoism studies showed that the O——O line of the PO_2-group makes an angle of about 70° and the bisector of < OPO an angle of about 40°, both with the helix axis. These parameters can be compared with those of the 10- and 11-fold models derived by Arnott et al. (25).

These workers have pointed out that the apparently meridional tenth layer line reflection, which is the main reason why Langridge and Gomatos (19) and Sato et al. (22) proposed that the RNA was 10-fold, can not, in fact, be truly meridional in either the α or β patterns unless the RNA is distorted from a regular helix or the reflection is the result of diffraction from ions or water in the structure which possesses a different crystal symmetry from the DNA. This is because the intensity distributions on the other layer lines in these patterns show that the three RNA molecules in the unit cell must be either 10-fold, and related in position by a left-handed screw triad, or 11-fold and related by a triad. The presence of the 3-fold screw axis (which must be present if the helix is ten-fold) results in all the meridional reflections except those on layer lines for which $l = 3n$ being systematically zero. If the intensity of the tenth layer line reflection is assigned to off-meridional reflections rather than (0, 0, 10), it can, after allowing for the degree of disorientation in the fibre, be accounted for by 10- or 11-fold models. The radius of the phosphorus in the 10- and 11-fold models built by Arnott et al. (25) is 8.96 Å and 8.84 Å respectively; separation of the two sugar-phosphate chains along the helix axis is about 13 Å in both models. In the 10-fold model the O ⎯⎯ O line of the PO_2-group makes an angle of 63° with the helix axis and the bisector of < OPO makes an angle of 30°. The corresponding values for the 11-fold model are 64° and 37° (calculations by S. D. Dover). Therefore all the structural parameters described above for both the 10- and 11-fold models agree within the limits of experimental error with those determined by Sato et al. (22). The remainder of this section is devoted to a summary of the principal conclusions drawn from a detailed analysis of the α and β patterns from reovirus RNA and the semi-crystalline patterns from yeast RNA by Arnott, Hutchinson, Spencer Wilkins, Fuller and Langridge (25).

The molecular conformation does not differ significantly in the two crystal forms. The position of the molecules in the α and β unit cells (Figure 5) is defined by X which was determined for both α and β by least squares analysis of the X-ray data. There are, in fact, within each unit cell, three triads formed by joining together points defining the molecular positions. Because the lattice is trigonal and one molecule has its centre on the b axis, the dimensions of two of these triads are identical. The angle ω between these two triads depends on X and it is particularly interesting that the experimentally determined values of X for both α and β are such as to make ω a multiple of 360°/11. Therefore for these values of X the contact which molecule 1 makes with molecule 2 is identical with that which it makes with molecule 3. This economy in the number of types of molecular contact is aesthetically attractive, particularly since by appropriate choice of the molecular orientation within the unit cell the contacts in α can be made identical with those in β. It should be stressed that this equivalence can only occur with these values of X if the helix is 11-fold. By calculating the agreement between observed and calculated diffraction as a function of the molecular orientation in the unit cell it is possible to estimate this parameter. For the 10-fold model in both α and β the agreement varied little with the molecular orientation. For the 11-fold model the best orientation obtained from the β diffraction data was within about 3° of an orientation which gave the maximum equivalence in intermolecular contact. For the 11-fold

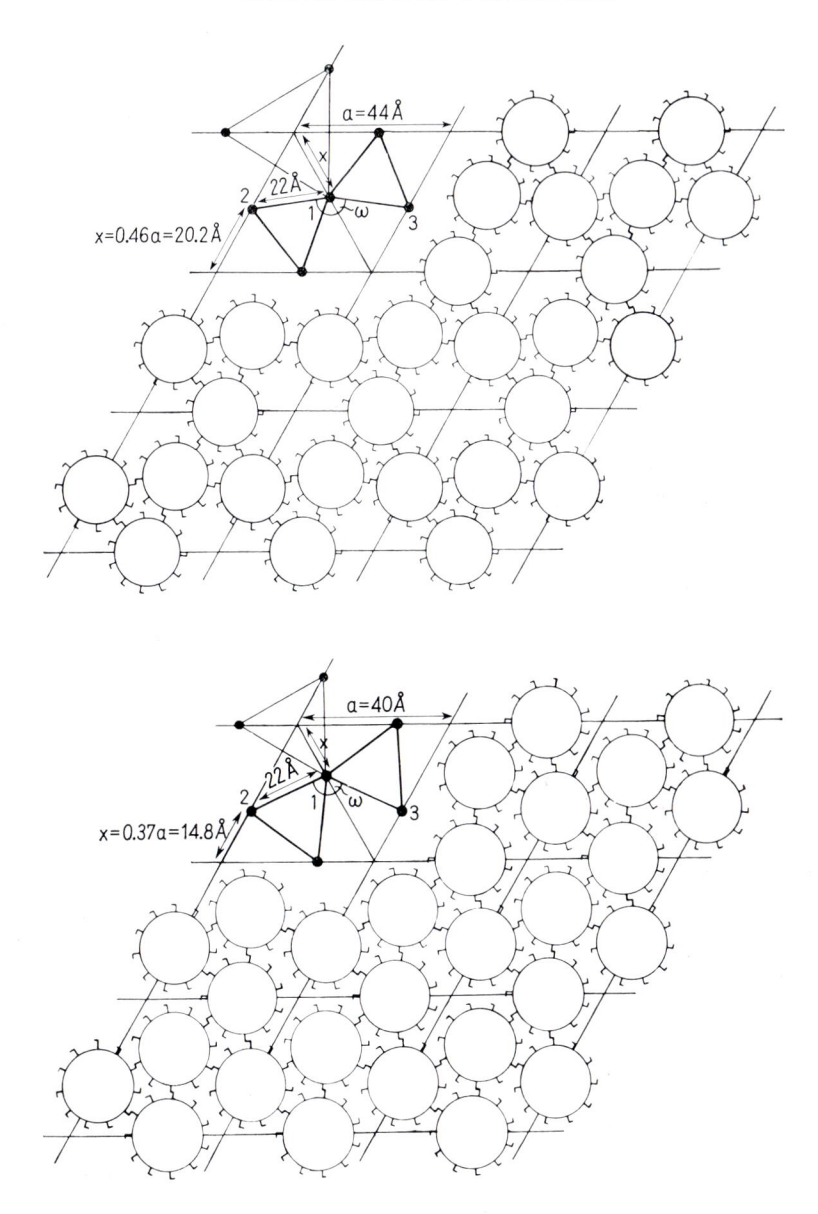

FIGURE 5. Molecular packing in the α (top) and β (bottom) crystalline RNA structures. The structures are projected onto a plane perpendicular to the helix axis. In this representation the RNA helices are assumed to be eleven-fold with each leg representing one residue. The alternative possibility that the helices are ten-fold does not allow such a high degree of equivalence in the intermolecular contacts. (Diagram by courtesy of Dr. Struther Arnott.)

model in α the orientation parameter was less well defined. The intermolecular stereochemistry was also calculated as a function of molecular orientation for both 10- and 11-fold models in both α and β. For the orientations favoured from diffraction considerations, the intermolecular stereochemistry of the 11-fold model is preferable. In addition to there being no short C—C or C—O contacts there are, for each pair of nearest-neighbours, two O—O contacts per helix pitch of each molecule which are an acceptable length for an O—H...O H-bond. However only one of these contacts has an angle at the donor atom compatible with H-bond formation. Small distortions of the helical structure are being investigated (with Dr J. Venable) to reduce this deviation. For the 10-fold model less intermolecular hydrogen bonding is possible.

The existence of the α form may be taken as support for the existence of intermolecular hydrogen bonds since it is, unlike β, an open structure with large holes filled by water and ions. Such open structures are often a consequence of highly directional intermolecular forces of which the hydrogen bond is the most common example. Such hydrogen bonding might be important in stabilizing the interaction of transfer RNA and ribosomal RNA during protein synthesis. Whilst it seems reasonable that the codon-anticodon interaction through Watson–Crick base-pairing should provide the specificity in the incorporation of the amino acid bound to the transfer RNA into the growing peptide chain, it seems doubtful if the two or three anticodon–codon hydrogen bonds could provide adequate stabilization of the complex. If the ribosomal RNA contained helical regions so arranged that a helical region of the transfer RNA molecule could be held there by intermolecular hydrogen bonds, this interaction could provide the additional stabilization. Interaction between RNA helices will also depend on the ionic strength of their environment since this determines how much the phosphate groups in different helices repel each other.

SINGLE STRANDED POLYNUCLEOTIDES

As was noted in a previous section, under normal conditions (i.e. near neutral pH, moderate ionic strength, room temperature) a single polynucleotide chain might be expected to fold up so that a number of "double stranded" loops are formed. In this way the polymer will obtain stabilization from interbase hydrogen bonding, dipole–dipole interactions between sucessive bases and hydrophobic interactions. If the loops form a significant fraction of the chain length the molecule will behave, from the point of view of many techniques, as a double-stranded molecule. If the material is recovered from solution in the looped form and made into a fibre, the X-ray diffraction pattern will resemble that of an irregular double stranded structure. The most characteristic feature in the diffraction pattern from a two stranded polynucleotide structure is the strong diffraction with a spacing of about 12 Å. If this is compared with the diffraction pattern to be expected from a structure identical to the double helical one, except that one chain has been removed (Figure 6), it can be seen that the strong 12 Å diffraction peak disappears and a strong peak appears at 22 Å. The strong peak at 12 Å is due to the X-ray scattering, from the two strands in the double helix which are separated along the c axis by approximately 12 Å, interfering constructively for this angle of diffraction. This particular single-stranded model is more regular than any structure which might be expected to occur naturally, but is adequate to illustrate the effect on the diffraction pattern of a second chain. (These calculations were done with S. D. Dover and M. H. F. Wilkins).

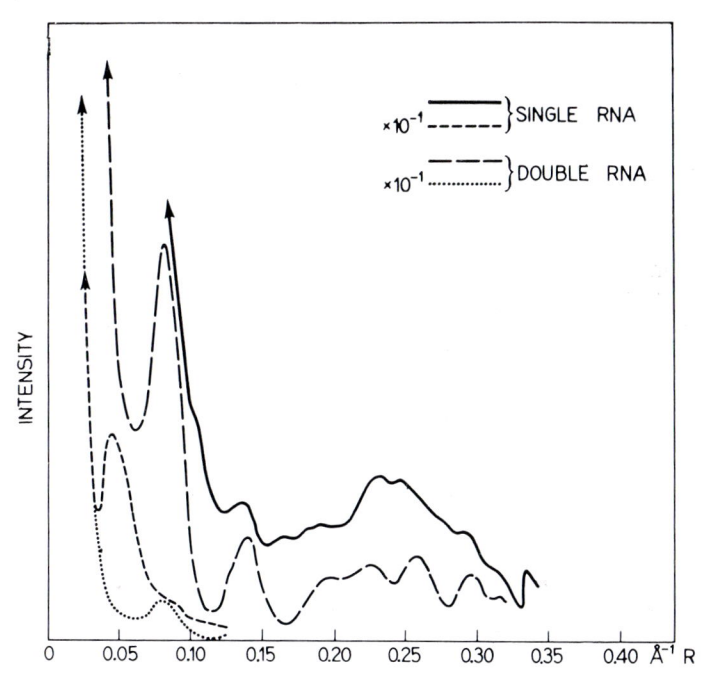

FIGURE 6. Comparison of the calculated diffraction from completely unoriented arrays of (a) single-stranded and (b) double-stranded RNA molecules (see text).

If diffraction patterns obtained from transfer RNA are compared with these calculated patterns it can be seen that the transfer RNA is in the "double helix" category. However, the X-ray technique alone is unable to distinguish between models where there are genuinely two strands and where there is a single molecule folded back on itself.

INTERACTION OF DEOXYRIBONUCLEIC ACID WITH ANTIMETABOLITES

There is considerable evidence that a number of antibiotics, carcinogens, drugs and dyes act by forming complexes with nucleic acids. It is natural therefore to attempt to understand the effects of these molecules in terms of their structure and properties and the structure and function of the nucleic acids. Furthermore, since some of these molecules interfere in a very specific way with biological function, it is possible that a study of their interaction with nucleic acids will result in an improved understanding of nucleic acid function. Moreover, from a consideration of the changes produced in the DNA structure and its stability by these molecules, it may be possible to learn something about the forces determining the DNA structure.

Often the physical and chemical techniques available for studying the complex require much higher concentrations of the antimetabolite than are needed for biological activity. In such circumstances it is important to determine whether the binding sites vary markedly in character as a function of the number of sites which are occupied. For drugs with characteristic spectra in the visible region, the existence of an isobestic point may provide useful information on this point.

One possible interaction mechanism which has excited considerable interest over the past few years is intercalation. The distance between corresponding points on a fully extended sugar-phosphate chain is about 7.2 Å, whereas the thickness of a DNA base-pair is only about 3.4 Å. Apart from an environment of very high ionic strength, mutual repulsion between successive phosphates will keep the sugar-phosphate chain in an almost completely extended form. Since base-pair stacking is energetically favoured, the slack in the DNA sugar-phosphate chains is taken up by the molecule assuming a helical conformation. However, the possibility remains of uncoiling the helix so that this slack is available for the insertion of a planar molecule of up to 3.8 Å thick between successive base-pairs. This insertion mechanism is called intercalation and if the inserted molecule contains unsaturated rings like the base-pairs, we might expect the insertion to be stabilized by stacking forces similar to those between DNA base-pairs. The increase in helix pitch following intercalation is given by

$$P' = \frac{P + \frac{N}{M} t}{1 - \frac{\theta N}{360 M}}$$

where P is the original pitch of the helix with N base-pairs per pitch and P' is the new pitch after 1 molecule, thickness t, has been inserted for every M base-pairs; we assume the uncoiling at the point of insertion is $\theta°$. If the intercalated molecule has approximately the dimensions of a base-pair with no bulky side-groups, then extensive intercalation will result in a DNA molecule with reduced radius. If the number of base-pairs per intercalation is less than about ten, then the changes in helix pitch should be detectable in fibre diffraction studies. If this number is as low as 4 the change in radius should be detectable.

However intercalation is not the only interaction mechanism which will result in an increase in the pitch of the DNA. Attachment of a molecule to the outside of the nucleic acid helix could conceivably distort the helix structure so that the pitch was increased. If the original helix had N nucleotide-pairs in pitch P, and if there is 1 attachment for every M base-pairs, then the new pitch of the helix, P', will be

$$P' = \frac{P}{1 - \dfrac{\theta N}{360 M}}$$

where θ is the degree of uncoiling of the helix at the point of attachment. However, this type of interaction would, unlike intercalation, result in an increased, rather than decreased, molecular diameter. The variation in helix pitch with the number of base-pairs per attached dye molecule for various degrees of uncoiling of the helix is plotted in Figure 7 for both this type of attachment and the intercalation interaction.

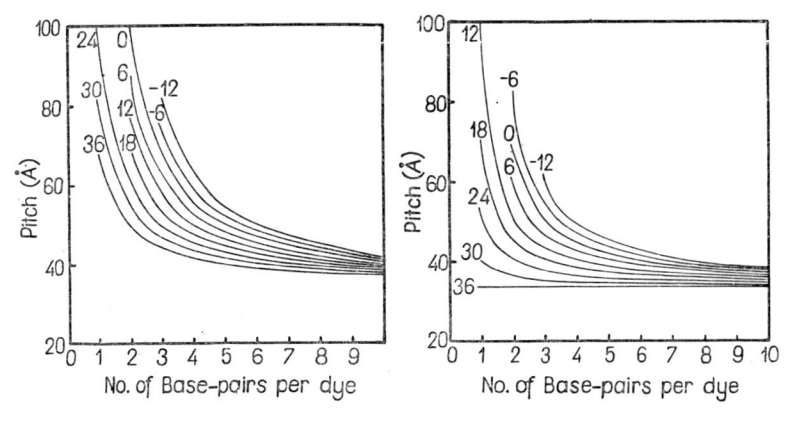

FIGURE 7. Variation of the helix pitch with the amount of dye bound to DNA. Left: When the dye intercalates. Right: When the dye is attached to the outside of the DNA. In both cases curves are plotted for various degrees of uncoiling of the DNA helix at the point of attachment. The number next to each curve denotes the rotation per residue of the DNA helix at the point of attachment of the dye. In the unperturbed helix this is 36°.

Lerman (27) postulated the intercalation mechanism to account for his X-ray diffraction, optical and solution studies of the interaction of DNA and acridines. Previous models for this interaction had involved binding of the acridine to the charged phosphate groups (28). Lerman's X-ray diffraction patterns indicated that the complex with 1 drug molecule per every 5 nucleotides had a diameter approximately 6 Å less than that of native DNA and that the original helical character had been almost entirely lost. Dichroism measurements indicated the acridine was approximately perpendicular to the length of the molecule. Viscosity and sedimentation studies indicated that the molecule was stiffer with a decreased mass per unit length. All these observations were consistent with intercalation and Lerman proposed a model for the interaction in which there was an uncoiling of 45° at the point of insertion.

Subsequent studies by Luzzatti and Lerman (29, 30) on the low angle X-ray scattering and flow birefringence of the complex support this model. Some observations of particular interest were made by Tubbs, Ditmars and Van Winkle (31) who showed the fluorescence quenching of acriflavine by DNA indicated a heterogeneity of binding sites with relative strength

$$\frac{AT}{AT} > \frac{AT}{GC} > \frac{GC}{GC}$$

($\frac{AT}{GC}$ indicates an intercalation site where the base-pair above the dye is AT and that below it is GC. These studies did not distinguish between $\frac{GC}{AT}$, $\frac{GC}{TA}$, $\frac{CG}{AT}$, $\frac{CG}{TA}$, $\frac{AT}{GC}$, $\frac{TA}{GC}$, $\frac{AT}{CG}$ and $\frac{TA}{CG}$ and similarly for $\frac{AT}{AT}$ and $\frac{GC}{GC}$. Gersch and Jordan (32) have extended the calculations of DeVoe and Tinoco (8) to estimate the binding energy of an acridine molecule intercalated between the sixteen different base-pairs. In agreement with Tubbs et al. (31) they find the strongest binding occurs when the acridine is sandwiched between two AT pairs and weakest when sandwiched between two GC pairs. Therefore those regions of the DNA which have the least stable helical structure (i.e. according to DeVoe and Tinoco, regions with successive AT pairs), and hence the most likely to accept an intercalated molecule, are, according to Gersch and Jordan (32), just the regions where the binding is strongest. This may be an important point in the study of acridine mutants. In support of these ideas Kleinwachter and Koudelka (33) have found that the melting point of the DNA–acridine complex increases with the adenine–thymine content of the DNA.

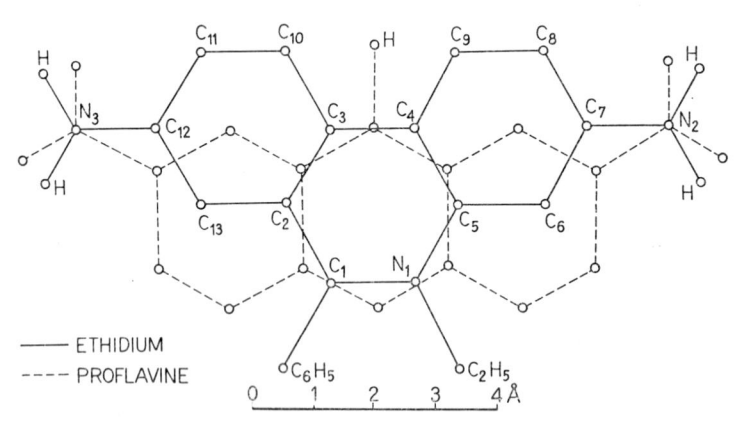

FIGURE 8. The chemical structure of ethidium (full line) and proflavine (dashed line).

Fuller and Waring (34) have postulated an intercalation model for the interaction of DNA and the trypanocidal drug ethidium bromide (Figure 8). This drug inhibits DNA dependent nucleic acid synthesis (35, 36) and its effect on the sedimentation constant, viscosity and melting temperature of DNA are similar to those of proflavine (36). The X-ray diffraction patterns obtained from the complex were similar to those obtained by Lerman with acridine except that the position of the equatorial reflection indicated a molecular diamater of 23.9 Å, which is intermediate between the value of 26.6 Å for native DNA and 20.1 Å for the DNA-proflavine complex. This intermediate value can be attributed to the effect of the bulky phenyl group. Fuller and Waring (34) dissociated the DNA-ethidium complex and showed that the original DNA structure had been regained. Their model building

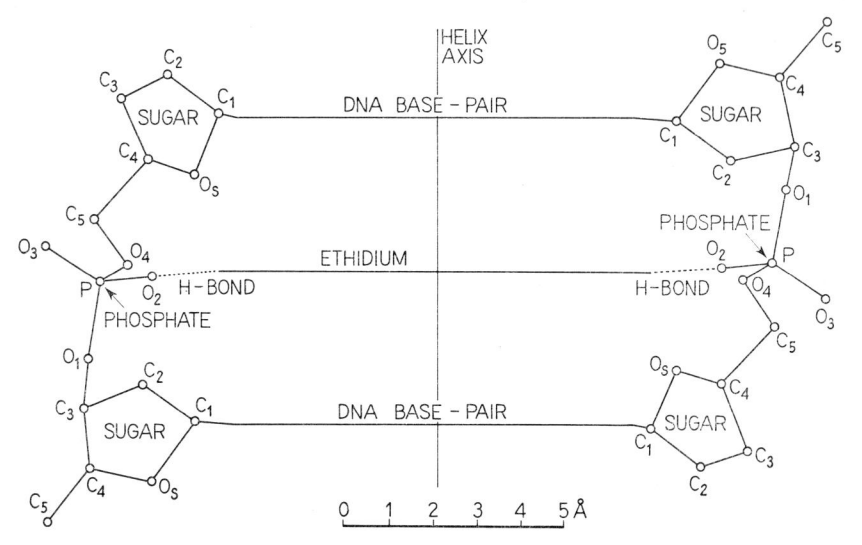

FIGURE 9. Projection onto a plane containing the helix axis of a model for the intercalation of a molecule of ethidium between successive deoxyribonucleic acid base-pairs.

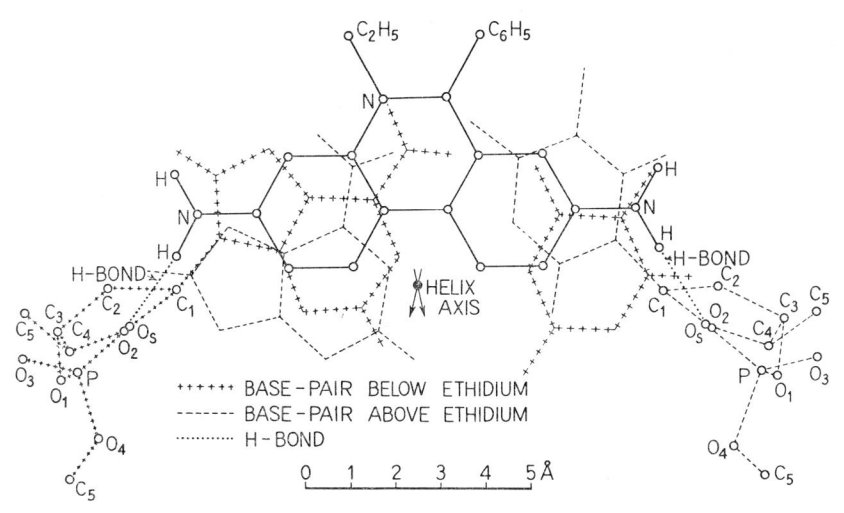

FIGURE 10. Projection onto a plane perpendicular to the helix axis of a model for the intercalation of a molecule of ethidium between successive deoxyribonucleic acid base-pairs.

studies (Figures 9, 10) showed that it was only necessary for the DNA molecule to uncoil by 12° at the point of insertion as compared with 45° proposed by Lerman. This difference may have relevance to studies on the kinetics of complex formation.

In the two interactions which have been discussed, the amount of drug bound to the DNA was so great that practically all the helical character in the structure had been lost. In studies on the interaction of DNA and the antibiotic daunomycin (37–40) (Figure 11), Fuller, Pigram and Hamilton (41) have studied the variation in helix pitch of the complex as a function of the amount of antibiotic complexed to the DNA. With 1 drug molecule per 4.5 base-pairs the pitch increased from 34 Å (native DNA) to 48 Å, whereas when the concentration was 1 per 8 base-pairs it increased to 41 Å. Using the curves in Figure 7 these observations are consistent with an uncoiling of 12° per insertion if the daunomycin is intercalated into the DNA. An

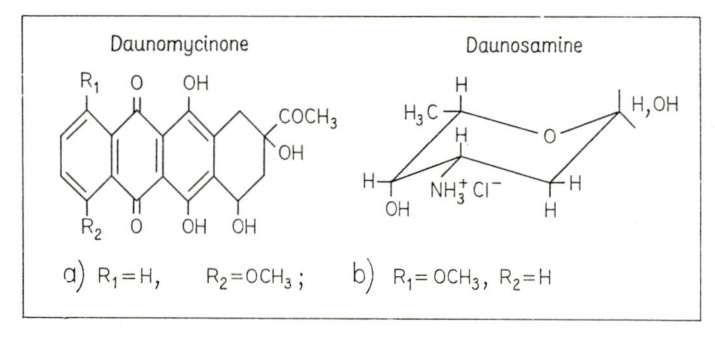

FIGURE 11. Chemical structure of daunomycin.

approximate model has been built showing that a daunomycin molecule can be intercalated into the DNA helix with an uncoiling of about twelve degrees. These studies were complicated because the primary structure of the antibiotic has not yet been completely determined. It is still not clear whether the bulk of the molecule will project into the large, or small, DNA groove. This may be important since Reich (42) has pointed out that, since RNA polymerase is more sensitive to daunomycin than DNA polymerase, then, by analogy with actinomycin (43), the drug may obstruct the small, rather than large, groove. Other physical and chemical studies of the daunomycin-DNA interaction (44) also (by analogy with similar studies on the DNA-acridine interaction) support the idea that when the concentration of daunomycin does not exceed one molecule for every 5 or 6 nucleotides, the drug interacts with the DNA through intercalation between successive base-pairs.

These studies strongly support the conception that the planar, unsaturated, triple ring grouping of proflavine, ethidium bromide and daunomycin may interact with the DNA by intercalation of the triple ring system between successive base-pairs. All the work described here refers to DNA–dye complexes with less than one drug molecule per two or three base-pairs. Clearly

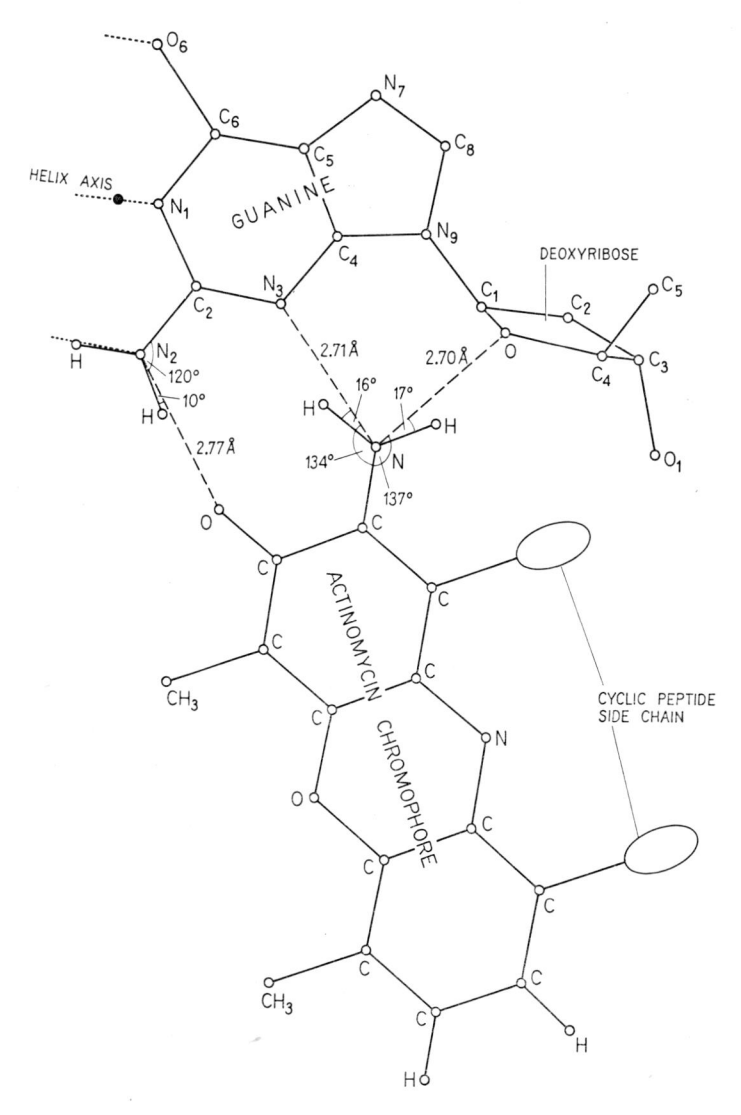

FIGURE 13. Proposed model for the binding of actinomycin C_1 to deoxyguanosine of DNA in the B conformation. Hydrogen bonds between actinomycin and DNA indicated by — — — — and between guanine and cytosine in DNA by · · · · · · · ·

not more than one drug molecule per base-pair can interact through intercalation with the DNA. Many years ago Peacocke and Skerrett (45) showed that there were two types of proflavine binding sites on a DNA molecule—a strong one, which saturated at about 1 proflavine per two base-pairs; and a weaker one, which was responsible for binding up to 1 dye molecule per DNA base. The intercalation reaction has been identified with the strong binding site, but as yet no detailed molecular model has been proposed for the weak interaction. The mutations produced in bacteriophage by proflavine can be accounted for if it is assumed that a base-pair is either deleted or inserted at the site of mutation (46). It may well be that such copying errors could be produced by an intercalated proflavine molecule.

However it appears that not all molecules containing an unsaturated triple ring system complex with DNA via intercalation. Extensive physical and chemical studies of the DNA–actinomycin complex (see Reich (42) for review) support the model proposed by Hamilton, Fuller and Reich (43), in which the antibiotic (Figure 12) is hydrogen bonded to a guanosine re-

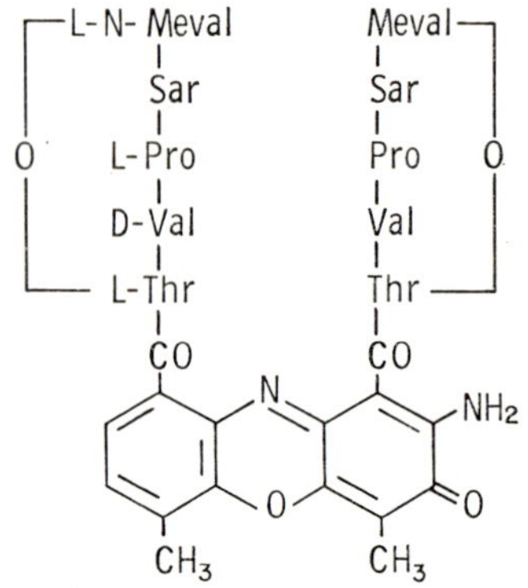

FIGURE 12. Structure of actinomycin C_1, sometimes referred to as actinomycin D.

sidue in the DNA *B* conformation (Figure 13), rather than intercalated. The large cyclic peptides can be packed into the small groove of the DNA. Molecular model building studies lend some support to the idea that intercalation does not occur when the molecule interacts with DNA because of steric hindrance from the cyclic peptides.

ACKNOWLEDGEMENTS

I would like to thank Professor Sir John Randall, F.R.S. for provision of facilities and encouragement, Professor M. H. F. Wilkins, F.R.S., and Dr. S. Arnott for helpful criticisms on an early draft of the manuscript, Professor M. H. F. Wilkins, Dr. S. Arnott, Dr. L. D. Hamilton, Mr. W. Pigram, Dr. J. Venable and Mr. Chang Yushang for discussion, and Miss A. Kernaghan and Mr. Z. Gabor for technical assistance. The daunomycin was a gift from Professor A. Di Marco.

REFERENCES

1. Wilkins, M. H. F., Stokes, A. R. and Wilson, H. R., *Nature* **171**, 738 (1953).
2. Franklin, R. E. and Gosling, R. G., *Nature* **171**, 740 (1953).
3. Chargaff, E., *The Nucleic Acids I* (Edited Chargaff, E. and Davidson, J. N.), p. 307, New York (Academic Press) (1955).
4. Watson, J. D. and Crick, F. H. C., *Nature* **171**, 737 (1953).
5. Langridge, R., Wilson, H. R., Hooper, C. W., Wilkins, M. H. F. and Hamilton, L. D., *J. Mol. Biol.* **2**, 19 (1960).
6. Langridge, R., Marvin, D. A., Seeds, W. E., Wilson, H. R., Hooper, C. W., Wilkins, M. H. F. and Hamilton, L. D., *J. Mol. Biol.* **2**, 38 (1960).
7. Fuller, W., Wilkins, M. H. F., Wilson, H. R. and Hamilton, L. D., *J. Mol. Biol.* **12**, 60 (1965).
8. DeVoe, H. and Tinoco, I. *J. Mol. Biol.* **4**, 500 (1962).
9. Cooper, P. J. and Hamilton, L. D., *J. Mol. Biol.* **16**, 562(1966).
10. Marvin, D. A., Spencer, M., Wilkins, M. H. F. and Hamilton, L. D., *J. Mol. Biol.* **3**, 547 (1961).
11. Wilkins, M. H. F., *Cold Spring Harbor Symposium on Quant. Biol.* **XXI**, 75 (1956).
12. Feughelman, M., Langridge, R., Seeds, W. E., Stokes, A. R., Wilson, H. R., Hooper, C. W., Wilkins, M. H. F., Barclay, R. K. and Hamilton, L. D., *Nature* **175**, 834 (1955).
13. Fuller, W., *University of London Thesis*, p. 169 (1961).
14. Pardon, J., Richards, B. M. and Wilkins, M. H. F., (In preparation).
15. Rich, A. and Watson, J. D., *Proc. Nat. Acad. Sci. Wash.* **40**, 759 (1954).
16. Spencer, M., Fuller, W., Wilkins, M. H. F. and Brown, G. L., *Nature* **194**, 1014 (1962).
17. Spencer, M., *Cold Spring Harbor Symp. Quant. Biol.* **XXVIII**, 77 (1963).
18. Spencer, M. and Poole, F., *J. Mol. Biol.* **11**, 314 (1965).
19. Langridge, R. and Gomatos, P. J., *Science* **141**, 694 (1963).
20. Tomita, K. I. and Rich, A., *Nature* **201**, 1160 (1964).
21. Langridge, R., Billeter, M. A., Borst, P., Burdon, R. H. and Weissman, C., *Proc. Nat. Acad. Sci.* **52**, 114 (1964).
22. Sato, T., Kyogoku, Y., Higuchi, S., Iitaka, Y., Tsuboi, M. and Mitsui, Y., *J. Mol. Biol.* **16**, 180 (1966).
23. Black, L. M. and Markham, R., *Neth. J. Plant Path.* **61**, 715 (1965).
24. Gomatos, P. J. and Tamm, I., *Proc. Nat. Acad. Sci. Wash.* **49**, 707 (1963).
25. Arnott, S., Hutchinson, F., Spencer, M., Wilkins, M. H. F., Fuller W., and Langridge, R., *Nature* **211**, 227 (1966).
26. Kaerner, H. C. and Hoffman–Berling, Z., *Z. Naturforschg.* **19b**, 593 (1964).
27. Lerman, L. S., *J. Mol. Biol.* **3**, 18 (1961).
28. Bradley, D. F. and Wolf, M. K., *Proc. Nat. Acad. Sci.* **45**, 944 (1959).
29. Luzzatti, V., Masson, F. and Lerman, L. S., *J. Mol. Biol.* **3**, 634 (1961).
30. Lerman, L. S., *Proc. Nat. Acad. Sci.* **49**, 94 (1963).
31. Tubbs, R. K., Ditmars, W. E. and Van Winkle, Q., *J. Mol. Biol.* **9**, 545 (1964).
32. Gersch, N. F. and Jordan, D. O., *J. Mol. Biol.* **13**, 138 (1965).
33. Kleinwachter, V. and Koudelka, J., *Biochim. Biophys. Acta* **91**, 539 (1964).
34. Fuller, W. and Waring, M. J., Bunsengesellochaft für physikalische Chemie **68**, 805 (1963).

35. Elliott, W. H., *Biochem. J.* **86**, 562 (1963).
36. Waring, M. J., *Biochim. Biophys. Acta* (Amsterdam) **87**, 358 (1964).
37. Arcamone, F., Francesche, G., Orezzi, P., Cassinelli, G., Barbieri, W. and Mondelli, R., *J. Amer. Chem. Soc.* **86**, 5334 (1964).
38. Arcamone, F., Cassinelli, G., Orezzi, P., Franceschi, G. and Mondelli, R., *Ibid* **86**, 5335 (1964).
39. Di Marco, A., Gaetani, M., Orezzi, P., Scarpinato, B. M., Silverstrini, R., Soldati, M., Dasdia, T. and Valentini, L., *Nature* **201**, 706 (1964).
40. Dorigotti, L., *Ibid* **50**, 117 (1964). Di Marco, A., Soldati, M., Fioretti, A. and Dasdia, T., *Ibid* **49**, 235 (1963). Di Marco, A., Gaetani, M., Dorigotti, L., Soldati, M. and Bellini, O., *Tumori* **49**, 203 (1963). Grein, A., Spalla, C., Di Marco, A., Canevazzi, G. and Giorn, G., *Microbiol.* **11**, 109 (1963).
41. Fuller, W., Pigram, W. and Hamilton, L. D., (In preparation).
42. Reich, E., *Science* **143**, 684 (1964).
43. Hamilton, L. D., Fuller, W. and Reich, E., *Nature* **198**, 538 (1963).
44. Calendi, E., Di Marco, A., Reggiani, M., Scarpinato, B. and Valentini, L., *Biochim. Biophys. Acta* **103**, 25 (1965).
45. Peacocke, A. R. and Skerrett, J. N. H., *Trans. Faraday Soc.* **52**, 261 (1956).
46. Brenner, S., Barnett L., Crick F. H. C. and Orgel A., *J. Mol. Biol.* **3**, 121 (1961).

SUMMARY

Recent X-ray diffraction and molecular model building studies of the structure of nucleic acids in the form of fibres are reviewed. Particular reference is made to the forces considered to be important in stabilizing nucleic acid conformation, and an account is presented of the influence of the ionic content of the fibre on the conformation of DNA. Models of nucleoprotamine and nucleohistone are described. The structure of regular and irregular two-stranded RNA molecules is discussed. The possible role of the ribose-2-hydroxyl in stabilizing the conformation of RNA, and defining its interaction with other RNA molecules, is discussed with particular reference to the interaction of ribosomal and transfer RNA molecules during protein synthesis. The mechanism of intercalation is discussed and models for the interaction of DNA with acridines, ethidium bromide, daunomycin and actinomycin are described.

DISCUSSION

(Chairman, V. ENGELHARDT)

U. Z. LITTAUER: Have you made any studies on intact ribosomal RNA and what would be your conclusions as to its structure?

W. FULLER: X-ray diffraction studies indicate that ribosomal RNA has a two-stranded helical structure with a conformation not unlike that of DNA. The structure is less regular than that of DNA but the X-ray studies have not so far been able to distinguish between two possible alternatives: (1) The molecule contains helical regions with a regularity comparable to that of

DNA. These are linked by non-helical regions and so prevented from packing regularly in the fibres; (2) The helical regions are much less regular than DNA, due perhaps to base-pairing of other than the Watson-Crick type. This would result in an irregular sugar-phosphate chain but the average conformation of the molecule could still be similar to that of the A conformation of DNA.

F. SOKOL: You said that the structure of double-stranded RNA of reovirus and of the so-called replicative form of MS2 bacteriophage RNA found in virus-infected cells on the one hand, and that of the double-helical regions of ribosomal RNA formed by folding-back on the other, are similar. How would you explain that the former RNA's are completely refractory to degradation by ribonuclease, whereas in the case of the latter even the double-helical regions are degraded by this enzyme?

W. FULLER: There is considerable evidence from studies on the enzymatic attack of ribosomal RNA that there are regions with a molecular weight of about 10,000 which exhibit high resistance to attack by ribonuclease. Whilst it is true that prolonged treatment with ribonuclease does completely degrade these fragments, this could be due to their relatively short length making them less stable, rather than to any intrinsic difference from the reovirus helix.

The Structure of Nucleic Acids in Solution, as Determined by X-ray Scattering Techniques: DNA, RNA, poly A

V. Luzzati, J. Witz and A. Mathis

Laboratoire de Génétique Physiologique, C.N.R.S., 91—Gif-sur-Yvette and Centre de Recherches sur les Macromolécules, 6, Rue Boussingault, 67—Strasbourg, France

Introduction

Some effort was devoted in our laboratory, over the last few years, to the study of the structure of nucleic acids in solution, by X-ray diffraction techniques. The structure analysis of these macromolecules is usually carried out on oriented fibers, by crystallographic methods. Only indirect evidence is generally available about the conformation in solution, mainly based upon hydrodynamic and spectroscopic techniques. Nevertheless, even in the case of solutions, X-ray scattering is likely to provide one of the most reliable methods for structure analysis, although the wealth of information is much smaller than in the case of fibers, as a consequence of the lack of orientation. An important technical advantage of studying macromolecules in solution is the ease with which a variety of parameters (temperature, nature of the solvent, concentration, pH, ionic strength) can be kept under control, and thus polymorphic transitions can be induced and analyzed.

We shall summarize here some of the results that we have obtained; most of these have been described elsewhere.

Techniques

A. SMALL-ANGLE X-RAY SCATTERING

The small-angle X-ray scattering techniques will only be briefly summarized here (1,2).

The characteristic feature of the X-ray diffraction pattern of the isotropic solutions used in this work, is the presence of a diffuse scattered halo around the incident beam. The distribution of the intensity in the halo, as a function of the angle of diffraction, is the experimental information to be recorded.

The apparatus makes use of a monochromatic X-ray beam, focussed on a long and straight line. The sample is mounted in a flat cell, covered by thin mica foils; the experiments are carried out at controlled temperature, in the range -30 to $+150°C$. The result of an experiment, after the elimination of the blank scattering (due to air, cell windows and solvent), is a normalized function $j_n(s)$, that is a function only of the structure of the sample.

If the sample is a perfect solution of long and rigid rods, the form of $j_n(s)$ is:

$$j_n(s) = Af(\pi R_c s) \tag{1}$$

$$f(x) = 1/2 \exp(-x^2) K_0(x^2) \tag{2}$$

$$A = c_e \mu (1 - \varrho_0 \psi)^2 \tag{3}$$

$s = 2\sin\theta/\lambda$ (2θ is the diffraction angle, λ is the wavelength), R_c is the axial radius of gyration and μ is the mass per unit length of the rods (see precise definition in ref. 1), c_e and ψ are the electron concentration and partial specific volume of the solute, ϱ_0 is the electron density of the solvent.

If the sample contains rather short rods, either isolated from each other, or somehow linked together into a compact particle, the form of $j_n(s)$, for s large, remains that of Eq. 1, but as s decreases $j_n(s)$ departs from Eq. 1. A difference by excess ($j_n(s)$ larger than Eq. 1) indicates that the molecules are more compact than a rigid rod of the same length; this may mean, for

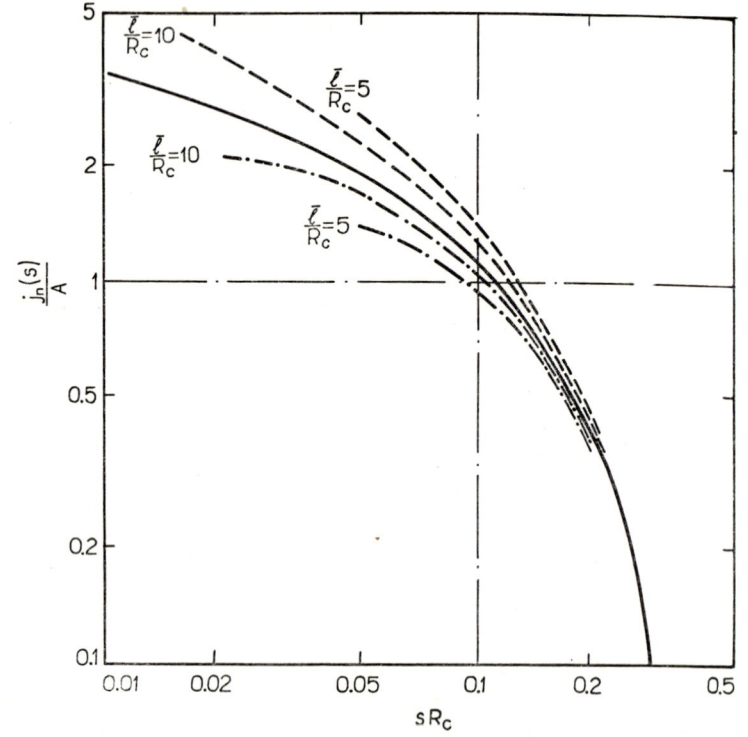

FIGURE 1. Theoretical curves for rod-like particles. ——— infinitely long rods (Eq. 1); ——— Gaussian coil of rods (Eqs. 4 and 5); —·—· independent rods of finite length (Eqs. 4 and 5, $n = 1$). $\bar{l}$ is the average length of the rods. (See Ref. 9).

example, that the rods are made up of rigid segments, linked together by flexible joints. A difference by defect is associated with the presence of free ends. The mathematical treatment of this model is given elsewhere (3,4,5). The result is the following asymptotic expression of $j_n(s)$:

$$j_n(s) \to A[f(\pi R_c s) + BR_c g(\pi R_c s)]. \qquad (4)$$

The expression of $g(x)$ is given by Witz et al. (5); $f(x)$ is defined in Eq. 2. The meaning of the parameter B depends on the choice of the model. Assuming for example that the molecule has the form of a random coil of n rods, of average length $\bar{l}$, the expression of B is:

$$B \doteq (2.94 - 4.94/n)/\bar{l}. \qquad (5)$$

Various curves corresponding to Eq. 4 are plotted in Figure 1, for different values of n and $\bar{l}$.

When the experimental points may be aligned on one of the curves of Eq. 1, the shape of the molecules of the solute can be assimilated to long and rigid rods, and the parameters R_c and μ can be determined (if c_e, ψ and ϱ_0 are known). If the coincidence of the experimental points with the curve for a rod is satisfactory at large s, and a small divergence is observed in approaching the origin of s, the model of a system of short rods may be adopted, and the value of B can be estimated, in addition to determining R_c and μ.

B. UV ABSORPTION

In order to correlate the X-ray scattering observations with some other physical parameter, we carried out UV absorption experiments as a function of temperature, on the same samples used for the X-ray experiments. A special cell was designed for this purpose, which allows the study of UV absorption of thin films (thickness of the order of 0.01 mm). This cell is mounted on a temperature-controlled support. The UV absorption was measured as a function of temperature on a Zeiss spectrophotometer.

RESULTS

A. DNA—WATER

DNA solutions have been studied extensively, at room temperature, by small-angle X-ray scattering techniques (2, 6, 7). The molecules are rod-like in shape, with $R_c = 8.5$ Å and $A/c_e = 15.5$. Adopting $\bar{v} = 0.575$ cm³g⁻¹ ($\psi = 1.85$ Å per electron) for Na.DNA (2), $\mu = 105$ electrons per Å, in excellent agreement with the Watson–Crick model ($\mu = 101$ for the B form).

If the temperature is raised above room temperature (8), the X-ray curves remain the same over a fairly extended temperature range, then a change takes place, the more pronounced the higher the temperature. The shape of the X-ray curves remains compatible with the rod-model at all temperatures: the experimental results can thus be represented by the two parameters R_c

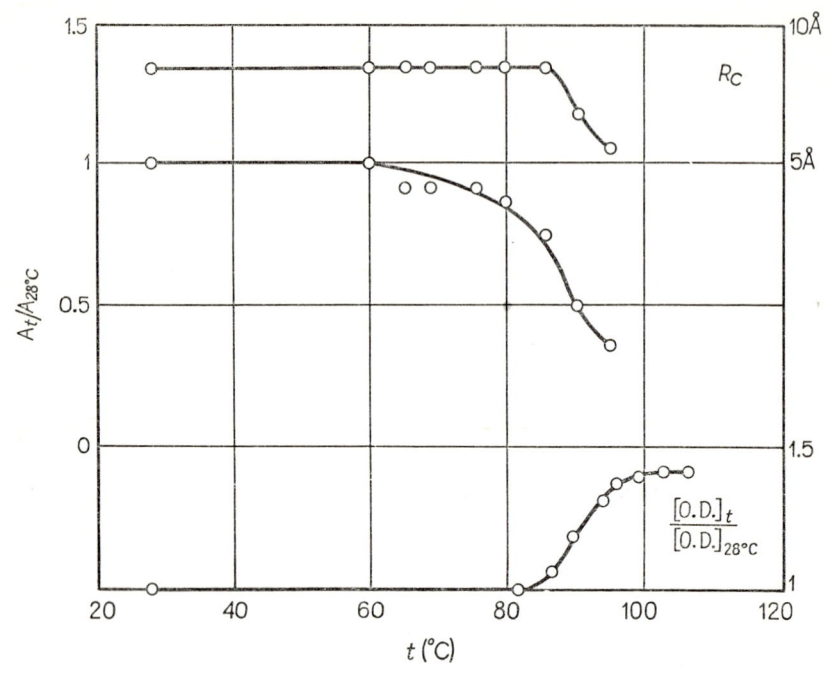

FIGURE 2. DNA (calf thymus) in saline (0.15 M NaCl; 0.015 M citrate), concentration 2%. Results of the small-angle X-ray scattering and ultraviolet absorption experiments, as a function of temperature. The X-ray curves are of the rod type from room temperature (28°C) up to 95°C: R_c and the ratio $A_t/A_{28°C}$ are plotted. Ultraviolet absorption is measured at 260 mμ. The ratio $(O.D.)_t/(O.D.)_{28°C}$ is plotted as a function of temperature. (See Ref. 8).

and A (Figure 2). It is important to note that from the onset of the transition (60°C in Figure 2) up to the highest temperature the parameters keep varying with temperature, without reaching any stable value. The variation of the O.D. is shown in Figure 2; the drop of R_c and A, and the rise of O.D. take place over the same temperature interval.

B. RNA–WATER

Various types of RNA have been studied: ribosomal (*E. coli*, yeast, Ascites tumor cells) (9), viral (turnip yellow mosaic, tobacco mosaic) (5). The small-angle X-ray scattering curves of all these preparations have common features, that can be visualized in the plot $j_n(s)/c_e$ versus s (Figure 3). At high angles ($s > 0.012$ Å^{-1}) all the curves are identical, and coincide with those of DNA. At smaller s the experimental curves depart from that of a rod: the difference increases as s decreases. The amplitude of the deviation varies in the different types of RNAs, and is dependent on the experimental conditions (especially ionic strength) (9). The effect of temperature, from 3 to 28°C, is negligible on the high angle portion of the scattering curves, and becomes more and more pronounced as s decreases (9).

These results may be interpreted as follows. The structure of RNA is, at high resolution, very similar to that of DNA: rod-like, with the same radius of gyration and mass per unit length. But rods of RNA are much

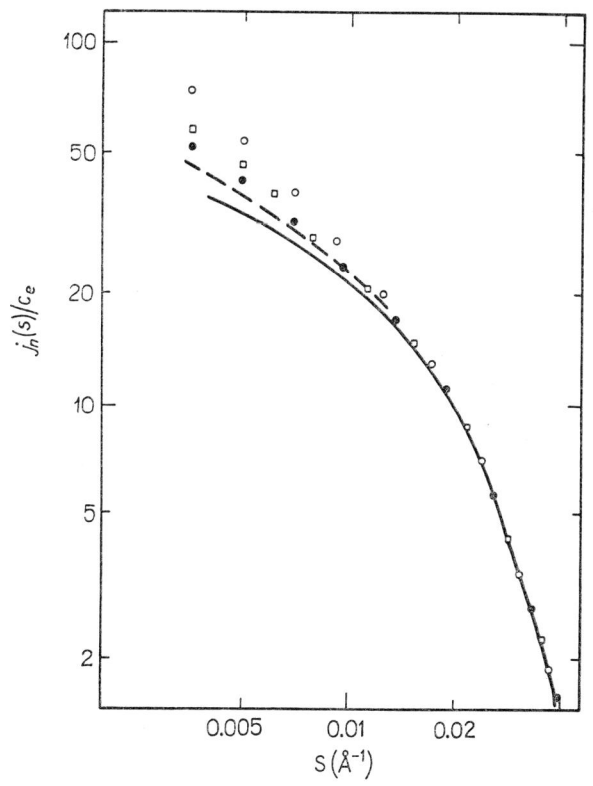

FIGURE 3. Experimental curves for RNAs of various sources in saline, 0.15 M NaCl: ○—TYMV; □—TMV; ●—ribosomal. Full line: theoretical curve for an infinitely long rod with $A/c_e = 16$, $R_c = 8.0$ Å. Dotted line: Gaussian coil of rods, with $A/c_e = 16$, $R_c = 8.0$ Å, $\bar{l} = 96$ Å. (See Ref. 5).

shorter than those of DNA, and are interconnected in such a way that the overall particle is more compact than DNA. Adopting the model of a random coil of rods, the average length of the rigid segments can be estimated in each case (the values are included from 80 to 200 Å).

C. POLY A–WATER

The conformation of this polymer in solution is known to be pH dependent. The molecule is a rigid rod at low pH, and becomes flexible (coil-like) at high pH (reviewed in Ref. 10).

The form of the small-angle X-ray scattering curves, recorded with various poly A preparations, is consistent with the rod model, in the pH range from 5.0 to 7.2 (11). R_c and A can thus be determined: the results are plotted in Figure 4, as a function of pH. The parameters remain constant over two fairly extended ranges of pH: two stable forms of the polymer are thus defined. The mass per unit length of the two forms can be determined, if $\bar{v}$ is known: at acidic pH the formation of an internal salt between phosphate and adenine has to be taken into account (4).

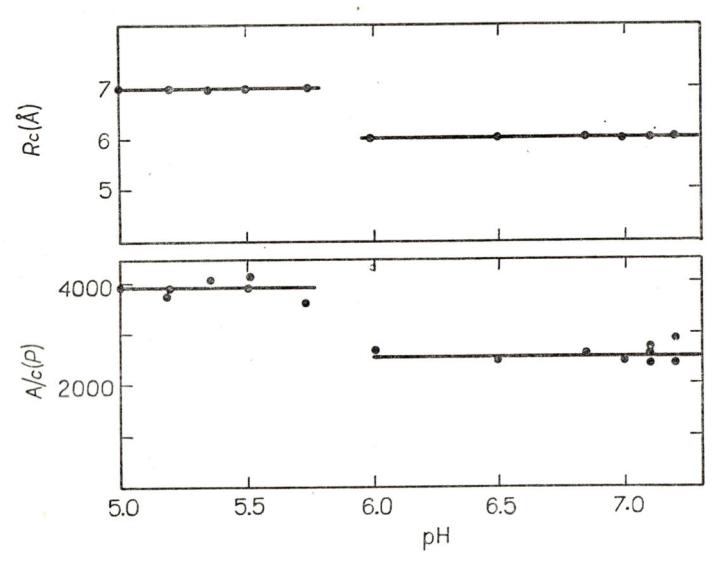

FIGURE 4. Poly A: $A/c(P)$ and R_c as a function of pH ($c(P)$ is the concentration, in moles of phosphate per gram of solution). (See Ref. 11).

The linear density of the two forms turns out to be two nucleotides per 3.8 Å at acidic pH, and one nucleotide per 3.5 Å at neutral pH.

D. DNA-WATER-FORMALDEHYDE

Several experiments were carried out on DNA solutions in saline, in the presence of variable amounts of formaldehyde, as a function of temperature (8). An example is shown in Figure 5. From room temperature to 50°C the X-ray diffraction curves remain the same, identical to those observed in the absence of formaldehyde (rod-type, with $R_c = 8.5$ Å, $A/c_e = 15.5$). From 50° to 65°C the curves change, but their shape remains typical of rod-like particles, A and R_c decreasing with rising temperature. Then, from 65 to about 90°C the curves remain stable, the value of A becoming close to one half of that at 28°C, and R_c becoming 5.8 Å. Beyond 90°C another transition is observed: the X-ray curves drop dramatically, especially at low angles. No stable form is observed after the beginning of this transition;

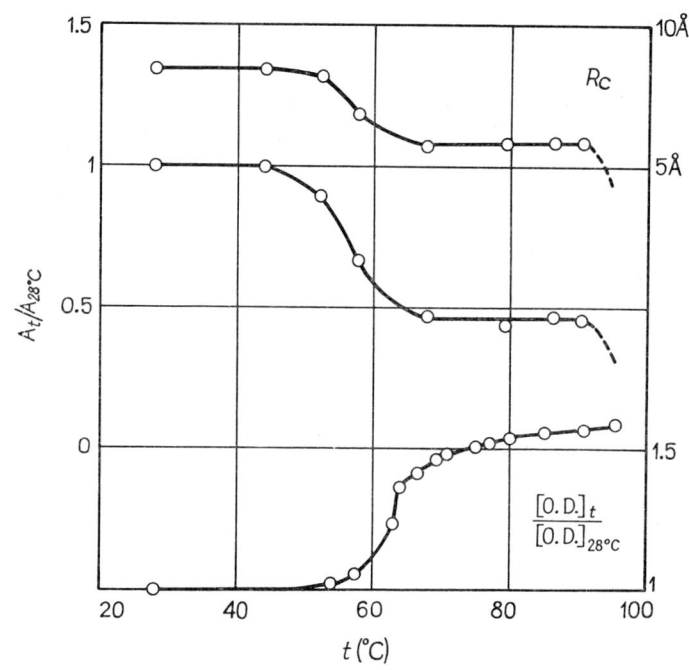

FIGURE 5. Same system as in Figure 2, plus 2% formaldehyde. The dotted line at high temperature shows the onset of the second transition, after which the X-ray curves are no longer of the rod type. (See Ref. 8).

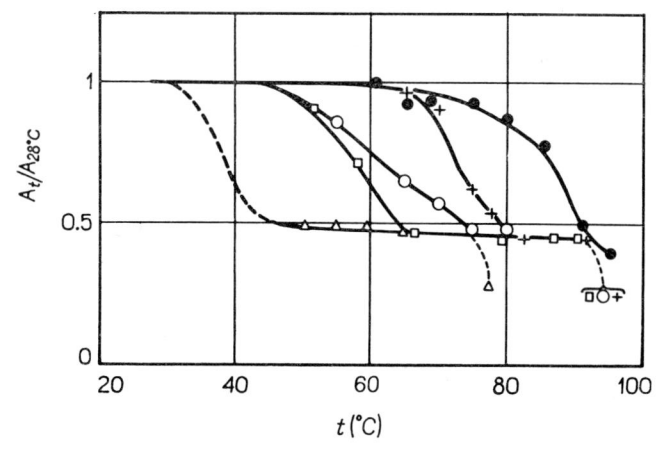

FIGURE 6. DNA in saline, with different amounts of formaldehyde. The ratio $A_t/A_{28°C}$ is plotted, as a function of temperature, for all the experiments in which the X-ray curves are of the rod type. The onset of the second transition is shown as in Figure 5; DNA concentration approximately 2%. -●-●- 0.15 M NaCl, 0.015 M citrate; -+-+- 0.15 M NaCl, 0.5% formaldehyde; -○-○- 0.15 M NaCl, 1.16% formaldehyde; -□-□- 0.15 M NaCl, 0.015 M citrate, 2% formaldehyde; -△-△- 0.15 M NaCl, 10% formaldehyde. (See Ref. 8).

the curves change with temperature, and change with time if the temperature is kept constant. Furthermore the form of the curves becomes incompatible with a rod model.

Results obtained with different formaldehyde concentrations are shown in Figure 6. In all the cases two transitions are observed, the first from the native to another apparently rod-like form, with an approximate halving of A and a drop of R_c from 8.5 to 5.8 Å, the second to an ill-defined form.

When formaldehyde is present the X-ray curves observed at any temperature remain the same after the sample is brought back to a lower temperature.

The variation of the O.D. is shown in Figure 5. A fairly steep rise is observed in the temperature range of the first transition, followed by a gradual increase at higher temperature.

E. DNA–ETHYLENE GLYCOL

DNA solutions in anhydrous ethylene glycol, containing some NaCl, were studied as a function of temperature (8). The solutions were made at low temperature (4°C) in order to preserve the native conformation (see below).

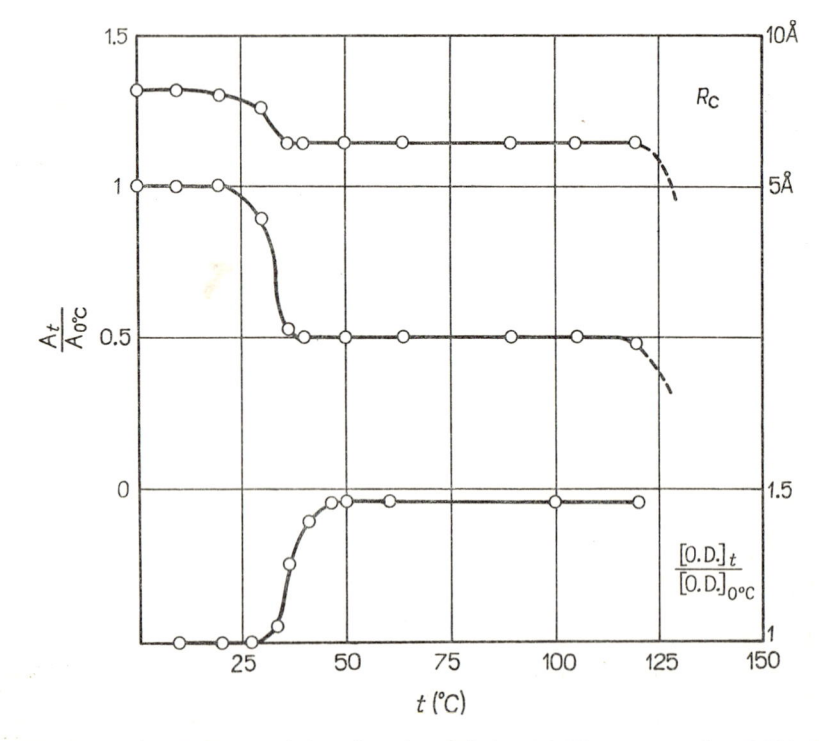

FIGURE 7. DNA (calf thymus), in saline-glycol (0.1 M NaCl), concentration 2.2%. Legends as in Figures 2 and 5. (See Ref. 8).

From −10 to 120°C the X-ray curves are of the rod-type, and can thus be represented by R_c and A (Figure 7). Two temperature ranges can be clearly distinguished, over each of which R_c and A remain constant. One, at low temperature, in which $R_c = 8$ Å and $A/c_e = 11.3$; the second, at high temperature, in which $R_c = 6.5$ Å and $A/c_e = 5.6$. Assuming that $\bar{v}$ is the same in water and in glycol, $\mu = 110$ electrons per Å in the low temperature form, in good agreement with the Watson–Crick model.

Beyond 120°C another transition is observed: the X-ray curves become time- and temperature-dependent, and their form becomes incompatible with the rod-model, as happens at high temperature in water-formaldehyde solutions (see above).

CONCLUSIONS

A. REMARKS ON THE X-RAY TECHNIQUES

Some aspects of the accuracy and the reliability of the experimental results should be discussed.

Some recent results obtained with polypeptides (12) have shown that the correlation effects are sometimes more serious than we had thought in the past. The ideal conditions, to which the theoretical functions (Eqs. 1 and 4) apply have to be approached by extrapolating, to infinite dilution, a series of experiments carried out at different concentrations. We have performed recently a few experiments with DNA over a fairly extended concentration range, in order to carry out a reliable extrapolation to infinite dilution. It turned out that the values of R_c and A, obtained previously, are only slightly modified by the extrapolation.

The determination of μ is critically dependent upon the partial specific volume ψ (see Eq. 3). The partial specific volumes of the nucleic acids are not known with great accuracy; we have discussed this problem elsewhere (2,4), and have justified the choice made here. We make the hypothesis here that ψ is the same in water and in ethylene glycol, and that the product $\varrho_0\psi$ is independent of temperature. Furthermore, we assume that the system contains only two components (solute and solvent), and we neglect possible specific associations between the solute and the various components of the solvent (water or glycol, electrolyte, formaldehyde). We have shown in fact (2,7), that at low salt concentration no serious correction need be taken into account. Finally it should be noted that most of the conclusions of this work are based upon the comparison of X-ray scattering curves obtained with the same preparation under different conditions (temperature, pH); the effects of the errors thus tend to cancel out.

The interpretation of the experimental curves is based upon the rod model. It should be remarked that a linear object behaves like a rod, so far as the X-ray small-angle scattering techniques are concerned, if the average radius of curvature and the length are much larger that the diameter. More precisely, in the conditions of our experiments, the length and the radius of curvature should not be smaller than about 300 Å. It is clear, nevertheless, that the fact that the X-ray curves are of the rod type is not sufficient to prove that the conformation of the solute indeed is rod-like: other lines of evidence have to be taken into account to confirm the model (see below).

B. DISCUSSION OF THE EXPERIMENTAL RESULTS

The experimental results can be briefly recapitulated.

In all the systems it is found that over one or two extended ranges of one variable (temperature or pH) the small-angle X-ray scattering curves remain unchanged. This is strong evidence in favour of the existence of one stable form over each of those intervals. This situation is met in the following cases:
  (a) DNA–water, at low temperature (28 to 60°C, Figure 2);
  (b) RNA–water, 2 to 20°C;
  (c) poly A–water, at low pH (pH 5 to 5.8, Figure 4);
  (d) poly A–water, at high pH (pH 6 to 7.2, Figure 4);
  (e) DNA–water—formaldehyde, at low temperature (28 to 45°C, Figure 5);
  (f) DNA–water—formaldehyde, at high temperature (65 to 90°C, Figure 5);
  (g) DNA–glycol, at low temperature (0 to 20°C, Figure 7);
  (h) DNA–glycol, at high temperature (40 to 120°C, Figure 7).

In all these cases the X-ray curves are of the rod type.

No stable form is found at the highest temperatures reached in our experiments; beyond the second high temperature transition ($t > 65°C$ in Figure 2; $t > 90°C$ in Figure 5; $t > 120°C$ in Figure 7) the X-ray curves change with temperature and with time of heating, and the form of the curves often becomes incompatible with the rod model.

The stable forms can be classified into three categories, according to the small-angle X-ray scattering curves:
  (1) two-stranded, fully ordered,
  (2) two-stranded, partially ordered,
  (3) single-stranded, ordered.

The structure of these will now be discussed.

1. Two-stranded, fully ordered

This conformation is characterized by rod-type diffraction curves, the linear density of the rods being close to two nucleotides per 3.4 Å (Table).

In the case of DNA solutions in saline, in saline plus formaldehyde, or in glycol, this conformation is that of the low temperature form, before the first transition. The dimensions of the rods (μ and R_c) are in excellent agreement with the Watson–Crick model of DNA, form B (2,7).

In the case of poly A, at acidic pH, the linear density of the rods agrees with the fiber structure, determined at acidic pH by Rich, Davies, Crick and Watson (13): two-stranded helix, with two nucleotides per 3.8 Å (Table).

TABLE

Two-stranded fully ordered

	DNA	DNA	DNA	poly A
Solute	DNA	DNA	DNA	poly A
Solvent	water	water formaldehyde	glycol	water
Experimental conditions	28–60°C (Fig. 2)	28–45°C (Fig. 5)	0–20°C (Fig. 7)	pH 5.0 (Fig. 4)
Number of nucleotides per rod length	2 per 3.2 Å	2 per 3.2 Å	2 per 3.1 Å	2 per 3.8 Å
Axial radius of gyration	8.5 Å	8.5 Å	8.0 Å	6.9 Å

Two-stranded partially ordered   Single stranded ordered

	RNA	DNA	DNA	poly A
Solute	RNA	DNA	DNA	poly A
Solvent	water	water formaldehyde	glycol	water
Experimental conditions	2–20°C	65–90°C (Fig. 5)	100–120°C (Fig. 7)	pH 6.0–7.2 (Fig. 4)
Number of nucleotides per rod length	ca 2 per 3.4 Å	1 per 3.5 Å	1 per 3.2 Å	1 per 3.5 Å
Axial radius of gyration	8.0 Å	5.8 Å	6.5 Å	6.0 Å

2. Two-stranded, partially ordered

The X-ray curves are consistent with molecules formed by fairly short rod-like segments, linked together by flexible joints. The parameters μ and R_c of these rods are similar to those of native DNA.

The analogy of dimensions suggests that the conformation of the rod-like segments is similar in RNA and DNA, namely helical and two-stranded, with the base planes stacked and perpendicular to the rod axis. A corollary of this model is that the base pairing is irregular in the helical segments, since the base composition of RNA is far from complementary, especially in the case of TYMV (5).

The agreement of the experimental curves with the rod-like model is quite satisfactory, and does not leave much room for a "non-helical" portion; furthermore the "helical content" appears to be the same for different types of RNAs, in spite of the difference in base composition (5). The existence of loops, as suggested by Fresco, Alberts and Doty (14), seems hard to reconcile with the small-angle X-ray scattering observations.

3. Single-stranded, ordered

The X-ray curves of this form are of the rod type, the mass per unit length of the rods being close to one half of that of the two-stranded native form (see Table). The most obvious structure consistent with those data is one strand of the double helix, separated from its complementary strand, in which the stacking and the orientation of the base planes is preserved. A two-stranded structure could also be constructed (8), but various recent observations indicate that the structure is more likely to be single-stranded (see below).

The existence of a single-stranded ordered conformation is confirmed by several observations.

Poly A, neutral pH. Circular dichroism (15) and optical rotatory dispersion (16,17) studies have shown that the conformation of poly A at neutral pH is ordered. Those data are consistent with a single-stranded helical structure, with stacked base planes. Similar optical studies have been carried out with other polymers (poly C, Ref. 18; poly U, Refs. 19, 20, 21), and agree with our observations on poly U (11).

DNA–water–formaldehyde. The "thermal denaturation" of DNA in this solvent has been studied by a variety of techniques. Freifelder and Davison (22) have presented data suggesting that the strands of the DNA separate at the transition defined by the rise of O.D.: this corresponds to the first transition in the X-ray study (near 55°C in Figure 5). Tikchonenko, Perevertajlo and Dobrov (23) have observed a two-steps transition in the viscosity versus temperature curve of T2 DNA, in the presence of 2% formaldehyde; the temperature gap between the two transitions is 30°C, thus similar to that of Figure 5. The most direct confirmation of the single-stranded ordered form is provided by the electron microscope study of ØXI74 DNA, by Freifelder, Kleinschmidt and Sinsheimer (24). These authors fixed the circular form of single-stranded DNA of this virus with formaldehyde. In the electron microscope each molecule appears as a ring, the length of which is very close to that of the two-stranded rings of the replicative form. Since the molecular weight of the single-stranded form is known to be one half of that of the replicative form, the conclusion can be drawn that the single stranded molecule is rod-like, and that the linear density of the rods is approximately one nucleotide per 3.4 Å.

DNA–glycol. Less information is available on this system. Several properties (viscosity and transforming activity, Ref. 25; UV absorption, Mathis, in preparation) display an irreversible change at the first transition (near 35°C, Figure 7). Strand separation is known to occur just above that temperature (26). Poly C (18) and transfer RNA (27) do not seem to possess an ordered conformation in glycol, when studied by ORD methods: no ORD analysis seems to have been carried out on DNA in glycol. The situation is thus still unclear, as the X-ray evidence points to the existence of an ordered form, whilst the optical properties, admittedly *not* on DNA, are in favour of a disordered conformation. Parallel X-ray and optical experiments on DNA, now in progress in our laboratory, should remove the uncertainty.

C. FINAL COMMENTS

The very existence of the single stranded ordered form (named at first "intermediate" form), that could well be questioned at the time we put forward our model (8), is now amply confirmed by other experimental evidence. It is perhaps surprising that in no case so far has a stable disordered form been observed, the X-ray curves of which are *not* of the rod type; it is not clear yet whether such a form does not exist, or has merely escaped our notice.

The stability of the structure of the three forms could hardly be visualized without assuming that the major stabilizing forces are provided by the stacking of the base planes, and that the hydrogen bonds play only a secondary rôle. This idea is now widely accepted; it was quite controversial at the time we were led to formulate it, in the interpretation of our X-ray experiments on RNA in solution (28).

A discussion of the biological significance of the ordered single stranded form remains in the realm of speculation, at least in the absence of evidence in favour of its existence *in vivo*. It is clear that if single-stranded polynucleotide chains could take up a fairly rigid ordered structure, strand separation and replication could be different events, in time and space, and some topological problems could be avoided (for example, the simultaneous unwinding of the double helix and the replication of the strands).

ACKNOWLEDGEMENTS

Most of the work described here was carried out at the Centre de Recherches sur les Macromolécules in Strasbourg. We wish to express our gratitude to Professors C. Sadron and H. Benoît for constant interest and encouragement. We are glad to acknowledge the contributions of our colleagues Drs. F. Masson, A. Nicolaïeff and S. N. Timasheff.

This work was supported by a grant from the "Délégation Générale à la Recherche Scientifique et Technique".

REFERENCES

1. Luzzati, V., *Acta Cryst.* **13**, 939 (1960).
2. Luzzati, V., Nicolaïeff, A. and Masson, F., *J. Mol. Biol.* **3**, 185 (1961).
3. Luzzati, V. and Benoît, H., *Acta Cryst.* **14**, 297 (1961).
4. Witz, J., *Thèse*, Université de Strasbourg (1964).
5. Witz, J., Hirth, L. and Luzzati, V., *J. Mol. Biol.* **11**, 613 (1965).
6. Luzzati, V., *Progress in Nucleic Acids Research* **1**, 347 (1963).
7. Masson, F., *Thèse*, Université de Strasbourg (1965).
8. Luzzati, V., Mathis, A., Masson, F. and Witz, J., *J. Mol. Biol.* **10**, 28 (1964).
9. Timasheff, S. N., Witz, J. and Luzzati, V., *Biophys. J.* **1**, 525 (1961).
10. Steiner, R. F. and Beers, R. F., *Polynucleotides*, Amsterdam, Elsevier (1961).
11. Witz, J. and Luzzati, V., *J. Mol. Biol.* **11**, 620 (1965).
12. Saludjian, P. and Luzzati, V., *J. Mol. Biol.* **15**, 681 (1966).
13. Rich. A., Davies, D. R., Crick, F. H. C. and Watson, J. D., *J. Mol. Biol.* **3**, 71 (1961).
14. Fresco, J. R., Alberts, B. M. and Doty, P., *Nature* (London) **188**, 98 (1960).
15. Brahms, J., *Nature* (London) **202**, 797 (1964).
16. Sarkar, P. K. and Yang Jen Tsi, *J. Biol. Chem.* **240**, 2088 (1965).
17. Holcomb, D. N. and Tinoco, I., *Biopolymers* **3**, 121 (1965).
18. Fasman, G. D., Lindblow, C. and Grossman, L., *Biochemistry* **3**, 1015 (1964).
19. Lipsett, M. N., *Proc. Nat. Acad. Sci.*, Wash. **46**, 445 (1960).
20. Brahms, J., *Journ. Am. Chem. Soc.* **85**, 3298 (1963).
21. Richards, E. G., Flessel, C. P. and Fresco, J. R., *Biopolymers* **1**, 431 (1963).
22. Freifelder, D. and Davison, P. F., *Biophys. J.* **3**, 49 (1963).
23. Tikchonenko, T. I., Perevertajlo, G. A. and Dobrov, E., *Biochim. Biophys. Acta* **68**, 500 (1963).
24. Freifelder, D., Kleinschmidt, A. K. and Sinsheimer, R. L., *Science* **146**, 254 (1964).
25. Eliasson, R., Hammarston, E., Lindhal, T., Björk, I. and Laurent, T. C., *Biochim. Biophys. Acta* **68**, 234 (1963).
26. Herskovits, T. T., *Biochemistry* **2**, 335 (1963).
27. Fasman, G. D., Lindblow, C. and Seaman, E., *J. Mol. Biol.* **12**, 630 (1965).
28. Luzzati, V., *Journ. de Chimie Phys.* **58**, 899 (1961).

SUMMARY

Different types of nucleic acids have been studied, in solution, by X-ray scattering techniques. All the structures observed belong to one of the following classes;

(a) The *fully ordered, two-stranded helix,* observed with native DNA, and with poly A at acidic pH. This conformation is characterized by long and rigid rods, whose mass per unit length agrees with a linear density of two nucleotides per 3.4 Å.

(b) The *two-stranded, partially ordered form,* found in ribosomal and viral RNA. The structure is formed by short rods, linked together by flexible connections; the internal structure of the rods is very similar to that of DNA, although the base pairing is less specific.

(c) The *single stranded ordered conformation,* observed with DNA, at high temperature, either in anhydrous ethylene glycol or in water-formaldehyde, and in poly A at neutral pH and at low temperature. The structure is rod-like, the mass per unit length of the rods is one half that of native DNA. The rods are probably formed by a single polynucleotide strand, and its rigidity is due to the stacking of the base planes.

Active Sites of RNA's

G. L. Brown, Sheila Lee and D. H. Metz

Medical Research Council Biophysics Research Unit, Biophysics Department, King's College, University of London

Four types of RNA are known to be involved in protein synthesis, sRNA, messenger RNA and the RNA's of the two ribosomal sub-units. A fifth type of RNA, the 5S RNA attached to ribosomes, observed by Rosset, Monier and Julien (1) may also have an, as yet unknown, role in protein synthesis. The primary sequences of several yeast sRNA's are now known (2, 3), and current interest lies mainly in the determination of the secondary and tertiary structure of sRNA's and the location of the active sites in the molecules. The determination of the primary sequence of the 5S RNA may also lie within the range of present methods. In the case of the other three types of RNA's, the detailed determination of their primary structures constitutes a daunting problem even using automated analytical procedures (5, 6) and computor-aided sequence analysis (7), as they contain more than 1,000 nucleotides per molecule.

However, there are certain limited regions of these large molecules which are of special interest, either because they are the regions involved in specific interactions between RNA's in protein synthesis or because they have a dominating role in the initiation and control of protein synthesis. These regions may be considered as the "active sites" of the RNA's, analogous to the "active sites" in proteins, and the determination of the primary sequences of these regions and of the conformation of their constituent nucleotides is a necessary step in the formulation of a detailed mechanism for protein synthesis.

The primary structure of limited regions of the high molecular weight RNA's could be determined by the present methods of nucleotide sequence analysis if these regions could be labelled, marked or specifically isolated from enzymic digests in some way.

The nucleotides in RNA's may exist in four general types of conformation; single-stranded with randomly oriented bases, single-stranded with stacked bases (8, 9), double-stranded helices as in the RNA of reovirus (10), and double-stranded helices with only partial Watson–Crick type pairing and possibly including other types of base-pairs (11), looped bases (12) or unpaired bases (13, 14). A single-stranded RNA molecule may contain regions with nucleotides in any of these types of conformation and an "active site" may consist of one or more of these conformational regions. A method of labelling and isolating sequences of nucleotides in a particular type of conformation, especially in the single-stranded conformations, would facilitate the study of the structure of the active sites. Before describing one experimental approach to the problem of locating and obtaining the information required about the structure of the "active sites" in transfer, ribosomal and viral messenger RNA's, it is convenient to first specify the nature of the "active sites" in these RNA's and discuss our present knowledge of their structures.

ACTIVE SITES IN SRNA

The active sites in sRNA which have been studied most intensively are the transfer coding site and the acceptor recognition site.

The transfer coding site is defined as the region in a specific sRNA which interacts with the coding triplet, in the ribosome-bound messenger RNA, that corresponds to the amino acid attached to the sRNA. The hypothesis due to Crick (15) states that this site is a non-hydrogen bonded triplet, known as the anti-codon, in a single-stranded region of the sRNA molecule, complementary in sequence to the corresponding coding triplet and that the correct aminoacyl–sRNA is selected by the formation of hydrogen bonds between the triplets. The codons for all the amino acids have now been determined (16, 17) and the complete sequences of several specific sRNA's have been determined so that the problem of this active site may shortly be solved by inspection of a number of different primary structures.

The acceptor recognition site of a specific sRNA is the region recognized by the corresponding aminoacyl–sRNA synthetase and therefore determines the specificity of the amino acid attachment reaction.

There are two main problems concerning these two active sites in sRNA. The first is the position of the two sites in the sRNA molecule and in particular whether the two sites are identical, separate or partially overlapping. The second is the sequence and the conformation of the nucleotides comprising these sites. The main approach adopted by the workers in this field has been to modify the primary or secondary structures of the sRNA's in some known way and to observe the effects on the acceptor and transfer activities. Information about the nucleotide composition of the active sites has been sought by studying the fractional inactivation by chemical reagents which react with specific nucleotides. These reagents and treatments include formaldehyde (18, 19), hydroxylamine (20, 21), aqueous bromination (22), bromination in dimethylformamide (23), methylation (23), nitrous acid (24, 25, 26). Similar inactivation studies using exposure to ultraviolet light have been made (27, 28, 29). In general, it has proved difficult to interpret the results unambiguously in terms of the composition of the acceptor site because of secondary reactions and because of the possibility of crosslinking or of inactivation due to chemical modifications of the pCpCpA end-group. Weil et al. (30) have overcome this objection by replacing the pCpCpA group after chemical modification of the rest of the molecule. They interpret their results as indicating that in some of the cases studied the hypothetical anti-codon participates in the acceptor activity of the corresponding transfer RNA. This interpretation is partly supported by studies of the inactivation of acceptor activities by ultraviolet irradiation (29). All the chemical procedures mentioned above, with the exception of formaldehyde, do not discriminate between nucleotides in different types of conformation. The difficulties in interpreting the effects of formaldehyde are that the modifications produced by it are largely reversible under the conditions of measurement of activities and that it forms cross links between different parts of the RNA chains (31).

Recently Burton and Riley (32) have found that osmium tetroxide oxidizes pyrimidines in single-stranded nucleic acids and oligonucleotides but not in double-stranded nucleic acids. Burton, Varney and Zamecnik (33) have observed correlations between the genetic code and the sensitivity of acceptor activities of sRNA to oxidation by osmium tetroxide, which suggests participation of the anti-codon in the acceptor activity. Seidel and Cramer (34) have found that perphthalic acid oxidizes adenine preferentially in single-stranded regions of RNA's and polynucleotides. These reagents together with the carbodiimide reagent to be described later, are capable of discriminating between nucleotides in different conformations and will clearly be of great value in exploring further the nature of active sites in RNA's.

Hayashi and Miura (35) have studied the inhibition of acceptor activity produced by adding oligonucleotides to the incubation mixture. They found that the attachment of phenylalanine, lysine and proline to sRNA was inhibited by oligo-A, oligo-U and oligo-G respectively. As the corresponding hypothetical anti-codons are AAA, UUU and GGG respectively, they concluded that the anti-codon was part of the acceptor recognition site.

The complexity of the cross-reactions between aminoacyl sRNA synthetases and sRNA's from different species, combined with the apparent universality of the genetic code, suggests that the complete acceptor recognition site must consist of a larger region than the anti-codon, or more than one region. Possibly tertiary folding of the sRNA molecule results in distinct regions of the primary structure being brought together to form the active acceptor site.

Arca et al. (36) attempted to obtain information about the conformation of the acceptor recognition site by measuring the acceptor activities of sRNA at temperatures where part of the secondary structure has melted out, using thermostable aminoacyl sRNA synthetaseses. They concluded that some of the secondary structure was necessary for acceptor activity.

Most of the methods described so far can yield only qualitative results and correlations and cannot be easily used to determine the sequence and conformations of the nucleotides in the active sites.

ACTIVE SITES OF MESSENGER RNA'S

At the moment, the only messenger RNA's available in sufficient quantities for detailed chemical studies are the polycistronic single-stranded viral RNA's, the simplest examples being the RNA-containing coliphages f-2, R17, MS2, and μ2. These phages contain a single-stranded RNA comprising about 3000 nucleotides coding for between 3 and 6 virus-associated proteins (37). One of these is the coat protein and one, or possibly two, others are the enzymes concerned with RNA replication. There are several features of *in vivo* and *in vitro* viral protein synthesis which are relevant to the consideration of active sites in the viral messenger RNA molecules.

In the RNA phage-infected cell, the cistron for the coat protein appears to be read at a much higher frequency than the other cistron, as most of the virus-associated protein formed in the infected cell is coat protein. This is also found in the cell-free system with isolated viral RNA as messenger where more than 90% of the protein formed appears to be coat protein (38). The RNA phage-infected cell also exhibits an ordered sequence of synthesis of different virus-associated proteins with time after infection, as Lodish Cooper and Zinder (39) have demonstrated with *f*-2 infected *E. coli*. They found that the synthesis of the virus-associated RNA polymerase ceases halfway through the infectious cycle, but that a host-dependent mutant of the "amber" type had lost this facility of switching off the RNA polymerase synthesis.

Spiegelman and Hayashi (40), using a cell-free system with MS2 as messenger were able to demonstrate a temporal sequence of synthesis of different proteins *in vitro*, coat protein being synthesized first, followed a few minutes later by the synthesis of other proteins. Larger single-stranded RNA viruses show a more complicated temporal sequence of synthesis of different proteins *in vivo* (41). The synthesis of a particular polypeptide chain must be initiated and terminated, and therefore the beginning and end of each cistron may be considered as active sites of the messenger RNA. As the coat protein cistron appears to be translated at a much higher frequency than the other cistrons, this whole region may also be considered as an active site of the viral messenger RNA. The problem as to how the temporal sequence of reading of a polycistronic message can be controlled is an intriguing one, as it may be related to the more general problem of the control of DNA-directed protein synthesis.

In the case of the RNA phages it appears from recent work that the phage RNA-directed synthesis of protein chains may be initiated by the attachment of N-formyl methionyl sRNA to the phage RNA-ribosomal complex at an AUG or UUG codon at the beginning of the cistron (42, 43, 44). Any particular cistron could be "switched off" during the infectious cycle by the chain initiating codon being inactivated by a change in one of its bases or by it becoming unavailable to a ribosome due to its participation in secondary structure. At the present moment, the presence of minor nucleotides in defined sequences in viral or other messenger RNA's has not been established although, by the use of a specific colour reaction and chromatography of digests of heavily labelled P-32 phage RNA, we have obtained evidence suggesting the existence of an average of a few molecules of dihydrouridylic acid per molecule in μ2 bacteriophage RNA (45). The possibility that the chain-initiating codon might be reversibly inactivated by a U$\rightleftharpoons$diHU transition catalyzed by a specific dehydrogenase cannot be ruled out, although it seems more likely that the secondary structure of the viral RNA is involved in the light of the following observations: Takanami (46) has found that ƒ-2 RNA appears to have only one binding site for ribosomes but on denaturation with formaldehyde a large number of ribosomes can attach to one RNA molecule. He also found that the bound ribosome protects 25-30 nucleotides of the messenger from digestion with ribonuclease (47). As double-stranded nucleic acids and polynucleotides cannot bind to ribosomes (48), an obvious explanation for the single binding site is that it is one end of the RNA molecule or an internal site consisting of a single-stranded region of 25-30 nucleotides containing a polypeptide chain initiation site.

On infection of an *E. coli* culture with P32-labelled MS2, the infecting RNA in the cell has a sedimentation constant of 20S when reisolated, whereas the RNA isolated directly from the phage has a sedimentation constant of 28S (49). The compact form of the viral RNA appears to reversibly unfold to a less compact form in the cell, the relative amounts of the two forms changing during the infectious process. One explanation for the control of protein synthesis in terms of the secondary structure of the RNA can be formulated as follows: The viral RNA can exist in two or more forms in which different chain initiation sites are exposed in single-stranded regions of 25 to 30 nucleotides, providing binding sites for ribosomes at the beginning of cistrons. In at least one of these forms the section of the RNA coding for the coat protein has more regions of single-stranded conformations and possibly shorter and less stable double-stranded regions than those corresponding to the other cistrons, enabling the ribosomes to translate this cistron at a higher frequency than the others. At the moment there is little relevant information about the detailed features of the secondary structure of viral RNA's such as the frequency distribution of lengths of the RNA in single-stranded and double-stranded conformations and their distribution along the viral RNA molecule. A method by which the single stranded regions could be excised and isolated, would enable this hypothesis to be eventually tested in some detail, and would be a step forward in the elucidation of the complete nucleotide sequence of a viral RNA.

ACTIVE SITES IN RIBOSOMAL RNA

The active sites of the ribosomal particles are
  (i) the messenger binding site which has been located on the small ribosomal subunit (50),
  (ii) the subunit recognition sites on each ribosomal subunit which enables them to recombine specifically to give the complete ribosome,
  (iii) the binding sites for the incoming amino acyl-sRNA and the sRNA attached to the end of the growing polypeptide chain, both located on the large subunit (51),
  (iv) the polymerizing site on the large subunit.

There is little evidence yet that regions of the 23S ribosomal RNA are involved in the binding sites for sRNA's on the large ribosomal subunit. In the case of the subunit combination sites and the messenger binding site on the 30S particle, however, there is evidence that the ribosomal RNA is involved (52). At low Mg^{++} concentrations, 2.5×10^{-4}M, the 70S ribosomal particles are dissociated into 30S and 50S subunits which reassociate into 70S particles on raising the Mg^{++} concentration to 10^{-2}M. Asano and Moore (53) found that if the separated 50S and 30S subunits are treated lightly with formaldehyde, which reacts with the free amino groups of adenine, guanine and cytosine at low Mg^{++} concentrations, the 70S particle does not reform on raising the Mg^{++} concentration to 10^{-2}M. They interpreted this as indicating that the 30S and 50S subunits combine to form specific hydrogen bonded base pairs between free nucleotide bases on the 23S and 16S RNA's of the ribosomal subunits. Asano and Moore (53) also found that ribosomes treated with formaldehyde lost their ability to bind radioactive poly U. This, together with the observation of other investigators that treatment of ribosomes with T1 RNAase also abolishes their ability to bind poly U as messenger, suggests that regions of the 16S RNA of the small ribosomal subunit are involved in the messenger binding site and that hydrogen bonding between bases in these regions and in messenger RNA is involved in the attachment of the messenger to the active site in the ribosome. In this case again, the problem is one of labelling and isolating specific regions in an RNA and the present evidence suggests that these regions at least are partly single-stranded.

REACTIONS OF N-CYCLOHEXYL, N'-β (4-METHYLMORPHOLINIUM) ETHYL CARBODIIMIDE SALTS WITH NUCLEOTIDES, POLYNUCLEOTIDES AND NUCLEIC ACIDS

Gilham (54) first observed that N-cyclohexyl, N'-β (4-methylmorpholinium) ethyl carbodiimide *p*-toluene sulfonate (CMEC *p*-toluene sulfonate) reacts specifically with uridine and guanosine in nucleotides, oligonucleotides and polyribonucleotides at pH 8.0, forming adducts with structures of the type shown in Figure 1, in the case of uridine. More recently, Naylor, Ho and Gilham (55) have found that CMEC reacts with pseudouridine to form adducts with either one or two CMEC groups attached to the base at the 3 and 1

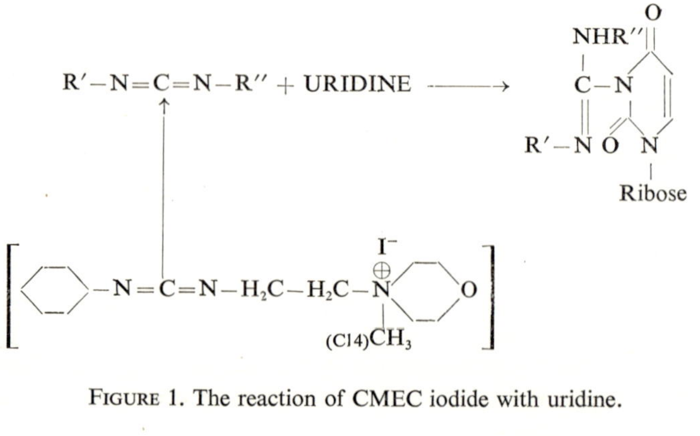

FIGURE 1. The reaction of CMEC iodide with uridine.

positions. Under mildly alkaline conditions, the CMEC groups were found to be removed by hydrolysis from guanosine, uridine and the 1 position of pseudouridine whereas the hydrolysis of the adduct in the 3 position required treatment with hot concentrated ammonia. CMEC does not appear to react appreciably with adenosine or cytidine but does form an adduct with thymidine. We have been studying the reactions of CMEC with nucleic acids and polyribonucleotides in order to develop methods for the labelling and study of the structure of active sites in ribonucleic acids and for the determination of nucleotide sequences in viral RNA's.

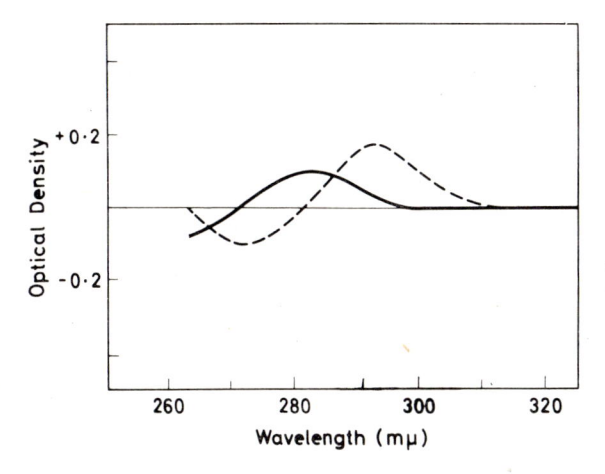

FIGURE 2. The difference spectra of (a) CMEC-uridine-5'-monophosphate versus uridine-5'-monophosphate ——————— at a nucleotide concentration of 7×10^{-5} M; (b) CMEC-guanosine-5'-monophosphate versus guanosine-5'-monophosphate — — — — — — at a nucleotide concentration of 7×10^{-5} M dissolved in 0.1 M sodium borate buffer, pH 8.0.

The reactions of CMEC with the bases in polynucleotides and nucleic acids can be followed by two methods, ultraviolet absorption spectrophotometry and the use of radioactively-labelled CMEC. The formation of CMEC-adducts produces shifts in the absorption spectra of the nucleosides and nucleotides, the difference spectra of the adduct-nucleotide pairs for guanosine-5'-phosphate and uridine-5-phosphate being shown in Figure 2. If pseudouridine is absent, the reactions of CMEC with uridine and guanosine in polynucleotides and nucleic acids can be followed separately by monitoring changes in the optical densities at 280 mμ and 300 mμ, in the absence of changes in optical density due to alterations in secondary structure. CMEC can be readily prepared with the 4-methyl group labelled with carbon-14 or tritium, and the amount of adduct formed with a CMEC-treated nucleic acid or polynucleotide estimated by the determinations of the radioactivity associated with the polyribonucleotide after removal of the unreacted CMEC salt by repeated precipitation from alcohol–salt solutions, by gel filtration on Sephadex or Biogel beads with porosities suitable for the exclusion of the nucleic acid or polynucleotide or by fractionation on DEAE–cellulose.

Using CMEC iodide radioactively labelled in the 4-methyl group, Tocco and Brown (56) obtained evidence that CMEC reacted rapidly with uridylic, guanylic and thymidylic residues in single-stranded polynucleotides but that the reagent appeared to react very slowly with these residues when they

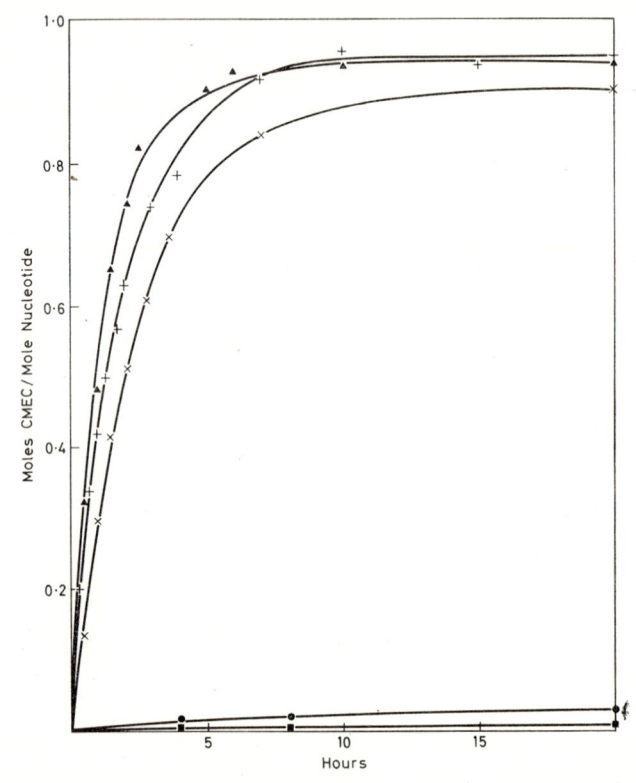

FIGURE 3. Reactions of nucleotides and synthetic polynucleotides with CMEC at pH 8.0 and 30°C as a function of time: (a) poly U, 0.05 mg/ml, CMEC, 19 mg/ml in 0.1 M sodium borate —▲—▲—▲—; (b) poly U (2mg/ml)+poly A (equimolar), CMEC, 8 mg/ml in 0.01 M sodium borate —■—■—■—; (c) poly G, 1 mg/ml, CMEC, 8 mg/ml in 0.01 M sodium borate —●—●—●—; (d) pU, 1,5 mg/ml, CMEC, 19 mg/ml in 0.1 M sodium borate —+—+—+— (e) pG, 0.05 mg/ml, CMEC, 19 mg/ml in 0.1 M sodium borate —×—×—×—.

were in hydrogen bonded multichain helices or in double-stranded helical regions of ribonucleic acids (57). The reactions of nucleotides and some polynucleotides with CMEC iodide are illustrated in Figure 3. From this diagram it can be seen that the rate of reaction with the uracil residues increases slightly from nucleotide to polynucleotide, but that the addition of poly A, which complexes with poly U, almost completely supresses the reaction of the uracil residues with the CMEC. In the case of guanine residues, in the free nucleotide they react slower than the uracil residues but in poly G, which is believed to have a multistranded helical structure, they react at an extremely slow rate. Figure 4 shows the reactions of CMEC-iodide with calf thymus DNA, de-

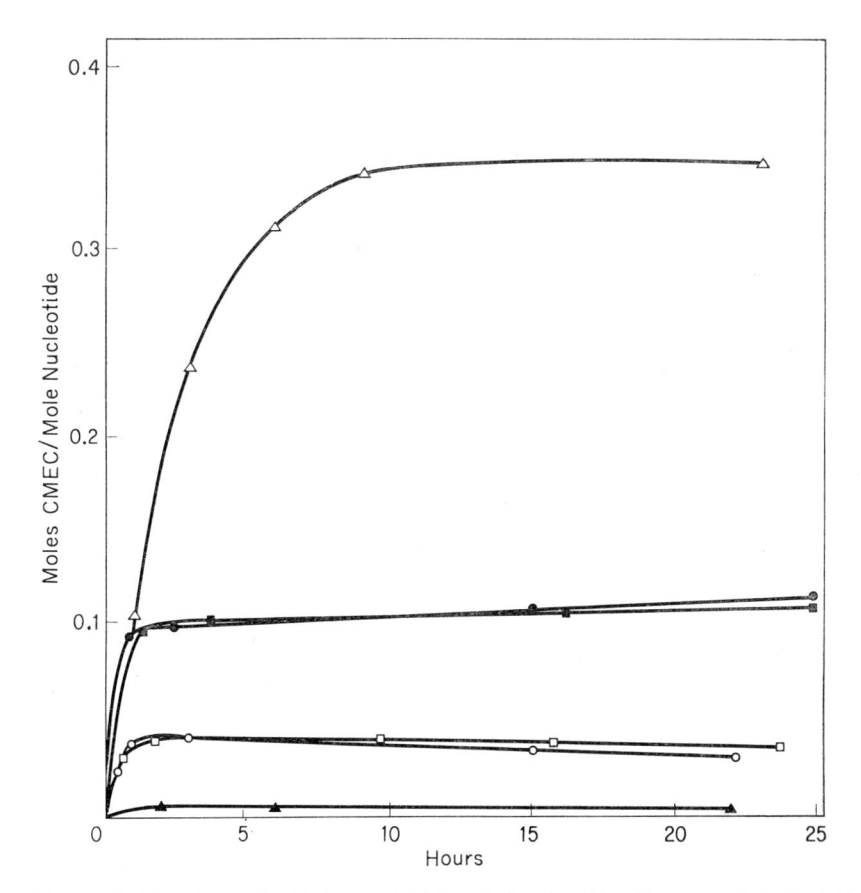

FIGURE 4. Reactions of calf thymus DNA and the sRNA's of yeast and *Escherichia coli* with CMEC (8 mg/ml) in 0.01 M sodium borate buffer, pH 8.0, as a functions of time: (a) calf thymus DNA 2 mg/ml ▲—▲—▲—▲—; (b) heat denatured calf thymus DNA (2 mg/ml) —△—△—△—△—; (c) yeast sRNA (4 mg/ml) ●—●—●—●—●—; (d) yeast sRNA (4 mg/ml) +0.01 M MgCl₂ ○—○—○—○—; (e) *Escherichia coli* sRNA (4 mgm/ml) —■—■—■—■—; (f) *Escherichia coli* sRNA (4 mgm/ml)+0.01 M MgCl₂ □—□—□—□

natured DNA, yeast and *Escherichia coli* sRNA's in presence and absence of magnesium at 30°C.

Most of the thymidine and deoxyguanosine residues of denatured DNA, which are known to be largely in single-stranded non-hydrogen bonded regions (59), react rapidly with CMEC, whereas in the native DNA they appear to be protected from the action of the reagent. From the extent of reaction with the polynucleotides and native and denatured DNA it is clear that uridine, guanosine and thymidine in hydrogen bonded multistranded helical conformations react extremely slowly with CMEC so that the rate of reaction of either of these residues in polynucleotides can be used to indicate whether or not it is in a hydrogen-bonded multistranded helical region or in a single-stranded region with unbonded bases. The question as to whether stacking of bases, as in neutral poly-A (9), possibly occurring in the case of a run of guanosine residues in a single-stranded region of an RNA, would affect the reaction of the residues with CMEC needs further study on model oligonucleotides. The same would apply to a uridine or guanosine residue in an imperfect helix. It seems clear then that the CMEC reagent discriminates strongly between residues in multistranded hydrogen bonded helical regions and single-stranded regions.

This interpretation is supported by the reactions of CMEC with sRNA, especially by the effect of Mg^{++} on the extent of the reaction, as shown in Figure 4. In the absence of Mg^{++} the sRNA's of both yeast and *Escherichia coli* react rapidly with CMEC iodide at 30°C, binding 8–10 moles of CMEC per 100 moles of nucleotide in 2–3 hours; this is followed by a slower reaction in which a further 2 moles is bound during the next 20 hours. In the presence of 0.01 M Mg^{++}, however, only about 3 moles of CMEC are bound per 100 moles of nucleotide in the initial fast reaction which, in this case, is not followed by a slow increase. Studies of the hyperchromic effects (58) produced in the ultraviolet spectra of solutions of sRNA by heating the solutions to temperatures from 0 to 100°C have been interpreted in terms of the "melting-out" of base-paired double-stranded helices in the secondary structure of sRNA, although recent work suggests that the unstacking with increase in temperature of base-stacked single-stranded regions could contribute to the effects observed. In the absence of Mg^{++}, raising the temperature of a solution of sRNA from 0 to 30°C produces a continuous increase in optical density at 260 mμ, which can be interpreted as a melting of hydrogen bonded helices and a corresponding increase of the fraction of nucleotides in single-stranded regions in the sRNA molecules. In the presence of 0.01 M Mg^{++} no such change in absorption spectra takes place until the temperature is raised to above 40°C, and the presence of Mg^{++} appears to stabilize the structure of sRNA against the effect of temperatures. The effects of Mg and temperature on the reaction of CMEC with sRNA are consistent with this interpretation, if it is assumed that CMEC reacts rapidly with the nucleotides in the single-stranded regions and that the slow reaction observed in the absence of Mg^{++} is due to a partial progressive denaturation of the remaining secondary structure.

Figure 5 shows the effect of concentration of CMEC on the amount of CMEC bound to sRNA in the presence and absence of Mg^{++}. In the presence of Mg^{++} a plateau region is observed when the reactive bases in single-stranded regions are presumably saturated with CMEC groups. The lack of a plateau region in the absence of Mg^{++}, and the increase in the amount of CMEC bound at high concentrations of CMEC iodide in the presence of

Mg^{++} are probably due to a progressive denaturation of the secondary structure of sRNA by CMEC.

The amount of CMEC bound at the plateau region obtained in the presence of 0.01 M Mg^{++} is about 17 moles/100 moles nucleotide, indicating that, on an average, about 14% of the nucleotides in the sRNA molecule are in single-stranded regions.

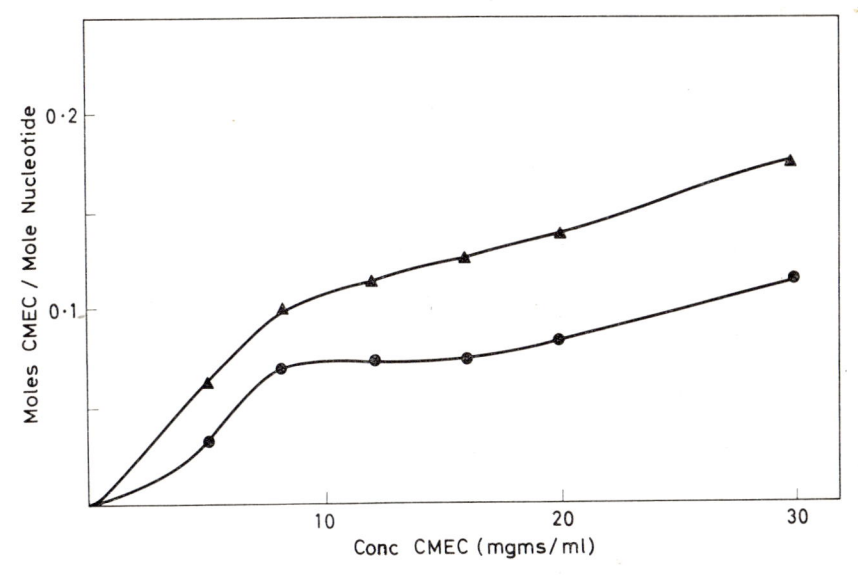

FIGURE 5. Moles of CMEC bound per mole of nucleotide to the sRNA of *Escherichia coli* (1 mg/ml) as a function of the concentration of CMEC iodide with a reaction time of 15 hours at 30°C in 0.1 M sodium borate buffer, pH 8.0: (a) in the absence of Mg^{++} ▲—▲—▲—▲—; (b) in the presence of 0.01 M Mg^{++}—●—●—●—●—.

The characteristic features of the reaction of CMEC with the viral messenger RNA from the RNA bacteriophage $\mu 2$ are shown in Figures 6 and 7. Figure 6 shows the time course of formation of CMEC adducts with the isolated RNA in solution and with the RNA in a virus suspension at the same RNA concentration. In both cases there is an initial rapid reaction followed by a very slow reaction. In the case of the complete virus particle, however, the extent of the reaction is very limited compared with the free RNA, indicating that confinement within the protein capsid protects a large fraction of the uridine and guanosine groups available in the free RNA from reaction with CMEC. This could be due to the RNA chain folding up within the protein capsid in such a way that nearly all the bases are hydrogen bonded or that most of those not base-paired are protected by interaction with the protein subunits of the capsid. Figure 7 shows the effect of increasing the concentration of CMEC on the amount of CMEC bound to $\mu 2$ RNA in the presence and absence of Mg^{++}. The phage RNA used was

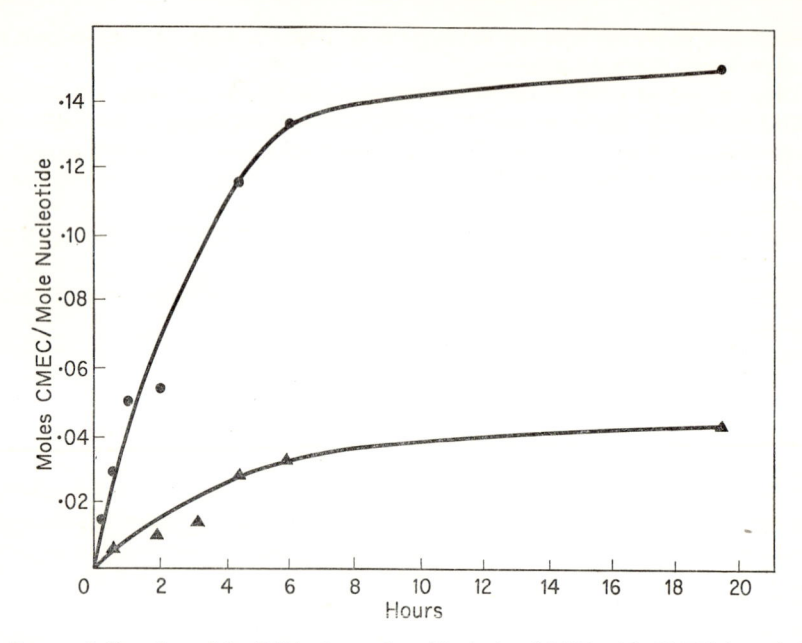

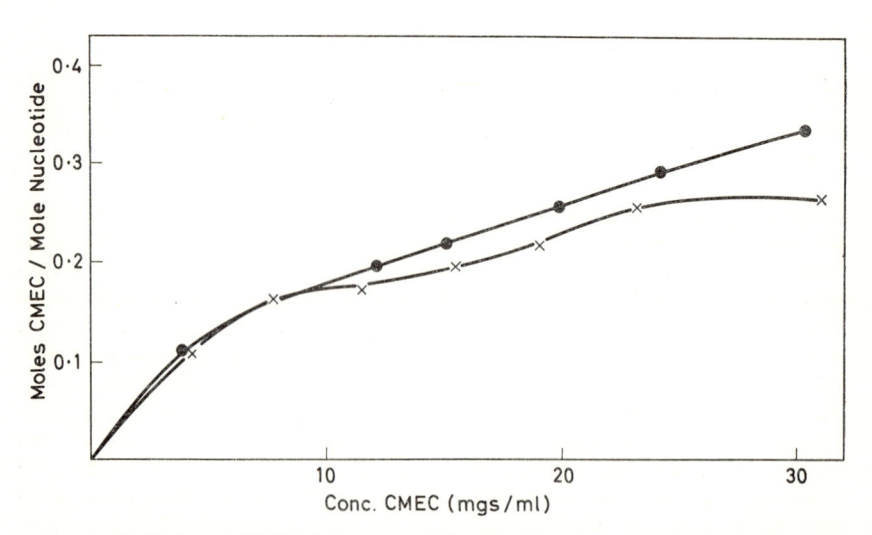

FIGURE 6. Reaction of the RNA phage $\mu 2$ and its isolated RNA with CMEC (8 mg/ml): (a) $\mu 2$ (1 mg RNA/ml) —▲—▲—▲—▲—; (b) $\mu 2$ RNA (1 mg/ml) —●—●—●—●—

FIGURE 7. Moles of CMEC bound per mole of nucleotide to the RNA of the phage $\mu 2$ (approx. 0.5 mg/ml) as a function of the concentration of CMEC iodide with a reaction time of 20 hours in 0.1 M sodium borate buffer, pH 8.0: (a) in the absence of Mg^{++} —●—●—●—; (b) in the presence of 0.01 M Mg^{++} —×—×—×—.

labelled with ^{32}P and passed through a column of Biogel P-300 to remove minor RNA components found in some phage RNA preparations. The shape of the saturation curve in the presence of Mg^{++} suggests that the RNA molecule contains two types of regions with nucleotides that react with CMEC, the first possibly single-stranded regions that are saturated with CMEC at 10-15 mg CMEC/ml and a second type which react only after denaturation with CMEC at concentrations of 20-30 mg CMEC/ml.

To confirm that the plateaus in the curves relating CMEC bound to RNA and CMEC concentrations do correspond to saturation with CMEC of the available nucleotide in the RNA's, we have carried out model experiments on the reactions of CMEC with nucleoside-5'-phosphates, at concentrations equivalent to those of the apparently reactive residues in the experiments with RNA, using difference spectrophotometry to follow the reactions. At CMEC concentrations of 10 mg/ml, over 90% of the guanylic residues and 95% of the uridylic residues were reacted at equilibrium. This contrasts with the reactions of formaldehyde where only 50% of the apparently available nucleotide residues react when the formaldehyde concentration is low enough to avoid denaturation of the sRNA (59). The evidence obtained so far strongly indicates that CMEC reacts predominantly with single-stranded regions in RNA and that under conditions where plateaus are obtained in the curves relating CMEC bound to RNA to the concentration of CMEC used, more than 90% of the uridylic and guanylic residues in the single-stranded regions form adducts with CMEC.

RESTISTANCE OF CMEC ADDUCTS TO NUCLEASES

Gilham (54) first demonstrated that the uridine-CMEC adduct in dinucleotides and ribonucleic acids is resistant to the action of pancreatic ribonuclease. Later Naylor *et al.* (55) extended this observation to include pseudouridine, the CMEC adducts of which are also resistant to pancreatic ribonuclease. They also examined the effect of the uridine and pseudouridine adducts on the digestion by spleen and snake venom phosphodiesterases of the dinucleoside phosphates UpA, ψpA, CpU, Cpψ and UpC. Snake venom phosphodiesterase, which is an exonuclease degrading polynucleotides by stepwise removal of nucleotides from the 3'-hydroxyl end, was found to be blocked by adducts formed with uridine or pseudouridine on 3'-hydroxyl ends and the rate of digestion was decreased considerably by adducts formed at the 5'-hydroxyl end. With spleen phosphodiesterase, which removes nucleotides in a stepwise fashion from the 5'-hydroxyl end, the CMEC group on the 5'-hydroxyl of uridine or pseudouridine does not block completely but does decrease the rate of digestion, whereas the adduct on the 3'-hydroxyl end blocks completely the removal of the nucleotide on the 5'-hydroxyl end.

We have studied the action of ribonuclease T1, spleen phosphodiesterase and snake venom phosphodiesterase on some other dinucleoside phosphates under conditions designed to minimize losses of CMEC from the adducts by hydrolysis, that is, short incubation time, high enzyme concentration and as low a pH as possible. In particular, the digests with T1 ribonuclease and snake venom diesterase were carried out at pH's below their optima as the CMEC adducts are hydrolyzed slowly at pH 8.0 and more rapidly above this pH but increase in stability as the pH is decreased. The results with three dinucleoside phosphates and the three enzymes are shown in Table I. It is clear that the addition of the CMEC group to guanosine makes the neighbouring 3′–5′ phosphodiester linkage substantially resistant to the action of ribonuclease T1 and, even when on the 5′-hydroxyl end, partially resistant to the action of spleen phosphodiesterase. ApŪ is resistant to the action of snake venom diesterase and largely resistant to the action of spleen phosphodiesterase.

Thus the addition of CMEC to the uridylic and guanylic residues of the single-stranded regions of RNA's will make the neighbouring phosphodiester linkages resistant both to the endonucleases, pancreatic ribonuclease and T1 ribonuclease, and to the exonucleases, spleen phosphodiesterase and snake venom phosphodiesterase. As the neighbouring double-stranded regions are susceptible to digestion by the endonucleases, theoretically the single-stranded regions can be isolated according to the schemes shown in Figure 8. An alternative scheme, using pancreatic ribonuclease instead of T1 ribonuclease, will give a different set of fragments in which the integrity of the single-stranded fragment is lost by scission at the cytidylic residues. The scheme illustrated can be used either to excise single-stranded loops as shown or to excise single-stranded regions linking neighbouring double stranded regions. In the case of sRNA, the simple scheme is likely to be complicated by the presence of minor nucleotides which might be resistant to the nucleases. The single-stranded regions can then be identified by using radioactively labelled-CMEC.

THE SINGLE-STRANDED REGIONS OF THE RNA OF $\mu 2$ PHAGES

One of the problems of investigating the role of secondary structure in the active sites of a viral messenger is that of studying the distribution of nucleotides in the single-stranded regions, as discussed in the introductory section defining active sites. A modification of the scheme described in the previous section has been applied by us to the problem, using disc electrophoresis in acrylamide gels (60) to separate the enzyme-resistant oligonucleotides.

^{32}P-labelled $\mu 2$ phage was grown in a casamino acids-glucose medium containing 10 mc ^{32}P-labelled phosphate per litre, on *Escherichia coli* K12, 58–161 F$^+\sigma^+$ as host, and purified by ammonium sulphate precipitation and by three CsCl density gradient centrifugation runs (61). ^{32}P-labelled phage RNA was prepared by 3 extractions with 90% phenol in the presence of bentonite and the phenol removed by six extractions with ether which was removed by air bubbling, and the RNA precipitated with alcohol. The specific radioactivity of the RNA was 840,000 cts per min/mg using counting samples, dried on filter paper discs, in a Packard Scintillation counter with a counting efficiency for ^{32}P of 84%. Approximately 0.5 mgm of RNA was dissolved in 1 ml of 0.1 M sodium borate buffer pH 8.0+0.01 M MgCl$_2$ containing 12 mgm of CMEC iodide and incubated at 30° for 3 hours. The

TABLE I. Percentage Hydrolysis of Dinucleoside Phosphates and their CMEC Adducts

	T1 RNAase	Spleen phosphodiesterase	Snake Venom phosphodiesterase
GpA	100	94	100
$\overline{G}$pA	9	24	92
ApU	—	100	100
Ap$\overline{U}$	—	18	12
GpC	100	98	97
$\overline{G}$pC	13	30	85

0.5 mg of each dinucleoside phosphate was incubated with T1 RNAase (0.1 mg) at 30°C for 2.5 hours in 1 ml of 0.05 M Tris acetate buffer pH 7.4, with spleen phosphodiesterase (0.5 unit) at 30°C for 2.5 hours in 1 ml of 0.05 M ammonium acetate buffer pH 6.0 and with snake venom phosphodiesterase (0.2 mg) at 30°C for 2.5 hours in 1 ml of 0.05 M Tris buffer pH 7.4. The products of digestion were analyzed by paper electrophoresis in 0.05 M ammonium acetate buffer pH 6.0. ($\overline{X}$ signifies a CMEC blocked nucleoside).

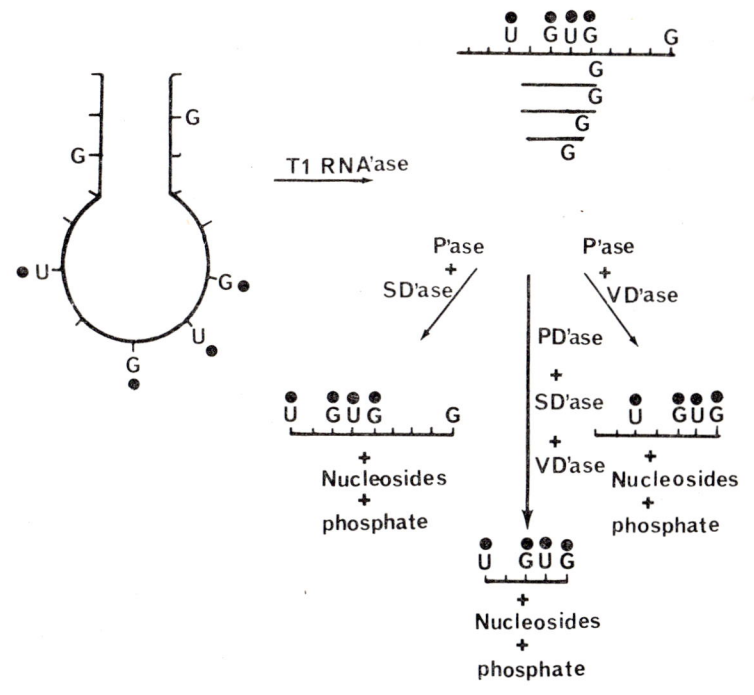

FIGURE 8. Schemes for the isolation of single-stranded segments of RNA's by CMEC and mixtures of exo- and endonucleases: ● — CMEC group; SD'ase — spleen phosphodiesterase; VD'ase — snake venom phosphodiesterase; P'ase — bacterial phosphatase.

sample was then passed through a column of Biogel P-300, washed with 0.01 M sodium phosphate buffer pH 7.0, and the sharp peak of radioactivity emerging first was freeze dried. The sample was then dissolved in 0.5 ml water and digested at 37° for 2 hours with 0.5 mg of ribonuclease T1 and 0.5 mg of snake venom diesterase after adjusting the pH to 7.0. A second example was treated in exactly the same way with the CMEC omitted.

Samples containing about 100,000 counts/min of ^{32}P were run on 10 cm long tubes of 15% acrylamide gels, using the creatinine-acetate pH 5.3 buffer

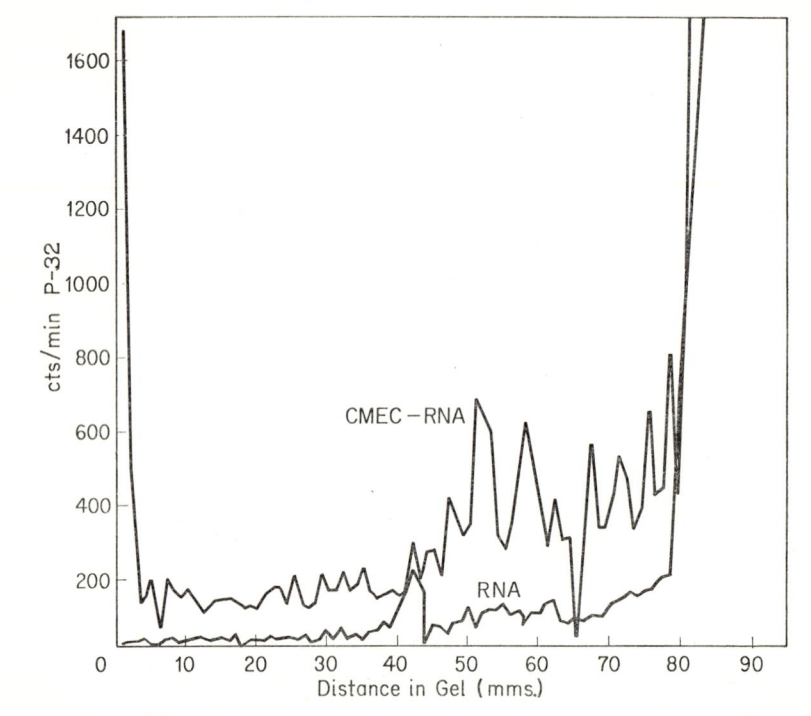

FIGURE 9. Disc electrophoresis in 15% acrylamide gels of ^{32}P-labelled CMEC-oligonucleotides from ^{32}P-labelled μ2 phage RNA, CMEC-treated (CMEC–RNA) and untreated (RNA), resistant to digestion by T1 ribonuclease and snake venom diesterase (see text for details). The radioactivity in the last six slices, which contain the mononucleotides, has been omitted.

system of Richards, Coll and Gratzer (60), for 3 hr at 150 volts at room temperature. At the end of the runs the gels were sliced into 1 mm slices which were dried on filter paper discs and counted in a Packard Scintillation counter. The distribution of ^{32}P activity in the gels is shown in Figure 9.

All the RNA untreated with CMEC should have been digested to mononucleotides which run in the large peak of activity beyond the 80 mm slice. In the untreated RNA control only about 2% of the total radioactivity is distributed between the origin and this mononucleotide peak, most of which is concentrated in a peak at 40 mm and must be due to an oligonucleotide sequence in the μ2 RNA molecule which is resistant to digestion by T1 ribonuclease and snake venom diesterase. In the case of the CMEC-treated RNA about 15% of the ^{32}P is distributed in a series of peaks between the mononucleotides peak and the origin. In particular 2% of the total counts have remained near the origin, a feature which is absent in the electrophoregram of the untreated RNA. Richards, Coll and Gratzer (60) have shown that the electrophoretic mobilities of RNA components in acrylamide gels depend on their molecular weights and are correlated with their sedimentation components. In the present case, the analysis is complicated by the amount of bound CMEC which affects both the size and the charge of the components and varies with the contents of uridylic and guanylic acids. As CMEC–uridylic and CMEC–guanylic acids travel with the wide mononucleotide band in this system, the size of the resistant oligonucleotides must be the predominant factor controlling their mobilities in this method. The 2% of the total radioactivity near the origin may then be derived from one or more long single-stranded regions in the μ2 RNA, comprising a total of about 60 nucleotides out of the 3000 nucleotides of the RNA molecule.

This result indicates that the finer details of the secondary structure of viral messenger RNA's may be investigated by these procedures and that some of the active sites of the messenger RNA's discussed previously may eventually be located in oligonucleotides separated by further refinements of the techniques described.

ACTIVE SITES OF SRNA'S.

The reactions of CMEC with sRNA have already been described and the effects of Mg^{++} illustrated in Figure 4 and 5. It is clear that only limited regions of the molecule are accessible to the reagent and that these are probably single-stranded. If these regions are active sites for the acceptor activity and the coding properties, then their reactions with CMEC might be expected to interfere with these activities. Figures 10 and 11 show the effect on some acceptor activities of *E. coli* sRNA of reaction with CMEC under the same conditions as for reactions illustrated in Figure 4. In the presence of 0.01 M Mg^{++} there is little decrease of acceptor activity for serine, arginine, lysine or tyrosine with up to 3 moles CMEC bound per molecule of sRNA. In the absence of Mg^{++} the acceptor activities decrease rapidly with the time of reaction, at the end of which 7 to 8 molecules of CMEC are bound per molecule of sRNA. Either the limited region of the molecule, which is exposed to the reagent on removal of Mg^{++}, is part of the acceptor recognition site, or possibly, the part of the secondary or tertiary structure in which this region is involved is essential for the active configuration of the site.

The effect of CMEC on the lysine-specific sRNA coding site has been studied by reacting sRNA from *E. coli* with CMEC in the presence of 0.01 M Mg^{++} at 30°C and pH 8.0, charging it with ^{14}C-lysine and then determining the amount of lysine transferred into protein in a ribosomal system, using poly A as a messenger (62). The attachment of up to 4 CMEC molecules

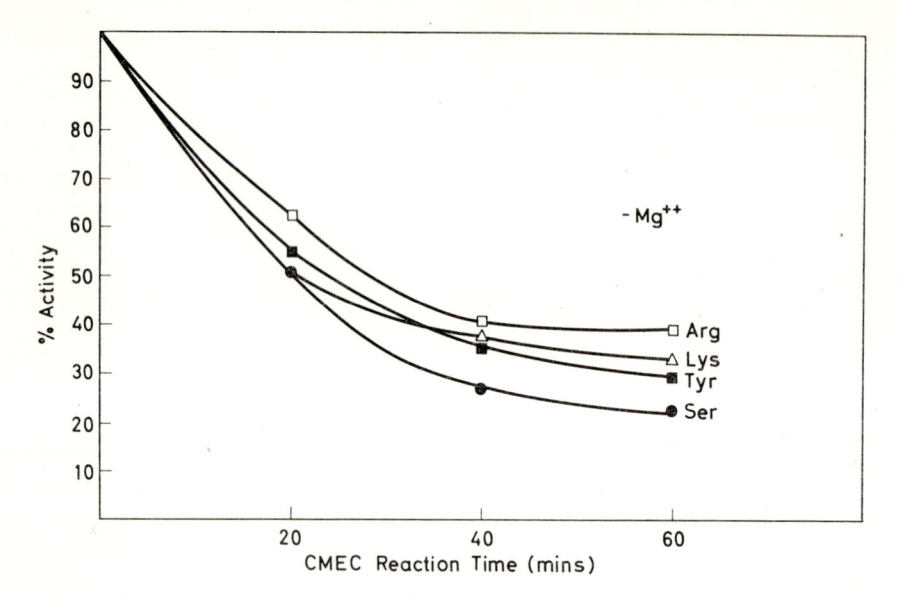

FIGURE 10. Effects of the reaction of CMEC with the sRNA of *Escherichia coli*, in the absence of Mg^{++}, on some amino acid acceptor activities. The conditions of the reaction were identical to those for Figure 4e. Arginine □—□—□—□, lysine △—△—△—△, tyrosine ■—■—■—■—, serine ●—●—●—.

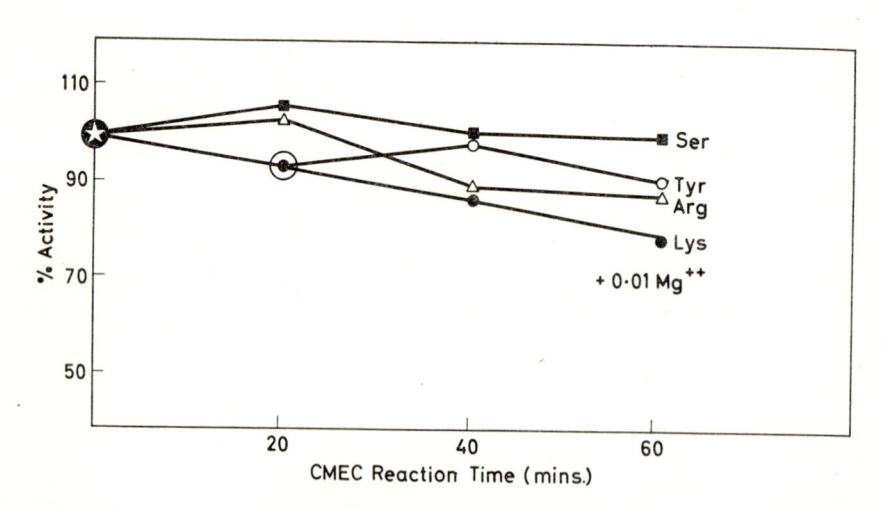

FIGURE 11. Effects of the reaction of CMEC with the sRNA of *Escherichia coli*, in the presence of Mg^{++}, on some amino acid acceptor activities. The conditions of the reaction were identical to those for Figure 4f. Arginine △—△—△—△, lysine ●—●—●—●—, tyrosine ○—○—○—○—, serine ■—■—■—■—.

per mole of the sRNA used in the test produced little inactivation in this test and even a degree of stimulation in some experiments. However, as CMEC was found to be slowly released from the sRNA during the incubation this negative result does not mean that the coding site is inaccessible to the reagent. The inactivation of the coding site of lysine-specific sRNA by CMEC has also been investigated by measuring the attachment to ribosomes (63) of ^{14}C-lysyl–sRNA samples, labelled with CMEC in the presence of 0.01 M Mg^{++} previous to charging the sRNA with lysine, using poly A as the messenger polynucleotide and untreated ^{14}C-lysyl–sRNA as a control. With equivalent amounts of ^{14}C-lysyl–sRNA added in each case, 10 $\mu\mu$moles, the amounts of CMEC-treated ^{14}C-lysyl–sRNA, with 3 CMEC groups per sRNA molecule, bound to ribosomes in the presence of poly A, at a concentration of 25 mμ moles of nucleotide per 50 μl incubation mixture, was 0.8 $\mu\mu$ moles compared with 2.6 $\mu\mu$ moles of untreated ^{14}C-lysyl–sRNA bound in a control experiment. (The concentrations of ribosomes and enzymes and the other experimental conditions were identical to those used by Bernfield and Nirenberg (64) except that the enzymes and ribosomes were prepared from a ribonuclease-free strain of *E. coli*, MRE 600.).

It appears from this last result that the attachment of CMEC to the region of the lysine-specific sRNA available to the reagent in the presence of 0.01 M Mg^{++} partly inactivates the attachment of the sRNA to the messenger-ribosome coding site. According to current theories, the anticodon in the lysine-specific sRNA should be a pUpUpU sequence in a single-stranded region of the sRNA molecule. As the inactivation of the coding attachment is not complete, it is possible that the anticodon site is partly protected from reacting with CMEC in the presence of Mg$^+$. Unfortunately, the instability of lysyl–sRNA at the pH's at which sRNA can react with CMEC, make it technically difficult to study the effect on its coding activity of reacting sRNA with CMEC in the absence of Mg^{++}.

CMEC AND ACTIVE SITES OF RIBOSOMES

The most interesting active site on the ribosomal particle is the messenger binding site situated on the 30S particle which must be directly involved in the mechanism by which the messenger is moved stepwise, triplet by triplet, relative to the 30S and 50S subunits. As discussed in an earlier section, there is evidence from the inactivation of this site by formaldehyde treatment and by treatment with T1 ribonuclease, that regions of the 16S RNA component must be involved. Therefore we have examined the effects of treatment with CMEC on the messenger-binding site in *E. coli* ribosomes, using the binding of ^{32}P-labelled μ2 RNA to ribosomes, measured by sucrose gradient centrifugation, as a test for activity.

Figure 12a and b shows that treatment of the ribosomes with CMEC inactivates the binding site as the amount of ^{32}P phage RNA binding to the ribosome is considerably reduced by a brief treatment with CMEC. If however this treatment is carried out in the presence of poly A or an oligo-A mixture produced by alkaline digestion of poly A, the binding site appears to be protected from the reaction with CMEC as the amount of $\mu 2$ RNA bound to the ribosomes is partly restored to its original value as shown in Figure 12c.

This result provides further evidence suggesting that single-stranded regions of ribosomal RNA are involved in the messenger binding site and also suggests a method for locating fragments from these regions in digests of

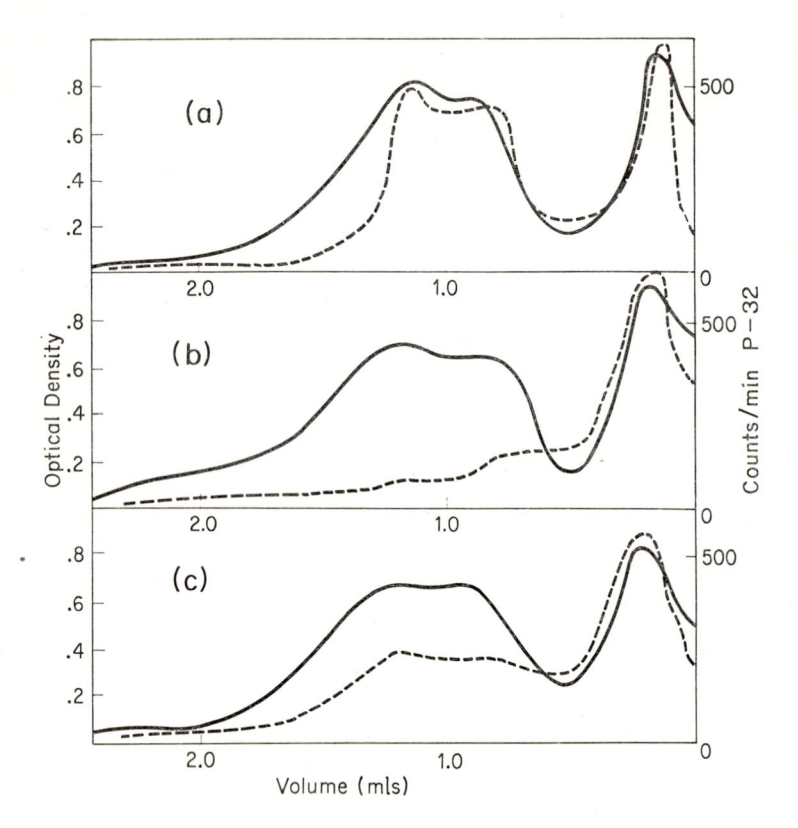

FIGURE 12. The inactivation of the messenger binding sites of the ribosomes of *E. coli* by treatment with CMEC and the partial protection given by the presence of oligo-A. Ribosomes were prepared from *E. coli* MRE 600 and, after freeing from endogenous messenger by preincubation (62) and washing, were treated at a concentration of 1 mg/ml in 0.01 M sodium borate buffer, pH 8.0, +0.01 M $MgCl_2$ with 2 mg/ml of CMEC iodide for 30 min at 25°C, and then repurified by centrifugation in 0.01 M magnesium acetate. Samples of treated and untreated ribosomes were incubated with a molar equivalent of ^{32}P-labelled $\mu 2$ phage RNA and run on a sucrose density gradient (5–20%) in a SW 39 head for 1.5 hours at 35,000 rpm. The tubes were analyzed with an Isco density gradient fractionator. Optical density ———, ^{32}P cts/min -------.

All incubations and runs were made in 0.05 M tris-acetate = 0.05 M magnesium acetate: (a) untreated ribosomes; (b) CMEC-treated ribosomes; (c) ribosomes treated with CMEC in the presence of 1 mg/ml of poly A that had been digested with 0.1 N KOH for 20 min at 20°C and neutralized with HCl.

the ribosomal RNA. If non-radioactive CMEC is used to saturate the exposed single-stranded regions of the RNA in the ribosomes, with oligo-A, say, blocking the messenger binding site, and radioactively-labelled CMEC used to label the messenger binding-site after removal of the oligo-A, the radioactivity-labelled CMEC oligonucleotides obtained in an enzymic digest can be identified as containing oligonucleotide fragments from the messenger binding-site.

SUMMARY

The reactions of CMEC salts with nucleotides, oligonucleotides, polynucleotides and nucleic acids and the action of nucleases on the products of the reactions described here demonstrate that the reagent can be used to label nucleotides in the single-stranded regions of RNA. By conferring resistance to digestion by nucleases, these regions can be excised almost intact for studies of chain lengths and sequences, thus linking the study of secondary structure to that of primary structure. In the case of small RNA's like sRNA, to which present sequence methods are applicable, the additional information obtainable by the use of the reagent will enable the secondary and tertiary structures to be deduced from the known primary structures. In the case of viral messenger RNA's, the single-stranded regions labelled with CMEC can be used as reference regions for the analysis and ordering of large oligonucleotide fragments obtained by partial enzymic digestion.

The active site of RNA's, which are either wholly or partially single-stranded, can be studied by the use of CMEC in procedures designed for a particular site as illustrated by the examples described here. In principle, methods similar to that used for the study of the messenger binding site could be used for locating regions of the 23S ribosomal RNA involved in the two sRNA binding sites or in the site or sites to which the 30S ribosomal subunit attaches.

ACKNOWLEDGEMENTS

We are indebted to Professor Sir John Randall and Professor M. H. F. Wilkins for their interest and encouragement and to Mr. H. Isenberg, H. Matthews, J. Kosinsky and Miss J. Deniset for supplying materials and help in carrying out these experiments. We also wish to thank the Director of the Microbiological Research Establishment, Porton, England and his collegues for the preparation of *E. coli* sRNA and growing μ2 bacteriophage.

REFERENCES

1. Rosset, R., Monier, R. and Julien, J., *Bull. Soc. Chim. Biol.* **46**, 87 (1964).
2. Holley, R. W., Apgar, J., Everett, G. A., Madison, J. T., Marquisee, N., Merrill, S. H., Penswick, J. R. and Zamir, A., *Science* **147**, 1462 (1965).
3. Zachau, H. G., Dutting, D. and Feldmann H. These Proceedings, p. 271.
4. Brownlee, G. G. and Sanger, F. Private Communication.
5. Lee, S., McMullen, D. W., Brown, G. L. and Stokes, A. R., *Biochem. J.* **94**, 314 (1965).

6. Brown, G. L., Lee, S., McMullen, D. W. and Stokes, A. R., *Biochem. J.* **92**, 4P (1964).
7. Bradley, D. F. Merrill, C. R. and Shapiro, M. B., *Biopolymers* **2**, 415 (1964).
8. Michelson, A. M., Ulbricht, J. L. V., Emerson, T. R. and Swan, R. J., *Nature* **209**, 873 (1966).
9. Brahms, J., Michelson, A. M. and van Holde, K. E., *J. Mol. Biol.* **15**, 467 (1966).
10. Langridge, R. and Gomatos, P. J., *Science* **141**, 694 (1963).
11. Donahue, J., *Proc. Nat. Acad. Sciences, U.S.* **42**, 60 (1956).
12. Fresco, J. R. and Alberts, B. M., *Proc. Nat. Acad. Sciences, U.S.* **46**, 311 (1960).
13. Marciello, R. and Zubay, G., *Biochem. Biophys. Res. Comm.* **14**, 272 (1964).
14. Kiselev, L. L., Frolova, L. Y., Boriosova, O. F. and Kukhanova, M. K., *Biokhimiya* **29**, 116 (1964).
15. Crick, F. H. C., *Biochem. Soc. Symp.* (Cambridge, Engl.) 14, 25 (1956).
16. Nirenberg, M., Leder, P., Bernfield, M., Brimacombe, R., Trupin, J., Rottman, F. and O'Neal, C., *Proc. Natl. Acad. Sciences, U.S.* **53**, 1161 (1965).
17. Söll, D., Ohtsuka, E., Jones, D. S., Lehrmann, P., Hayatsu, H., Nishimura, S. and Khorana, H. G., *Proc. Natl. Acad. Sciences, U.S.* **54**, 1378 (1965).
18. Kukhanova, M. K., Kiselev, L. L. and Frolova, L. Y., *Biokhimiya* **28**, 1053 (1963).
19. Penniston, J. T. and Doty, P., *Biopolymers* **1**, 209 (1963).
20. Takanami, M. and Miura, K., *Biochim. Biophys. Acta* **72**, 237 (1963).
21. Cerna, J., Rychlik, I. and Sorm, P., *Coll. Czechoslov. Chem. Comm.* **29**, 2832 (1964).
22. Yu, C. T. and Zamecnik, P. C., *Biochim. Biophys. Acta* **76**, 209 (1963).
23. Weil, J. H., Befort, N., Rether, B. and Ebel, J. P., *Biochem. Biophys. Res. Commun.* **15**, 447 (1960).
24. Zillig, W., Schactschnabel, D. and Krone, W., *Hoppe-Seyl. Z.* **318**, 100 (1960).
25. Carbon, J. A., *Biochem. Biophys. Res. Commun.* **15**, 1 (1964).
26. Ebel, J. P. and Lazzari, M., Fed. Europ. Bioch. Soc. 2nd. Meeting. Vienna (1966).
27. Fawaz-Estrup, F. and Setlow, R. B., *Biochim. Biophys. Acta* **87**, 28 (1964).
28. Swenson, P. A. and Nishimura, S., *Photochem. Photobiol.* 3, **85** (1964).
29. Zachau, H. G., *Hoppe-Seyl. Z.* **336**, 176 (1964).
30. Ebel, J. P., Weil, J. H., Rether, B. and Heinrich, J., *Bull. Soc. Chim. Biol.* **47**, 1599 (1965).
31. Fel'dman, M. Ya., *Biokhimiya* **30**, 203 (1965).
32. Burton, K. and Riley, W. J., *Biochem. J.* **98**, 70 (1966).
33. Burton, K., Varney, N. F. and Zamecnik, P. C., Private Communication.
34. Seidel, H. and Cramer, F., *Bioch im. Biophys. Acta* **108**, 367 (1965).
35. Hayashi, H. and Miura, K., *Nature* **209**, 376 (1966).
36. Arca, M., Calvori, C., Frontali, L. and Tecce, G., *Biochim. Biophys. Acta* **87**, 440 (1964).
37. Haywood, A. M. and Sinsheimer, R. L., *J. Mol. Biol.* **14**, 305 (1965).
38. Nathans, D., *J. Mol. Biol.* **13**, 521 (1965).
39. Lodish, H. F., Cooper, S. and Zinder, N., *Virology* **24**, 60 (1964).
40. Spiegelman, S. and Hayashi, M., Cold Spring Harbor Symp. Quant. Biol. **28**, 161 (1963).
41. Summers, D. F., Maizel, J. V. and Darnell, J. E., *Proc. Natl. Acad. Sciences. U.S.* **54**, 505 (1965).
42. Marcker, K. and Sanger, F., *J. Mol. Biol.* **8**, 835 (1964).
43. Clark, B. F.C. and Marcker, K. A. *Nature* **207**, 1038 (1965).
44. Adams, J. M. and Capecchi, M. R., *Proc. Nat. Acad. Sciences, U.S.* **55**, 147 (1966).
45. Brown, G. L., Lee, S. and Deniset, J. Unpublished.

46. Takanami, M., Yan, Y. and Jukes, T., *J. Mol. Biol.* **12**, 761 (1965).
47. Takanami, M. and Zubay, G., *Proc. Natl. Acad. Sci. U.S.* **51**, (1964).
48. Okamoto, T. and Takanami, M., *Biochim. Biophys. Acta* **76**, 266 (1963).
49. Kelly, R. B., Gould, J. L. and Sinsheimer, R. L., *J. Mol. Biol.* **11**, 562 (1965).
50. Cannon, M., Krug, R. and Gilbert, W., *J. Mol. Biol.* **7**, 360 (1963).
51. Matthaei, J. H., Amelunxen, F., Eckert, K. and Heller, G. Z., *Electrochem.* **68**, 735 (1964).
52. Watson, J. D., *Bull. Soc. Chim. Biol.* **46**, 1399 (1965).
53. Asano, K. and Moore, P., Quoted in (52).
54. Gilham, P. T., *J. Amer. Chem. Soc.* **84**, 687 (1962).
55. Naylor, R., Ho, N. W. Y. and Gilham, P. T., *J. Amer. Chem. Soc.* **87**, 4210 (1965).
56. Augusti-Tocco, G. and Brown, G. L., *Nature* **206**, 683 (1965).
57. Marmur, J., Rownd, R. and Schildkraut, C. L., *Progr. Nucleic Acid Res.* **I.**, 232, (1962).
58. Tissières, A., *J. Mol. Biol.* **1**, 365 (1960).
59. Penniston, J. T. and Doty, P., *Biopolymers* **1**, 145 (1963).
60. Richards, E. G., Coll, J.A. and Gratzer, W. B., *Analytical Bioch.* **12**, 452 (1965).
61. Cooper, S. and Zinder, N. D., *Virology* **20**, 605 (1963).
62. Gardner, R. S., Wahba, A. J., Basilio, C., Miller, R. S., Lengyel, P. and Speyer, J. F., *Proc. Nat. Acad. Sci. U.S.* **48**, 2087 (1962).
63. Nirenberg, M. and Leder, P., *Science* **145**, 1399 (1964).
64. Bernfield H. R. and Nirenberg M. W., *Science* 147, 479 (1965).

DISCUSSION

(Chairman V. ENGELHARDT)

G. M. BLACKBURN: You have assumed that reactivity of the di-imide reagent with the nucleic acid is governed solely by accessibility of the G and U bases. But the reagent is a very polar molecule and its dipole will be sensitive to the field induced by all the bases in the single-strand region, i.e. all G's or U's will not be equally reactive. Have you observed such a phenomenon?
G. L. BROWN: Not yet. We have aimed at saturating the reactive U's and G's in the single-stranded regions. The effects that you mention will probably show up in kinetic experiments that we are carrying out using the spectrophotometric procedure.
P. LEDER: It might be more useful to carry out your binding studies on treated ribosomes using sRNA in reaction mixtures. Hatfield and Nirenberg have shown that sRNA is required for specific binding of trinucleotide codons to ribosomes.
G. L. BROWN: Yes. I agree with you that the presence of sRNA will probably increase the protection of the messenger binding site by poly-A.

Mitochondrial DNA and other Forms of Cytoplasmic DNA

P. Borst, A.M. Kroon and G. J. C. M. Ruttenberg

Department of Medical Enzymology, Laboratory of Biochemistry, University of Amsterdam, Jan Swammerdam Institute, 1e Constantijn Huygensstraat 20, Amsterdam, The Netherlands, Laboratory of Biochemistry and Toxicology, University of Amsterdam, The Netherlands

Introduction

For years, the biochemical studies of the biogenesis of mitochondria have suffered from lack of respectability. Cytoplasmic inheritance was considered a sideline, if not a freak, by most geneticists. DNA in mitochondrial preparations was routinely used as an indicator for nuclear contamination. RNA in mitochondrial preparations was considered to reflect the close association of the mitochondria with the endoplasmic reticulum, and the well-known inhibition of mitochondrial protein synthesis by chloramphenicol was considered proof for the *bacterial* origin of the amino acid incorporation by mitochondrial preparations, a trivial explanation which appeared to derive support from the fact that all attempts to pin-point the product of mitochondrial protein synthesis *in vitro* met with failure.

In the last two years, however, this sceptical attitude was over-taken by a series of important advances in our knowledge of the biochemical aspects of mitochondrial biosynthesis. In two laboratories it was independently shown that stable genetic determinants which influence the structure and enzymatic make-up of the mitochondrion can be transferred from one cell to another by purified mitochondrial preparations (1-3). It was firmly established that DNA is an intrinsic, and probably indispensable component of all mitochondria (see following sections) and mitochondrial DNA from vertebrate tissues was identified as a unique circular DNA with a molecular weight of 11 million (4,5). Lastly, RNA synthesis was demonstrated to occur in isolated mitochondria (6-8) and our insight into the mechanism of mitochondrial protein synthesis was considerably increased (cf. review in Ref. 9).

In this paper some of these advances are reviewed. No attempt will be made to cover the literature completely and emphasis will be put on the identification and properties of mitochondrial DNA, the biochemical evidence for a genetic function of this DNA, and possible models for the biosynthesis of mitochondria which can accomodate the experimental facts available. Where possible, relevant data on other organelles, such as chloroplasts and kinetoplasts, will be included in the discussion.

Identification and Properties of Mitochondrial DNA

Although good evidence for the presence of DNA in cytoplasmic organelles can be found in the older literature (reviewed in Refs. 9 and 10), the path to a rapid biochemical attack on this problem was opened by the discovery that the DNA of certain plants and algae contains a minor component which is concentrated in the chloroplast fraction (11) and which is absent in aplastidic mutants (12). The minor component has a base composition different from that of nuclear DNA (11, 13–15) which made it possible to separate nuclear and chloroplast DNA in CsCl gradients. The important advantage of this approach was that it circumvented the difficulty of preparing cytoplasmic organelles free of nuclear contamination by establishing a criterion for the unambiguous recognition of organelle DNA in the presence of nuclear DNA. This has led to a rapid development of methods to obtain organelle DNA free of nuclear DNA and to the establishment of many additional criteria to distinguish organelle DNA from nuclear or bacterial DNA.

IDENTIFICATION OF MITOCHONDRIAL DNA

The buoyant density of DNA found in more or less purified mitochondria from a variety of organisms has been studied and the results obtained up till now are given in Tables I and II. Since the methods used to calculate the absolute densities in different laboratories are different*, only the relative values found for nuclear and mitochondrial DNA should be compared. In all cases given in these tables the band-forming species was identified as DNA by appropiate means. This is of importance because carbohydrate material may form sharp DNA-like bands in CsCl at a density of about 1.7 g/cm^3 (25, 28, 34). The buoyant density of mitochondrial DNA may be higher than, lower than, or about equal to that of the nuclear DNA. The third category is represented only by rodents and ruminants but this may be the fortuitous consequence of the tendency not to publish results considered negative. The maximal difference in buoyant density observed between mitochondrial and nuclear DNA is 0.018 g/cm^3, which is equivalent to a difference of about 18 mole per cent in G+C content (35), if the assumption is made that the difference in buoyant density is due to a difference in base composition and not to the presence of unusual bases or tightly bound protein or RNA. This assumption has been checked by independent methods— e.g. direct determination of base composition or melting curves —with mitochondrial DNA from yeast (17) and chicks (24, 28) and found to be correct.

There is an interesting tendency of the mitochondrial DNA of related organisms to have similar buoyant densities, notably in the case of mammals,

* The buoyant density in CsCl of an unknown DNA is usually calculated from its position in the gradient relative to a marker DNA of known absolute density and a published equation for the steepness of the gradient under the conditions used. Originally the equation derived by Sueoka (32) was used with a value for *E. Coli* DNA of 1.710 as an absolute reference. More recently Hearst and Vinograd (33) have derived a more refined equation for the density gradient under standard conditions which differs significantly from that of Sueoka. Moreover, other absolute reference values have come into use. Since several recent papers do not contain sufficient information as to the equation and absolute reference value chosen, it is impossible to recalculate all results reported on a common basis. All buoyant densities determined in our laboratory have been calculated according to Sueoka (32), with 1.710 for *E. Coli* DNA as reference.

TABLE I. Density of Mitochondrial DNA in CsCl (g/cm³)

	Nuclear DNA	Mitochondrial DNA	Reference
Unicellular organisms			
Saccharomyces cerevisiae	1.700	1.685	16, 17
Saccharomyces carlsbergensis	1.700	1.685	16
Neurospora crassa	1.712	1.701	6
Paramecium aurelia	1.689	1.702	18
Tetrahymena pyriformis	1.688	1.682	18
Tetrahymena pyriformis	1.685	1.671	19
Euglena gracilis	1.707	1.691	20, 21
Plants	1.688–1.692	1.706	22
Animals			
Ox	1.704, (1.715)†	1.703	9, 16, 23–25
Lamb	1.704, (1.714)†	1.714	16, 26
Rat	1.703	1.701	27
Mouse	1.701, (1.690)†	1.701	9, 16, 24
Guinea pig	1.700, (1.704)†	1.702	9, 16, 24
Rabbit	1.701	1.703	4
Chick	1.701	1.708	4, 16, 28
Pigeon	1.700	1.707	24
Duck	1.700, (1.708)†	1.711	25

† The values between brackets refer to satellite components found in nuclear DNA.

TABLE II. Density in CsCl of Organelle DNA of Plants and Algae (g/cm³)

	Nuclear DNA	Chloroplast DNA	Mitochondrial DNA	Reference
Plants				
Spinacia oleracia	1.695	1.719, 1.705		11
Beta vulgaris	1.695	1.719, 1.705		11
Beta vulgaris	1.689	1.700		29
Nicotiana tabacum	1.690	1.703		30
Phaseolus aureus	1.691		1.706	22
Brassica rapa	1.692, (1.700)	1.695	1.706	22
Ipomoea batatas	1.692		1.706	22
Allium cepa	1.688		1.706, (1.718)	22
Algae				
Euglena gracilis	1.707	1.686	1.691	15, 20, 21, 31
Chlamydomonas reinhardi	1.723	1.695		11, 13
Chlorella ellipsoida	1.716	1.695?		11

birds and plants. An exception to this generalization was noted by Suyama and Preer (18) who found that the mitochondrial DNA's of the related flagellates *Tetrahymena* and *Paramecium* have widely different buoyant densities although the buoyant densities of the nuclear DNA's are similar. A second, rather surprising, exception is the recent finding in two different laboratories that there is a large difference in the buoyant densities of the mitochondrial DNA of beef and sheep, which are very closely related (36). The possible implications of this are numerous. However, in a preliminary experiment in this laboratory this result could not be confirmed, and the buoyant density of sheep heart mitochondrial DNA was found to be 1.702 g/cm³. Therefore, it seems wise to withhold speculations for the moment. Lastly, we have observed a small, but reproducible difference in buoyant density between mitochondrial DNA from chicks and ducks. This indicates that the over-all base composition, and *a fortiori* the base sequence, may be useful additional criteria in the taxonomic classification of birds. Insufficient experimental evidence is available to decide whether the small differences in buoyant density observed for mitochondrial DNA of different rodents are significant.

Although the finding of a DNA with a buoyant density different from that of nuclear DNA, concentrated in the mitochondrial fraction, may be a very useful method for the rapid identification of mitochondrial DNA, this criterion obviously fails when the buoyant densities of nuclear and mitochondrial DNA are about the same. In addition, this criterion does not yield information on the exact subcellular localization of the DNA in the absence of extensive cell fractionation studies; it does not allow a decision as to whether there is DNA with the buoyant density of nuclear DNA present in the mitochondria in addition to the unique DNA studied; and, lastly, it does not enable the essential distinction to be made between mitochondrial and bacterial DNA.

Additional criteria are therefore required to prove that the mitochondrial DNA isolated is indeed mitochondrial, and these will now be discussed in turn:

(1) The presence of mitochondrial DNA within the semi-permeable mitochondrial membrane completely protects it against degradation by DNase under conditions where added DNA or DNA in intact nuclei or nuclear fragments is degraded (17, 24, 28). With chick-liver mitochondria we found that after mild swelling at 0° in 1 mM phosphate (pH 6.8)+approx. 0.025 M sucrose (a treatment that is known to lyse lysosomes (37), a quantitatively important contaminant of mitochondrial preparations from liver) followed by DNase treatment, we could still isolate 50% of the mitochondrial DNA present in a high-molecular weight form (25). After more drastic swelling, followed by DNase treatment, we were unable to isolate any mitochondrial DNA at all (25). The same result was obtained after treatment of the mitochondria with the detergent Triton X-100. In a similar way Suyama and Bonner (22) extracted DNA from isolated plant mitochondria with deoxycholate which did not lyse the bacteria present in their preparations.

The resistance of mitochondrial DNA to DNase so long as the mitochondrial membrane is intact is therefore an important identification criterion, especially in organisms which do not contain other organelles with swelling properties similar to those of mitochondria. In plants and algae, the presence of immature chloroplasts and proplastids poses additional problems which have not yet been studied in detail.

(2) The petite mutants of yeast, which are characterized by the absence of functional mitochondria, contain very little or no mitochondrial DNA

(16, 17). Unfortunately, this important criterion can only be used with facultative anaerobes.

(3) Mitochondrial DNA from vertebrate tissues has unique physicochemical properties which set it apart both from nuclear and bacterial DNA. This will be discussed below. It is likely that this criterion will be found to hold for mitochondria from all organisms.

Although none of these arguments is conclusive, taken together they prove that at least in some cases, notably chick liver, mouse liver, beef heart and yeast the mitochondrial DNA referred to is indeed an integral part of the mitochondrion. This conclusion is in agreement with independent qualitative histochemical evidence that all mitochondria studied up till now contain acid-insoluble DNase-sensitive material with the staining characteristics of DNA (38–41), while tritiated thymidine can be incorporated into material with similar properties which is closely associated with mitochondria (42).

PHYSICOCHEMICAL PROPERTIES OF INTACT MITOCHONDRIAL DNA

It was recently found in different laboratories that carefully isolated mitochondrial DNA from several organisms can be distinguished from nuclear DNA by its high degree of homogeneity (16, 17, 24, 27). This could already be inferred from the narrow band obtained for this DNA in a CsCl gradient (see Figure 1) and from the steep melting curve (24, 27), illustrated in Figure 2, but it was directly proven by the ability of mitochondrial DNA to renature after thermal or alkaline denaturation (4, 16, 17, 24). This is shown in the experiments given in Figures 1–3.

In the experiment with mouse-liver DNA of Figure 1, denaturation and renaturation is demonstrated qualitatively by band shifts in CsCl gradients. Since nuclear DNA does not renature at all under these conditions (16, 25, 43), it is clear that renaturation is a very useful criterion to distinguish mitochondrial from nuclear DNA in organisms where the buoyant densities of these DNA's are the same.

The experiment with chick-liver mitochondrial DNA (Figure 2) demonstrates that mitochondrial DNA, after thermal denaturation, completely recovers its original hypochromicity on cooling, while for nuclear DNA a curve is obtained typical for a mixture of single-stranded polynucleotides which do not renature at all (44). On reheating of the mitochondrial DNA the same melting curve is obtained (results not shown). Therefore, mitochondrial DNA renatures much faster than bacterial DNA, which indicates that it consists of a homogeneous population of molecules in the molecular weight range of the DNA viruses. This is further illustrated by the experiment given in Figure 3, in which the renaturation rates of DNA from chick-liver nuclei and mitochondria and *Vibrio el Tor* are compared. *Vibrio el Tor* DNA was chosen as a control because it has about the same melting point as DNA from chick-liver mitochondria (4). Under optimal conditions at 70°, renaturation of mitochondrial DNA was so rapid that the original hypochromicity had already been largely regained by the time the first reading could be made at 3 min. On reheating, the DNA melted out with the steep curve characteristic of a homogeneous double-helical polynucleotide. After 10 min at 100° the temperature was shifted back to 83°, which is an unfavourable temperature for renaturation, as shown by the fact that *Vibrio el Tor* does not renature at all (optimal = about 25°

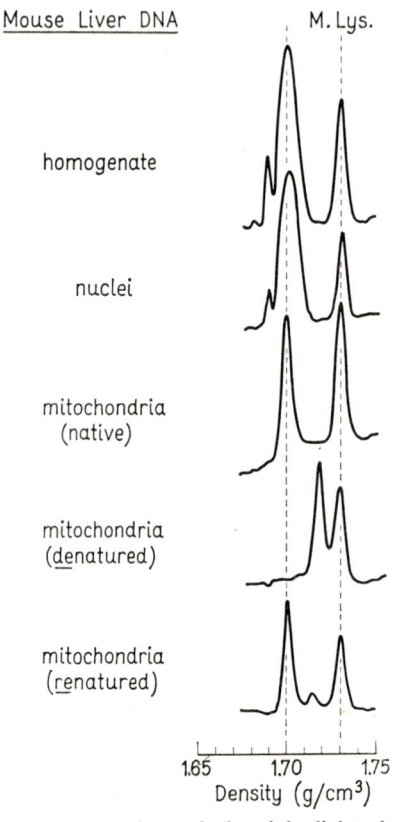

FIGURE 1. Microdensitometer tracings of ultraviolet-light photographs of mouse-liver DNA in a CsCl gradient at equilibrium. The tracings are matched at the *Micrococcus lysodeikticus* marker, banding at 1.731 g/cm³. Mitochondrial DNA in 0.15 M NaCl, 0.015 M sodium citrate (pH 7.0) was denatured by adding NaOH to 0.1 M final conc., followed after 15 min by 0.15 vol. of 1 M NaH₂PO₄ at 0°. The denatured DNA was renatured by heating for 2 hr at 70° (Reproduced with kind permission of the publishers from Ref. 4).

below the melting point; see Refs. 44–47). Again rapid renaturation of mitochondrial DNA occurred, apparently following the kinetics of a second order reaction. The high tendency to renaturation is therefore due to the high degree of homogeneity of the DNA and not to cross-links preventing complete separation of the strands.

The homogeneity of mitochondrial DNA demonstrated by these experiments, and the fact that the calculated amount of DNA per mitochondrion is around 40×10^6 Daltons (see following section), made it of interest to study the molecular weight of our DNA preparations. The results obtained were rather unexpected. In all preparations studied two components were found, sedimenting at $S_{20,w}$ values of about 39-42 and 27-29, with very little material sedimenting at intermediate S values or not sedimenting at all (4). The two components were present in about equal amounts in freshly-prepared DNA but extensive handling of the DNA apparently led to a conversion of compo-

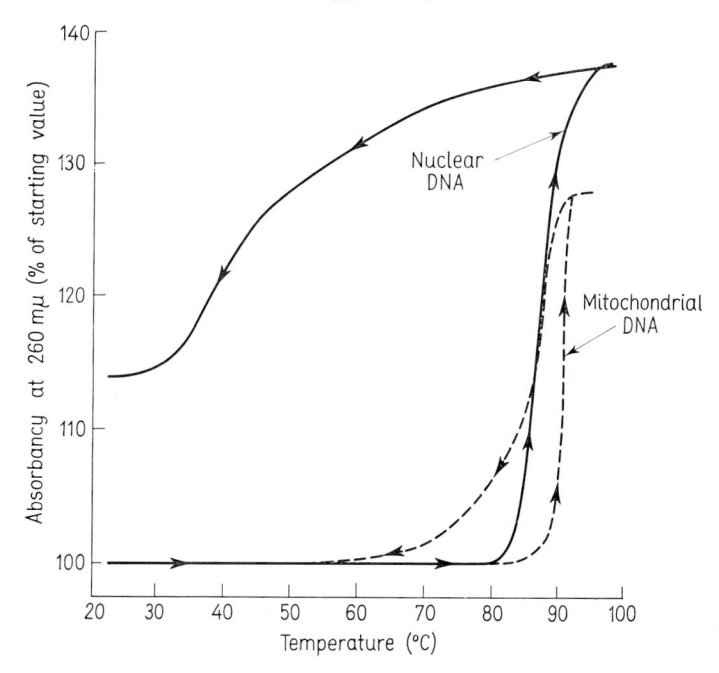

FIGURE 2. Thermal transition profiles of nuclear and mitochondrial DNA from chick liver in 0.15 M NaCl, 0.015 M sodium citrate (pH 7.0).

nent I into II, as illustrated in Figure 4. Similar results have not been obtained before with DNA from animal tissues, but two-component systems have been observed with linear viral DNA and its half-molecules (cf. 48) and with hypertwisted circular and open circular forms of viral DNA (cf. Table 3). Electron microscopy by Dr. van Bruggen (University of Groningen) made it possible to decide between these alternatives (4,5). Freshly prepared chick liver mitochondrial DNA contained predominantly highly-twisted DNA molecules in which no free ends could be distinguished, as illustrated in Figure 5 (a and b). In addition a minority of half-twisted, or coiled, circular molecules (Figure 5c) and open circles (Figure 5d) and an occasional linear molecule were present. In DNA preparations that contained mainly or exclusively component II, more open and half-twisted circles and less highly-twisted molecules were seen. The same result was obtained with freshly-

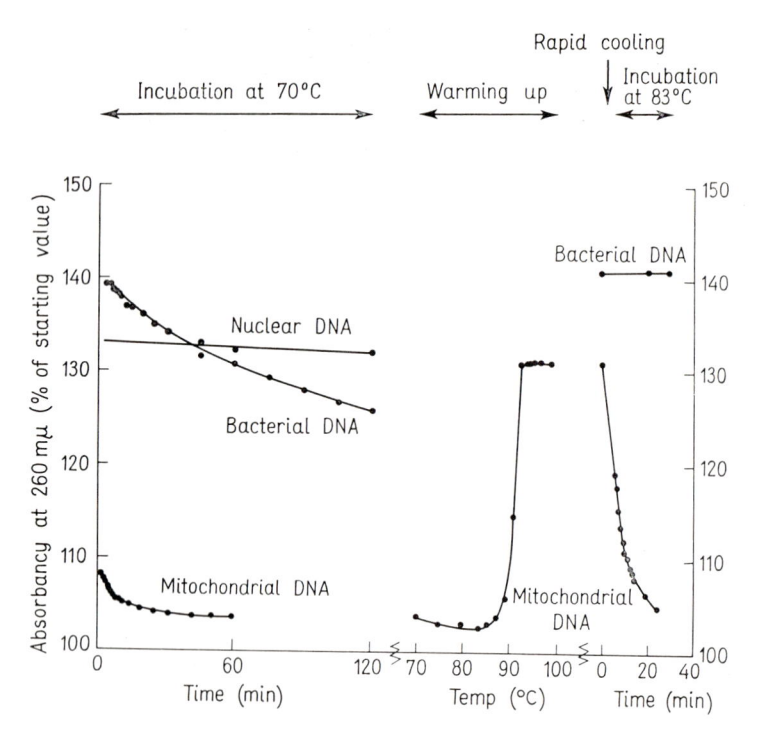

FIGURE 3. Renaturation of denatured DNA from chick-liver mitochondria, chick-liver nuclei and *Vibrio el tor*. The DNA was denatured by NaOH and neutralized as described in the legend to Figure 1. At the start of the experiment the denatured, reneutralized DNA was rapidly brought to 70° and put into the spectrophotometer cell, also thermostatted at 70°. Further details are described in the text.

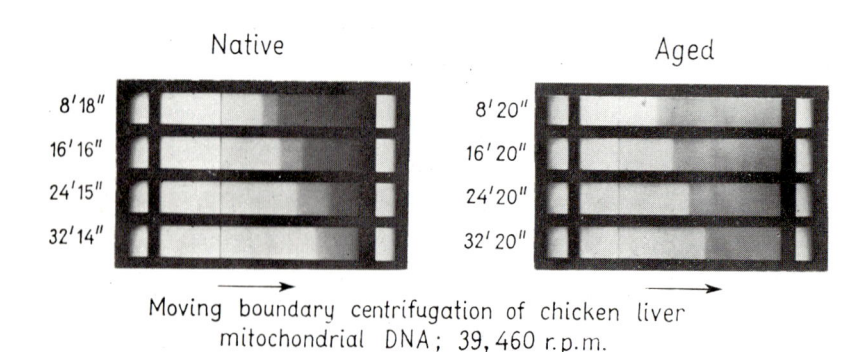

FIGURE 4. Ultraviolet light photographs of chick-liver mitochondrial DNA sedimenting in the analytical ultracentrifuge. The figures on the left side of the photographs refer to the time elapsed after reaching full speed. Solvent: 1.2 M NaCl, 0.05 M sodium phosphate, pH 6.8.

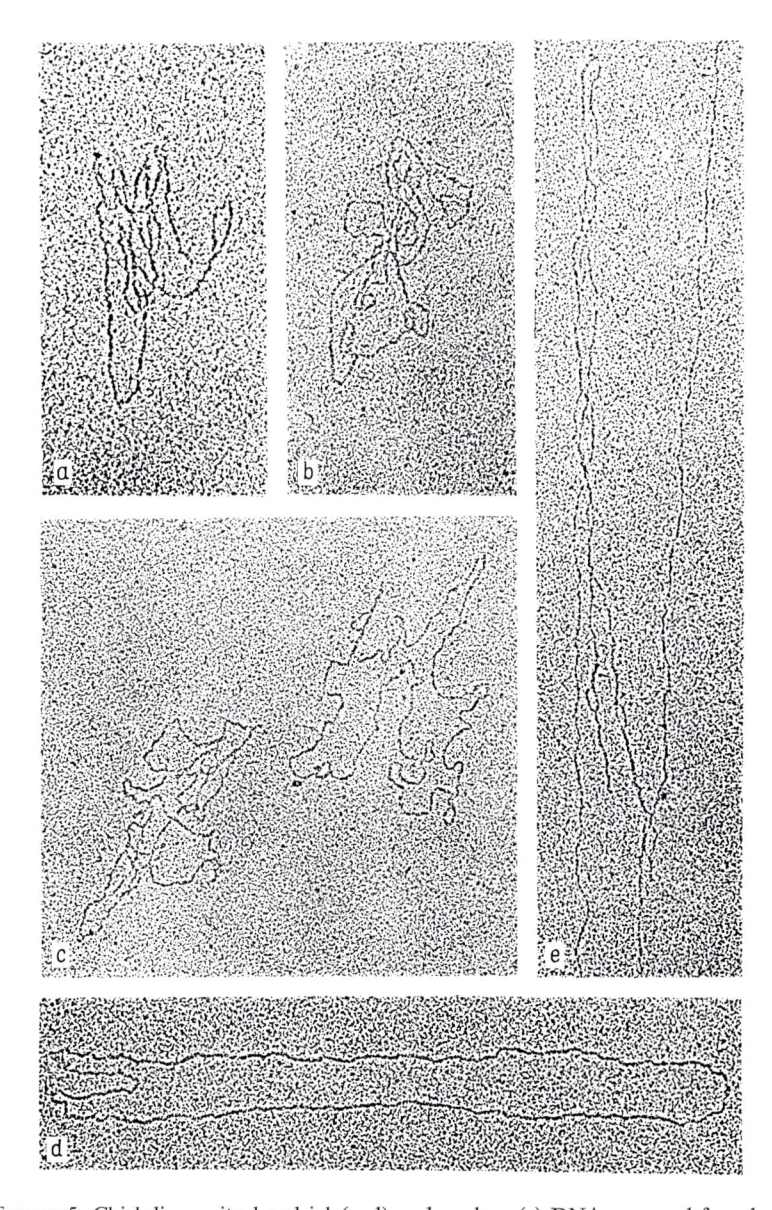

FIGURE 5. Chick-liver mitochondrial (a–d) and nuclear (e) DNA prepared for electron microscopy by the Kleinschmidt technique (55); DNA, 1–2 μg/ml; cytochrome c, 10 mg/ml; ammonium acetate, 1.3 M; rotatory shadowing with 6 mg platinum at an angle of 8° and a distance of 5 cm. Magnification 62,550 ×. The magnification was calibrated with a replica of a grating 2160 lines per mm. (Reproduced with kind permission of the publishers from Ref. 4).

prepared mitochondrial DNA which had been pretreated with hydroquinone, a treatment which converts the hypertwisted form of polyoma DNA into the open circular form (49).* No circular DNA molecules at all were found in nuclear DNA from chicken liver (Figure 5e).

The contour length of 21 circles which could be accurately measured is given in Figure 6. The contour length follows a normal frequency distribution with a mean of 5.45 μ and a Standard Error of 0.08 μ. Since the apparent circumference of the twisted molecules and of the occasional linear

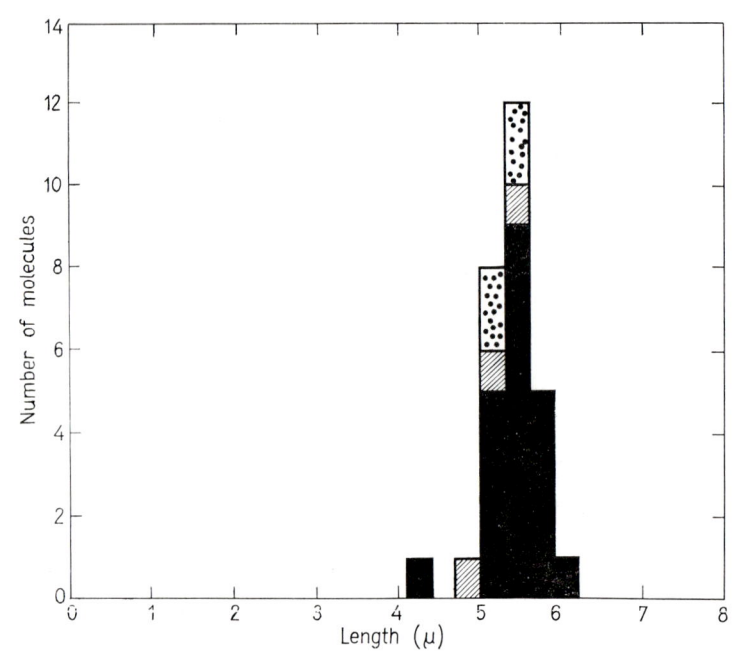

FIGURE 6. Length distribution of circular mitochondrial DNA from chick liver, mouse liver and beef heart. Black, chick-liver DNA; hatched, mouse-liver DNA; stippled, beef-heart DNA.

molecules present is compatible with this value, we concluded (4,5) that the molecular weight of mitochondrial DNA from chick liver is about 11×10^6 and that this is probably the size of all mitochondrial DNA molecules. Our further interpretation of these data is that the components I and II observed in the ultracentrifuge are the hypertwisted, circular form and the open circular form of mitochondrial DNA respectively, with structures similar to those proposed for several double-stranded DNA viruses (43, 49, 51, 53) and the Replicative Form of ϕX174 (49, 50). We believe that the quantitative interpretation of the electron micrographs of these large circles is compli-

* The assignments "predominantly hypertwisted", or "less twisted" were made without prior knowledge of the type of preparation examined, to avoid bias.

TABLE III. Circular DNA; Broken, Intact and Twisted

Source	Mol. wt.	Linear form	$S_{20,w}$ Values Open circle	Twisted circle	Reference
Polyoma virus	3.0×10^6	14.5	16	20	49
øX174 phage RF	3.4×10^6	?	17	21	50
Papilloma virus	5.3×10^6	18	20	28	43, 51
Lambda phage	33×10^6	32	37	48[†]/61[‡]	52, 53
Mitochondria	11×10^6	24[§]	27–29	39–42	4, 5

[†] In high salt.
[‡] In low salt.
[§] Calculated from the molecular weight with the Eigner–Doty equation (54).

cated by the fact that open circles tend to become entangled during the preparation of the grid which leads to many partly-twisted forms. This precludes even a semi-quantitative determination of the relative amounts of hypertwisted and open circles. This interpretation is supported by the comparison of the relative $S_{20,w}$ values obtained for the hypertwisted circular, open circular and linear DNA's of DNA viruses, given in Table III. It is clear from these results that the sedimentation coefficients that we have obtained for the two components of mitochondrial DNA fit the values expected for a hypertwisted circular and open circular molecule with a molecular weight of 11×10^6. Definite identification of components I and II as the hypertwisted and open forms of mitochondrial DNA will require direct evidence for the absence of strand separation in I under denaturing conditions and the demonstration that I can be converted into II by the scission of one phosphodiester bond. Technical difficulties with the analytical ultracentrifuge and with the scaling up of our procedure for the preparation of intact mitochondrial DNA have precluded such identification up till now.*

Although the experimental evidence available is very limited as yet, there are indications that the results obtained with mitochondrial DNA from chick liver will hold for mitochondria in general. Renaturation has been observed with mitochondrial DNA from mouse (4,16), guinea pig (16) and rat liver (16), beef heart (16) and yeast (16,17). In the latter case it was also shown that renaturation occurs more readily than could be expected for bacterial DNA (17). The various forms of circular DNA were also found as the predominant species in mitochondrial DNA from beef heart and mouse liver (4,5). This is illustrated in Figures 7 and 8. The circumference of 3 mouse and 4 beef circles was measured. The values obtained were very close to the mean value found for chick-liver DNA as shown by the histogram of Figure 6. In addition, purified mitochondrial DNA from duck liver also sediments in two components in the ultracentrifuge, with about the same $S_{20,w}$ values as components I and II of mitochondrial DNA from chick liver. Therefore, at least all vertebrates appear to have circular mitochondrial DNA with a circumference of about 5.45 μ.

It cannot be guessed at the moment whether mitochondria from lower organisms contain larger DNA molecules. Luck and Reich (6) reported that mitochondrial DNA from *Neurospora crassa* contained long strands of varying lengths in the micron range. No mention was made of circular molecules. An electron micrograph in their paper shows a linear DNA molecule 6.6 μ long, which is outside the range observed for vertebrate mitochondrial DNA. However, the mitochondrial DNA preparation studied by Luck and Reich was not free of contamination by nuclear DNA and it is

---

* The extraction of DNA from mitochondria which have a continuous turnover and in which DNA apparently replicates in a semi-conservative fashion (16) might be expected to lead to the extraction of branched molecules comparable to those detected in radioautograms of *E. Coli* (56). Such molecules were not found in any of the preparations studied. This may reflect the short time required for the replication of a DNA molecule of this size in relation to the turnover of the rest of the mitochondrion or it may be that the purification procedures employed eliminate branched molecules.

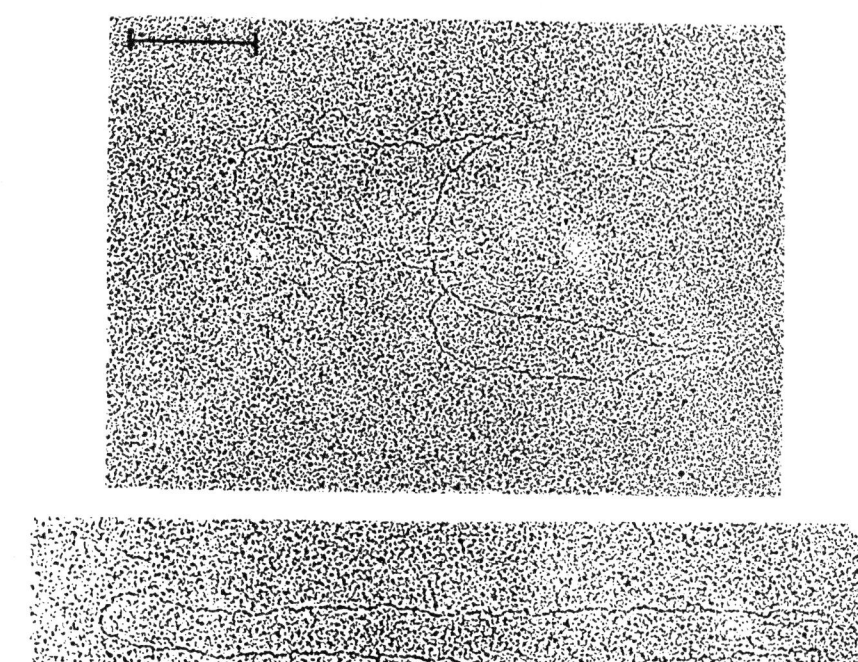

FIGURE 7. Two circular molecules in beef-heart mitochondrial DNA, prepared and photographed as described in the legend to Figure 5.

therefore premature to conclude from their data that mitochondrial DNA of *Neurospora crassa* has a higher molecular weight than that of vertebrates.

Sedimentation coefficients for purified mitochondrial DNA from yeast have been determined by Tewari *et al.* (17). A mean $S_{20,w}$ of 33 was obtained. No indication for two components was found and no conclusions can therefore be drawn from this value.

The Information Content of Mitochondria

In recent years much effort has been spent in different laboratories to determine the so-called true amount of mitochondrial DNA per mitochondrion, in order to estimate the information content of the mitochondrial population. This approach is based on the assumption (9), which is rarely stated explicitly, that all mitochondria in a cell contain the same amount of identical DNA and that this DNA is not redundant, i.e. that there are not several molecules of the same base sequence in one mitochondrion. Unfortunately, there is

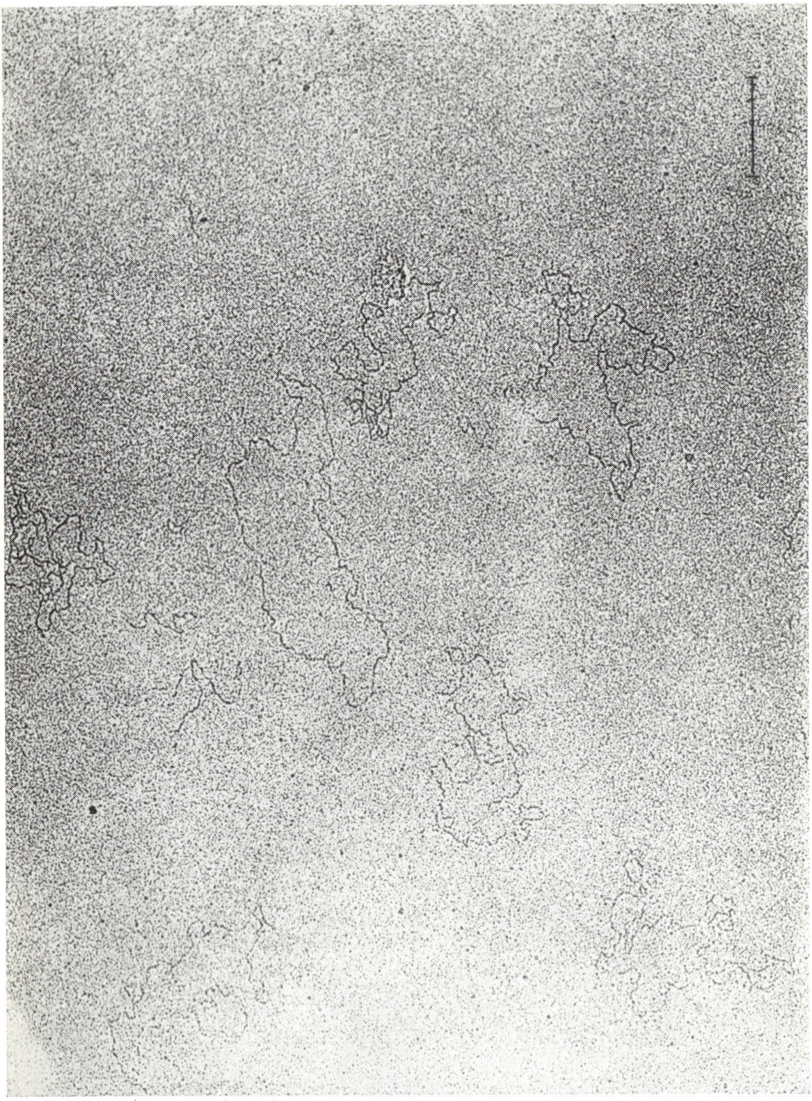

FIGURE 8. Different types of circular DNA molecules in a field of mouse-liver mitochondrial DNA, prepared and photographed as described in the legend to Figure 5.

no evidence to support this assumption. Histochemical techniques are only able to demonstrate the presence of DNA in an occasional mitochondrion (Refs. 38–41) and quantitative studies on the distribution of mitochondrial DNA in mitochondrial sub-fractions have not even been started. The only direct evidence on the information content of the total mitochondrial population is therefore our estimate of the degree of homogeneity of chick-liver mitochondrial DNA obtained from quantitative renaturation studies (4, 25), as shown in Figures 2 and 3. Our results indicate that the information content of the mitochondrial population is much less than that of a bacterium and it could well be the amount contained in a DNA molecule with a molecular weight of 11×10^6. Experiments are in progress to refine the renaturation criterion in order to get a more precise estimate.

It is of interest to compare some recent values for the DNA content of mitochondrial preparations, as given in Table IV. The results of two methods are compared: the first two columns give total acid-insoluble deoxyribose in mitochondrial preparations with and without pretreatment with DNase. In the absence of DNase pretreatment very high values may be found which have been ascribed to contamination of the mitochondrial fraction by nuclear fragments (24, 26). In the third column values are given for the specific mitochondrial DNA which can be extracted from mitochondrial preparations and which is recognized by its specific buoyant density or melting behaviour. In this case corrections have to be made for losses during purification and this was done in most cases (22, 24) by adding a known amount of marker DNA to the lysate and assuming that the recoveries of marker DNA and mitochondrial DNA are the same. Both methods are based on unproven assumptions and have obvious drawbacks (see discussion in Ref. 9), but it is comforting to see that they yield similar values, and that the values found for mitochondria from different organisms are similar. Recalculation of these values to obtain the amount of DNA per mitochondrion leads to variable results because the methods used to determine the amount of mitochondria per mg of mitochondrial protein are not very accurate (cf. Ref. 60) and the values used by different workers for mitochondria from the same tissue, e.g. rat liver, may differ by a factor of 2 (Refs. 27, 57, 61, 62). If we take an average value of 7×10^{-17} g per mitochondrion, this is equivalent to 40×10^6 Daltons per mitochondrion or sufficient to code for 100 proteins containing 200 amino acids. As pointed out before (4,9), this amount would probably be insufficient to code for all mitochondrial proteins and for a complete protein-synthesizing machinery, including ribosomes, sRNA, aminoacyl-sRNA synthetases, transfer enzymes, methylases, etc. As mentioned before, this calculation is based on the assumption that all mitochondria contain the same amount of identical DNA without redundancy. Since mitochondrial DNA in vertebrate mitochondria is packaged in molecules with a molecular weight of 11×10^6, there are three possibilities:

(a) The estimates based on the amount of DNA per mitochondrion given in Table IV are too high by a factor of 4.
(b) There are 3–4 circles of different base sequence per mitochondrion.
(c) There are several identical circles per mitochondrion.

TABLE IV. DNA in Purified Mitochondrial Preparations

	Acid-insoluble deoxyribose		Mitochondrial DNA	Mitochondrial DNA per mitochondrion
	No pretreatment (μg DNA/mg prot.)	DNase pretreatment[†] (μg DNA/mg prot.)	(μg/mg prot.)	(g)
Rat liver	0.46–0.65 (27, 57, 58)[‡]			1–2×10^{-16} (27, 57)
Chick liver	0.5 (25)		0.5 (24)	7×10^{-17} (24)
Beef heart	0.24 (58)			3×10^{-17} (58)
Lamb heart	3.4–5.6 (59)		0.8 (22)	10^{-16} (22)
Mung bean			0.7 (17)	
Yeast	1.0–2.5 (17)	0.7–1.0 (17)		6×10^{-17} (17)

[†] This refers to pretreatment of the intact mitochondria, prior to DNA extraction.
[‡] The reference numbers are given between brackets.

An argument in favour of the last possibility is that mitochondria are replicating organelles and it is known that, in growing bacteria, cell division may lag behind DNA duplication (63). A second argument is that there is morphological evidence that the mitochondrial population of a cell is continuously mixing, mitochondria fusing, or sub-dividing into smaller mitochondria (64, 65). If this fusion is indeed complete fusion and not temporary attachment, the average number of DNA molecules per mitochondrion would be expected to be above one, assuming that mitochondrial DNA has an indispensable function in mitochondrial replication.

For these reasons we propose as a working hypothesis that the total information contained in the mitochondrial population of a vertebrate cell is that present in a DNA molecule of molecular weight 11×10^6. This could maximally code for 5000 amino acids, which would be a rather modest contribution to the synthesis of a complete mitochondrion. It is clear from the reservations outlined above that this working hypothesis is based partly on unproven, but reasonable assumptions.

IDENTIFICATION AND PROPERTIES OF DNA FROM CHLOROPLASTS AND OTHER ORGANELLES

DNA has been detected in chloroplasts by histochemical and autoradiographic techniques (66–69) and, as mentioned above, satellite DNA's with a base composition different from the bulk of the cellular DNA were found to be concentrated in the chloroplast fraction in plants and algae (see Table II: Refs. 11, 13–15, 21, 22, 29, 30). The predominant localization of these satellite DNA's in the chloroplast is supported in the case of algae by the observation that mutants in which functional chloroplasts are missing do not contain the satellite DNA (12, 20, 21). Estimates on the amount of DNA per chloroplast vary from 10^{-16} to 4×10^{-15} g/chloroplast (10, 70, 71), which would be equivalent to 10^8 to 4×10^9 Daltons DNA per chloroplast. Attempts to assess the homogeneity of purified chloroplast DNA by renaturation studies have not been made and, although melting curves of chloroplast DNA were determined in several laboratories, no mention was made of renaturation on cooling. The molecular weight of chloroplast DNA was determined by Ray and Hanawalt (14, 21) by sedimenting purified radioactive chloroplast DNA through a sucrose gradient in the presence of [^{3}H]-λ-DNA as marker. The DNA sedimented in two distinct components for which the authors calculated molecular weights of 40 and 20×10^6, assuming that the DNA was linear. Ray and Hanawalt suggest (14) that the two components observed arose from larger molecules by shear-degradation during the isolation of the DNA. No electron micrographs have been made of chloroplast DNA.

At present nothing can therefore be said about the molecular weight and structure of *intact* chloroplast DNA. This point is under investigation in our laboratory.

Little work has been done on the DNA apparently associated with the kinetoplast of flagellates. DNA was shown to be present in kinetoplasts by histochemical techniques and autoradiography (72) and more recently it was shown that the satellite band present in DNA from *Leishmania enrietti* (see Table V) is concentrated in the kinetoplast fraction (73). Since kinetoplast DNA can easily be detected in DNA extracted from whole cells, and since these flagellates contain only one kinetoplast, it is probable that the amount of DNA per kinetoplast is high compared with the amount of DNA per chloroplast or mitochondrion. Nothing is known about the homogeneity or structure of kinetoplast DNA.

TABLE V. Density of Kinetoplast DNA in CsCl (g/cm³)

Organism	Nuclear DNA	Kinetoplast DNA	Reference
Crithidia fasciculata	1.717	1.698	74
Leishmania enrietti	1.721	1.699	73

To our knowledge no systematic study has been made of the possible presence of small amounts of DNA in cytoplasmic organelles other than mitochondria, chloroplasts or kinetoplasts, e.g. the protein bodies from wheat endosperm described by Morton and Raison (75).

RELATION BETWEEN MITOCHONDRIAL AND NUCLEAR DNA

The interesting question of whether the base sequence of mitochondrial DNA is partially or fully represented in nuclear DNA has not yet been answered. The obvious approach to this problem would be to look for DNA with the buoyant density of mitochondrial DNA in highly purified nuclei from organisms in which there is a large difference in buoyant density between mitochondrial and nuclear DNA. The latter requirement excludes the animal tissues studied up to now as starting material. The most suitable test object would probably be yeast, since a method of obtaining highly purified yeast nuclei is available (76). An added advantage of using yeast is that the amount of DNA per nucleus is only 5×10^{-14} g (79), which is two orders of magnitude lower than the amount present in higher plants (78) and animals (79). Since the amount of DNA per mitochondrion in yeast is not lower than that found in animal or plant mitochondria (Table 4), and since the number of mitochondria per cell in yeast (80, 81) is only one order of magnitude lower than that found in plants and animal tissues, the odds are favourable in yeast.

It was recently shown that there is very little (17) or no (16) mitochondrial DNA in cytoplasmic "petite" mutants of yeast. It is doubtful, however, whether the techniques employed were sufficiently sensitive to detect one molecule of mitochondrial DNA per cell, which might be equivalent to 1/50th of the normal amount of mitochondrial DNA or even less, if several copies were present in a normal mitochondrion. This objection is underlined by the results obtained with the induction of "petite" mutants by ultraviolet light (82). In cells cultured under anaerobic conditions, the increase in the number of mutants is linear with dose, indicating a single-hit process, while in aerobic cells there is a pronounced lag in the induction, indicating multiple-hit. "In other words, in anaerobically growing cells there appears to be only one determinant, whereas in cells under aerobic culture there are many (82)". Therefore the conclusion that mitochondrial DNA is absent from cytoplasmic "petite" mutants cannot be accepted before it has been demonstrated that the methods used are sensitive enough to detect mitochondrial DNA in anaerobic yeast. Unfortunately, this control experiment has not yet been done.

Although no attempt has been made to study the presence of chloroplast DNA in highly purified nuclei from plants or algae, this DNA was reported to be undetectable in the whole-cell DNA from aplastidic Euglena and Chlamydomonas mutants, studied by analytical CsCl gradient centrifugation (12, 20, 21). This was confirmed in subsequent work (21) in which whole-cell DNA from Euglena, prefractionated in a preparative CsCl density gradient, was used to increase resolution. Furthermore, it was shown that chloroplast DNA could easily be detected in dark-grown cells which did not contain mature chloroplasts. Although these results are suggestive, they are subject to reservations: as already pointed out by Gibor and Granick (10), the degree of redundancy in the chloroplast DNA of algae is not known. If every chloroplast contained 10 or 100 copies of the same DNA molecule, which were only represented once in the nucleus, this one nuclear copy might easily escape detection in the aplastidic mutants, even in organisms containing but one chloroplast, like Chlamydomonas. This would certainly be so if this nuclear copy were present as part of a much longer DNA molecule. This would break randomly during isolation, leading to a range of molecules with densities intermediate between those of nuclear and chloroplast DNA. Furthermore, one should not overlook the possibility that mutations leading to a complete disappearance of all cytoplasmic DNA could also affect mitochondrial DNA which might be present in the nucleus.

A second approach to the possible presence of organelle DNA in the nucleus is to look for base-sequence homology between nuclear and organelle DNA by hybridization techniques. Before embarking on this type of experiment the odds should be calculated. As we have shown above the amount of "mitochondrial information" per cell is below 300×10^3 Daltons and it may well be close to 11×10^6 Daltons. A haploid mammalian cell nucleus contains about 2×10^{-12} g DNA (79) which is equivalent to 10^{12} Daltons. Obviously, therefore, the odds are very unfavourable, unless a high degree of redundancy is present in nuclear DNA, an assumption not supported by the experimental evidence available. It is therefore necessary to choose a very sensitive hybridization technique and for practical reasons the only feasible approach is to make highly labelled RNA copies of mitochondrial DNA *in vitro*, using RNA polymerase, and hybridize these with DNA from highly purified nuclei. In this way even less than 0.01% hybridization should be detectable (cf. 83 and 84). Obviously, the DNA agar technique (85), which has a detection limit for DNA–DNA hybrids of about 0.5 per cent, is too insensitive for this purpose. Unfortunately this is the technique chosen in two recent papers (30, 86) in which the extent of base-sequence homology of nuclear and organelle DNA was studied. Shipp et al. (30) could not detect hybridization between labelled chloroplast DNA and total cell DNA from immature, nongreened leaves, but the authors point out themselves that less than 50 "chloroplast complements" per nucleus would not have been detected. Similar results have been reported by Humm and Humm (86) for undefined "mitochondrial" DNA from mouse liver. The authors conclude from their experiments, in the report of which essential controls are unfortunately not mentioned, that "the mitochondrial DNA is in part a unique DNA and not related to its nuclear counterpart". For the reasons outlined above, however, this conclusion does not follow from the experimental evidence. In addition Humm and Humm (86) have studied the hybridization with nuclear and mitochondrial DNA of mitochondrial RNA and nuclear RNA labelled *in vivo*. Their most striking finding is that non-radioactive mitochondrial RNA competes with radioactive nuclear RNA for sites on nuclear DNA as efficiently as non-radioactive nuclear RNA. They conclude from this finding that "a considerable portion of the mitochondrial RNA must be coded for by nuclear DNA" Unfortunately the experimental evidence on which this conclusion is based is unsatisfactory and several trivial alternative explanations are not excluded.

In conclusion, there is no convincing evidence as to whether the base sequence of mitochondrial DNA is represented in the nucleus or not. Since methods to settle this question are, in principle, available, an answer should be forthcoming soon.

Relation between "Satellite" DNA and Mitochondrial DNA in Animal Tissues

As discussed in previous sections, the chloroplast DNA of plants was originally detected because a minor satellite DNA, present in DNA from whole cells, was found to be concentrated in the chloroplast fraction. Since DNA from several vertebrates was already known to contain satellite components, it seemed possible that these might be associated with the mitochondria. Although the first experiments by Marmur (quoted in Ref. 6) appeared to provide support for this hypothesis, it was subsequently shown in this laboratory that the relative concentration of satellite DNA of beef, mouse and guinea-pig tissues was the same in DNA from homogenates and purified nuclei, while the satellite DNA was absent from, or at least not concentrated in, DNA from the mitochondrial fraction (4, 23, 24, 87). These results were confirmed in a recent paper from Marmur's laboratory (16).

In Table I we have summarized available data on the relation of satellite DNA to mitochondrial DNA. Only *major* satellites, which represent at least a few percent of the total cellular DNA, are included. It is clear that in general the major satellites of animal tissues do not have any relation to mitochondrial DNA, with the doubtful exception (see above) of mitochondrial DNA of sheep heart.

The intracellular localization of the satellites detected in salmon-sperm

DNA (35) and DNA from the tissues of Cancer species (88) is not known. We have attempted (89) to study the distribution of dAT in *Cancer pagurus* hepatopancreas and ovary. Although the results obtained indicate that in this case the satellite also has a nuclear localization, the problem of isolating well-defined cell fractions containing high-molecular weight DNA from *Cancer* was not satisfactorily solved, and no definite conclusions could be drawn.

Why nuclear satellites band at a density different from that of the bulk of nuclear DNA has been very little studied with the exception of experiments on the satellite of *Cancer* (cf. 88 and 90), and limited experiments on the nature of the satellites of beef (91, 92) and mouse (91, 93) tissues. The biological function of these intriguing components remains unknown and it is not fruitful to speculate on this point as long as their chemical nature has not been fully elucidated.

Evidence for a Genetic Function of Cytoplasmic DNA

Although the presence of DNA in cytoplasmic organelles seems established on the basis of the evidence presented in previous sections, this does not necessarily imply that such DNA has a genetic function. The DNA could represent a somewhat odd storage form of nucleotides, and the presence of very large amounts of cytoplasmic DNA in egg cells (94, 95), and the accumulation of large amounts of Feulgen-positive material in the cytoplasm of fibroplasts in tissue-culture under special conditions (96), could even be interpreted as support for this suggestion.

Evidence for a genetic function of cytoplasmic DNA has come both from genetic and biochemical studies. These will now be discussed in turn.

GENETIC EVIDENCE

The existence of cytoplasmic mutations concerning the ability to form functional mitochondria, kinetoplasts or chloroplasts is supported by a large body of evidence, to be found in several recent reviews and monographs (10, 82, 97–99). A very simple summary of this evidence is given in Table VI, taken (in modified form) from the excellent review by Gibor and Granick (10). From the biochemical standpoint the following points are of special interest:

(1) Very little or no DNA could be detected in all cytoplasmic mutants studied, indicating a close correlation between organelle DNA and the presence of functional organelles. As pointed out above, this finding does not exclude the presence of some organelle DNA in these mutants. Moreover, it can be predicted that cytoplasmic suppressive mutants should contain mitochondrial DNA, albeit in altered form.

(2) Some mitochondrial and chloroplast proteins can still be synthesized in all mutants, even those which behave as if the cytoplasmic determinant (organelle DNA?) has been completely lost (see 10, 82, 100–102).

(3) The mutagenic agents inducing cytoplasmic mutations are known to

TABLE VI. Properties of Cytoplasmic Mutations (after Gibor and Granick (10))

	Euglena	Yeast
Phenomenon	Bleaching; impaired photosynthetic apparatus.	"Petit" colony; impaired respiration.
Anatomy	Proplastids fail to differentiate on exposure to light; no chloroplast lamellae develop.	Promitochondria fail to differentiate on exposure to air; no cristae.
Mutation rate	High.	High.
Mutagenic agents	U.V. light, 260 mμ max. Basic drugs (streptomycin).	U.V. light, 260 mμ max. Basic drugs (acridines).
Variability of mutants	Difference in carotenoid content and in porphyrinsynthetic abilities.	Suppressive and neutral types; suppressives can give rise to neutrals.
Biochemistry	Partial (but not complete) absence of chloroplast proteins.	Partial (but not complete) absence of mitochondrial proteins.
Organelle DNA	Probably absent in mutants studied.	Probably absent in mutants studied.

interact with DNA. The mutiple-hit induction of mutations by ultraviolet light (82, 103) strongly suggests that there are many equivalent genetic units in the cytoplasm which must all be hit before the ability to synthesize functional organelles is lost. The extraordinarily high efficiency which is nonetheless observed for the induction of cytoplasmic mutations could either be due to a difference between nucleus and organelles in the organization of the nucleic acids (82), or to the absence of "repair enzymes" (see Ref. 104) in the organelles.

Recently, it was independently shown in two laboratories that cytoplasmically determined characters can be transmitted from one cell to another by purified mitochondrial preparations. Following an approach pioneered by Pittman and Loker (quoted in Ref. 82), Tuppy and Wildner (1) used the spheroplasts of a stable acriflavin-induced "petite" mutant of *Saccharomyces cerevisiae*, auxotrophic for adenine and thymine, and incubated these with purified mitochondrial preparations from a leucine auxotroph with normal respiration. Under optimal conditions 2.6% "recombinants" were obtained, with normal respiration and without leucine requirement, but still auxotrophic for adenine and thymine. Apparently mitochondria can be taken up by yeast spheroplasts without irreparable damage to either the mitochondria or the spheroplasts. The results of control experiments with other cell fractions were not reported.

Similar experiments were done by Diacumakos (2, 3) with *Neurospora crassa*, but in this case a micro-injection technique was employed to introduce the abnormal mitochondria of a mutant into a normal recipient, since it had been shown (2) that the cytoplasmic character became phenotypically dominant in a heterocaryon. A high percentage of the injected *Neurospora* cells acquired the growth characteristics and abnormal cytochrome spectrum of the mutant donor strain. In control experiments it was shown that this result could not be duplicated by micro-injection of purified *Neurospora* nuclei, nuclear DNA or purified mitochondria from a normal strain.

Although these important experiments demonstrate that cytoplasmic characters can be transmitted by mitochondrial preparations, it has not been shown that the transmitted mitochondria replicate in the acceptor cell, or that mitochondrial DNA is involved in this process. With the experimental techniques used it should be possible, however, to obtain definite answers to both questions.

BIOCHEMICAL EVIDENCE

DNA synthesis: Since organelle DNA does not disappear in normal dividing cells, nucleic acid precursors should be incorporated into this DNA. This process has been studied in different systems and several points of interest were noted. Parsons (42) studied pulse-labelled *Tetrahymena* by autoradiography and showed that mitochondrial DNA became labelled even when no nuclear label was present. These results prove that incorporation of [^{3}H]-thymidine into mitochondrial DNA in *Tetrahymena* is a continuous process which is not synchronized with DNA synthesis in the nucleus. This suggests that mitochondrial DNA is not synthesized in the nucleus. In view of the large difference in size between mitochondrial DNA and nuclear DNA, and the known rate of DNA synthesis in mammalian nuclei (105) or bacteria (56), the continuous labelling of mitochondrial DNA throughout the growth cycle must indicate that replication of mitochondrial DNA is relatively slow, or that DNA synthesis in different mitochondria of one cell is not synchronized.

More rapid incorporation of DNA precursors into "heavy particle" (probably chloroplast) DNA as compared to nuclear DNA has been observed in growing plants (106, 107) and algae (108, 109). For plants good evidence was presented that the organelle DNA actually had a higher turnover than nuclear DNA. More rapid labeling of mitochondrial, than of nuclear, DNA *in vivo* was also recently reported for rat liver (27, 57, 110).

We do not think that these experiments provide an argument against the proposed genetic function of organelle DNA. The half-life of rat-liver mitochondria has been estimated to be about 10 days (111). If the half-life of mitochondrial DNA is also 10 days, it should be expected to turn over more rapidly than nuclear DNA. Genetic continuity will not be endangered by this apparent instability of the genetic material, unless growth and death of all mitochondria in a cell were completely synchronized, which is improbable. Even then, however, the organelle genome could be preserved in dormant promitochondria with a very low turnover.

Recently, Grossman and Marmur have shown (see Ref. 16) by ^{15}N transfer experiments, that mitochondrial DNA in *Saccharomyces carlsbergensis* replicates semiconservatively. Heavy (^{15}N–^{15}N) mitochondrial DNA could still be seen after nuclear DNA had undergone one replication to a completely hybrid (^{15}N–^{14}N) form. This indicates, in our opinion, that replication of mitochondrial DNA is asynchronous in yeast and that in rapidly multiplying cells the turnover rate of mitochondrial DNA is low in relation to the rate of duplication.

Incorporation of deoxynucleotides into the DNA present in mitochondrial preparations was briefly reported by Helge and Neubert (110). It would be of interest to determine whether these nucleotides are incorporated into mitochondrial DNA, identified as such by applying the criteria outlined above.

In conclusion, these experiments contribute suggestive evidence that organelle DNA can be synthesized in the organelle itself, and they provide no arguments against the proposed genetic role for this DNA.

RNA and protein synthesis: The ability of both isolated chloroplasts and mitochondria to incorporate labelled nucleotides into RNA and labelled amino acids into protein is now well established. Directly relevant to the possible role of organelle DNA in these processes is the observation that Actinomycin D exerts an inhibitory effect on the *in vitro* incorporation of nucleotides into chloroplast RNA (112, 113) and mitochondrial RNA (6, 7, 8, 114) from different organisms, while protein synthesis is also inhibited (115–119). In all cases the inhibition was far from complete, especially if intact mitochondria were used, although quite large amounts of inhibitor were added.

The interpretation of these experiments is complicated by the recent finding (120, 121) that Actinomycin D may have inhibiting effects on protein synthesis which are unrelated to its effect on template RNA synthesis and which may be due to an interaction with cellular membrane systems (120). It was shown by Kroon (116, 25), however, that the concentration of Actinomycin D, required for inhibition of mitochondrial protein synthesis, had no effect on mitochondrial oxidative phosphorylation, a sensitive measure for damage to the mitochondrial membrane. Moreover, the fact that high Actinomycin concentrations are required for inhibition of protein and RNA synthesis can be satisfactorily explained by the poor permeability of the intact mitochondrial membrane to Actinomycin, since swollen mitochondria are far more sensitive to the drug (8, 11, 110).

Table VII compares the requirements and properties of protein synthesis in intact chloroplasts and mitochondria with microsomal systems on the one hand and intact bacteria on the other. The differences between microsomes and intact organelles are clear. The sensitivity to inhibition by RNase and cycloheximide, and the need for external ATP and added pH 5 enzymes of the former system, together with the sensitivity to Actinomycin D, acriflavin and chloramphenicol of the latter, exclude a contribution of microsomes to the incorporation measured with isolated chloroplasts and mitochondria. The differences between bacteria and chloroplasts or mitochondria are less numerous and less obvious. It is therefore not surprising that reports have appeared in which amino acid incorporation by chloroplasts (123) and mitochondria (124) was attributed to bacterial contamination. The evidence against this point of view was recently summarized (9, 125). The RNase sensitivity and the need for external ATP in the case of chloroplasts (126–128), and the inhibition by low concentrations of 2,4-dinitrophenol and by swelling under mild conditions in the case of mitochondria (cf. 9), should be stressed.

TABLE VII. Protein Synthesis; Properties in Different Systems

	Microsomal systems	Mitochondria (intact)	Chloro-plasts	Bacteria
Inhibition by:				
Actinomycin D	—	±	±	+
Acriflavin (2 µg/ml)	—	++		±
Chloramphenicol	—	++	+	++
Puromycin	++	+	++	++
Cycloheximide	++	—		—
Deoxyribonuclease	—	—	±	—
Ribonuclease	++	—	++	—
Dinitrophenol (5×10⁻⁵M)	—	++		—
Swelling		+	+	—
Requirement for:				
pH 5 enzymes	++	—	—?	—
External ATP	++	—	++	—
$S_{20,w}$ value ribosomes	80		70	70

Acriflavin has been reported to inhibit RNA and protein synthesis by isolated mitochondria from yeast (17, 114, 118) and liver (8). The concentrations used for inhibition in these studies ranged from 20 to 50 µg/ml. We have recently shown (25) that these concentrations of acriflavin severely inhibit respiration and phosphorylation of isolated rat-liver mitochondria. However, nearly complete inhibition of protein synthesis by these mitochondria could already be obtained at an acriflavin concentration of 2 µg/ml, which had no detectable effect on oxidative phosphorylation. Maximal inhibition of protein synthesis occurred after a lag phase of 15 min, which was abolished by preincubation of the mitochondria with acriflavin (25). This suggests that the inhibition may indeed be due to a more direct effect on protein synthesis, presumably at the level of messenger-RNA synthesis (cf. Ref. 129).

A question relevant to the problem of a possible template function for the DNA in isolated organelles is, furthermore, whether all factors known to be involved in RNA and protein synthesis are present in these organelles or not. The presence of RNA polymerase may be deduced from the fact that under certain conditions RNA synthesis by isolated organelles is inhibited by Actinomycin D and DNase and is to some extent dependent on the presence of all four nucleoside triphosphates (6, 8, 17, 112–114). Fractions comparable with pH 5 enzymes have been obtained from disrupted mitochondria (130, 131) as well as from the mobile phase of chloroplasts (127). The presence of ribosomes in chloroplasts is well established (115, 128, 132). For mitochondria the biochemical evidence is restricted to the observations of Wintersberger (7), who reported the isolation of two RNA fractions with sedimentation properties which suggest that the fractions are of ribosomal origin. Moreover, André and Marinozzi (133) have recently presented morphological evidence for the existence of mitochondrial ribosomes.

In the case of chloroplasts the ribosomes differ from the cytoplasmic ribosomes *sensu stricto* because they sediment with a $S_{20,w}$ value of 70 (128), hence like bacterial ribosomes (134). It has been shown that chloramphenicol is specifically bound to the 50S sub-units of bacterial ribosomes and this binding appears to be necessary for inhibition of protein synthesis (135). It is therefore tempting to speculate that a 70S-type ribosome is required for inhibition of protein synthesis by chloramphenicol and that mitochondrial ribosomes will also turn out to be of the "bacterial" (134) type.

Although microsomal protein synthesis of mammalian systems is insensitive to chloramphenicol in short-term experiments, long-term effects were noted which have been ascribed (136) to the inhibition by chloramphenicol of the attachment of messenger RNA to cytoplasmic 80S ribosomes. One may inquire, however, whether the effects observed could not be secondary and due to an insufficient supply of energy in rapidly dividing cells, in which the biosynthesis of new mitochondria is blocked because mitochondrial protein synthesis is inhibited. That this inhibition can occur in the intact cell was recently demonstrated by an important experiment of Huang *et al.* (137), who cultured yeast under aerobic conditions in a glucose-containing medium with and without chloramphenicol. Although the growth rate and the final yield of cells were not influenced by the drug, striking changes occurred in the treated cells: After 16 hr of growth only a few diffuse mitochondrial profiles without significant amounts of cristae were visible per cell; the rate of respiration had dropped to less than 5 per cent of the control value and the cytochrome spectrum of the cells had completely changed. Cytochrome oxidase had disappeared completely while the concentration of cytochrome c had not diminished, with the result that the spectrum resembled that obtained for the various types of "petite" mutants of yeast (101).

These experiments strongly suggest that mitochondrial protein synthesis can be specifically inhibited in intact yeast cells. It is of interest that this inhibition prevents the synthesis of some, but not all, mitochondrial proteins. This fits in well with the results obtained with mitochondria *in vitro*, since in all cases amino acids were only incorporated into membrane-bound proteins, while loosely bound dehydrogenases and cytochrome c were not labelled at all (138, 139).

Although the experimental evidence reviewed in this section supports the hypothesis that organelle DNA provides the information for the synthesis of part of the organelle proteins, direct proof for this hypothesis is still lacking. Further characterization of mitochondrial RNA and hybridization studies of this RNA with highly purified mitochondrial DNA might settle this matter.

The Biosynthesis of Mitochondria: Available Evidence

For decades biosynthesis of mitochondria has been a subject of intensive speculation by morphologists. The fact that these organelles are highly mobile in the cell makes it possible to obtain photographs of mitochondria which appear to evolve from every type of cellular membrane conceivable, so that each membrane has been implicated in turn in mitochondrial biosynthesis (for reviews see Refs. 9, 141 and 142). More recently, considerable advances have been made in our knowledge of the formation of mitochondria in lower organisms, which have obvious advantages as test objects. In an outstanding and detailed study of the formation of mitochondria in rapidly growing *Neurospora crassa* filaments, Luck (142–144) designed experiments to decide between two alternative hypotheses for the synthesis of mitochondria: *de novo* synthesis from submitochondrial precursors or other cell membranes; or synthesis due to growth and division of existing mitochondria. In the first series of experiments (142), a choline-requiring *Neurospora* mutant was grown in a medium containing radioactive choline. Following a shift to a medium containing unlabeled choline the decrease in the specific radioactivity of the purified mitochondrial fraction was followed for three doubling cycles by autoradiography. It was found that radioactivity remained randomly distributed over all mitochondria and that the mitochondrial population did not resolve into "old" mitochondria with the original high radioactivity and "new" mitochondria without radioactivity. Secondly, it was shown that after a 10 min pulse of radioactive choline the radioactivity was randomly distributed over all mitochondria and that fusion, followed by fission, (if occurring at all), was too slow to ascribe the results to a rapid equilibration of material between "old" and "new" mitochondria synthesized *de novo*. A second series of experiments (144) using different, but not independent, techniques gave similar results. These experiments provide good evidence that in rapidly growing *Neurospora crassa* mycelia mitochondria are predominantly, if not exclusively, synthesized by growth and division of pre-existing mitochondria. However, as already pointed out by Luck, these results cannot exclude the possibility that choline is incorporated into a macromolecular structure outside the mitochondrion, which subsequently fuses with the latter. We shall return to this possibility in the next section.

A second organism in which mitochondrial biosynthesis has been intensively studied in recent years is yeast. Although anaerobic yeast does not contain any mitochondria-like structures, work in several laboratories (145–148) has provided strong evidence for the existence of vesicular structures with the characteristics of promitochondria, from which the mitochondria appear to develop on switching to aerobiosis. It has also tentatively been shown that specific mitochondrial enzymes are associated with these promitochondria (149).

These two pathways for the biosynthesis of mitochondria, by growth and division and from promitochondria, are at the moment the only pathways for which adequate evidence exists. There is no convincing evidence in our opinion for the *de novo* derivation of mitochondria from other membrane systems (however, see Refs. 140, 141, 150 and 151 for different opinions) although it cannot be excluded that mitochondria can grow by fusing with other membrane structures.

It should be obvious that these two pathways for mitochondrial biosynthesis are not mutually exclusive and mitochondrial biosynthesis in yeast growing for generations under aerobic conditions may well be similar to that observed in *Neurospora crassa*. Whether the results obtained with yeast and *Neurospora* can be extended to higher organisms is not known.

The biosynthesis of chloroplasts has been studied only by morphological techniques but the study of this problem is simplified by the fact that formation of chloroplasts is inducible in many simple organisms. There is evidence that chloroplasts are formed both from proplastids and by growth and division of pre-existing chloroplasts, although the latter process is rarely seen in higher plants. The proplastids remain present in dark-grown algae and are apparently able to multiply by growth and division. No evidence is available to support a complete *de novo* synthesis or evolvement of chloroplasts from other cell structures than the proplastids. For an extensive discussion of this problem and references to the original literature see Ref. 152.

Nothing is known as to how the components of such complicated multi-component systems as the respiratory chain, with the associated phosphorylative machinery, are assembled in the correct relative proportions and in the right sequence. The complexity of this problem was discussed *in extenso* in a recent paper by Green and Hechter (153) who came to the conclusion that "The possibility of spontaneous self-assembly cannot be seriously entertained". Two alternative hypotheses were considered: In the first hypothesis "polysome units" are involved "as the directing element in the assembly of macromolecular arrays ...... via complementary interactions of nascent polypeptide chains still attached to one or more polysome units". Apparently Green and Hechter presume that the spatial sequence of nascent proteins on a polysome might somehow result in the formation of a multi-component respiratory chain with the correct spatial arrangement of subunits without the interference of self-assembly. This hypothesis, however, was not extended and the authors discarded it because the polysome array required would be "too gigantic to discuss seriously".

The alternative proposed by Green and Hechter—the template hypothesis—has the following features: The phosphorylating respiratory chain is built up stepwise on templates. First the single components, e.g. cytochromes, are assembled on a protein template I in a protected form to prevent aggregation. After exchange of the protecting protein for phospholipid, complex I can leave template I to be assembled into a larger unit on a second template etc. Stoichiometry and sequence are therefore built into the template and are not a reflection of the ability of the monomers to interact specifically with each other.

The proposal by Green and Hechter is essentially based on the arguments that a mitochondrion is extremely complex and that tne isolated protein components have a strong tendency to interact non-specifically with each other or to form aggregates which can only be solubilized with detergents. Neither argument is conclusive. Self-assembly is responsible for the proper assembly of Tobacco Mosaic Virus from 1 molecule of RNA and about 2000 molecules of coat protein (154, 155); and self-assembly is very likely to occur with other viruses, some of which are very complex indeed, or have coat-protein subunits which form intractable aggregates *in vitro* which can only be solubilized with detergents. Similar considerations hold for the biosynthesis of ribosomes. Moreover, in our opinion the template hypothesis of Green and Hechter only displaces the main difficulty: the templates required, especially in the later stages of assembly, must be very large macromolecules; and since the molecules must have a very distinctive template surface, they must be built up of an enormous array of non-identical peptide chains, which must be assembled by self-assembly unless one wants to extend the template hypothesis *ad infinitum*. Another objection against this template hypothesis is that it would make the mitochondrion more vulnerable to mutations, since cistrons specifying the template would provide an extra target for mutagenic agents, while it would also considerably decrease the evolutionary flexibility of the mitochondrion.

We therefore prefer to retain the simple idea that the phosphorylating respiratory chain is made by a process of self-assembly in which stoichiometry and sequence are determined only by the properties of the monomeric units. It is possible that stepwise assembly, preventing "incorrect" interactions, is achieved by a system similar to that proposed by Gruber and Campagne (156) for the regulation of enzyme induction, e.g. that the finished polypeptide chain in extended configuration blocks transcription and/or translation and that it can only fold in its proper tertiary structure (with consequent release from the ribosomes) after interaction with its specific site on the growing respiratory chain. Such a process could both control rate of synthesis of the individual components and prevent non-specific interactions. Finally recent advances in our knowledge of protein synthesis (157, 158) form a reminder of the possibility that the primary structure of peptide chains could be modified *after* their assembly in a macromolecular system. Therefore the properties of the monomers which are isolated from the completed mitochondrion might be different from those of the monomers as they are inserted into the membrane subunit.

The Contribution of Organelle DNA to the Biosynthesis of Organelles: Working Hypotheses

If one accepts the extensive circumstantial evidence, outlined in previous sections, that organelle DNA has a genetic function, it is of interest to consider briefly possible ways in which this genetic system might function in the intact cell. Possible models for the biosynthesis of the DNA-containing organelles can be divided into two main groups:

I. All structural genes for organelle proteins are represented on the organelle DNA.

II. Only some of these genes are represented on the organelle DNA; the remainder are represented in the nucleus only.

Two types of model II should be distinguished: according to model II A, nuclear messenger RNA's containing copies of the structural genes for organelle proteins are translated inside the organelle by the protein-synthesizing machinery of the organelle. According to model II B, these nuclear messengers are translated outside the mitochondrion.

A version of model I has been discussed in detail by Gibor and Granick (10). This model implies the presence in organelles of organelle-specific information equivalent to at least 100×10^6 Daltons DNA. Although this possibility cannot be excluded in the case of chloroplasts or kinetoplasts, it is very unlikely from the available evidence that this requirement will be met in the case of mitochondria.

For the biosynthesis of mitochondria we are therefore forced to invoke one of the versions of model II. An objection against model II A is that it is difficult to see how a highly charged macromolecule like RNA could be taken up by intact mitochondria, in view of the known properties of the mitochondrial membrane. No direct evidence on this point is available, however. A second difficulty is that the synthesis of low amounts of organelle-specific proteins in cytoplasmic mutants cannot be explained satisfactorily by model II A, unless some rather unattractive assumptions are made.

Two serious objections can be formulated against simple versions of model II B: In general mitochondrial enzymes are not found in the cytoplasm and vice versa (see 140). Probably exceptions to this rule are only apparent since, for instance, the mitochondrial and extra-mitochondrial species of malate dehydrogenase and aspartate transaminase have been shown to be not identical but iso-enzymes (159). Secondly, several mitochondrial enzymes are very loosely bound, or not bound at all, to the membrane, and they are only retained by the mitochondrion because its membranes are impermeable to these enzymes (140), just as they are impermeable to added DNase and RNase.

Although there are ways of getting around these objections, we would like to restrict ourselves to one version of model II B, which has not been considered before. This version is based on an analogy with the Virus-Synthesizing Bodies formed in mammalian cells under the direction of replicating RNA viruses (160). These bodies are deoxycholate-sensitive structures containing the bulk of the viral RNA synthetase (161) and the newly-synthesized viral RNA and coat protein present in the cell. Since the single-stranded RNA viruses are essentially messenger-RNA's, it is reasonable to assume that similar lipoprotein structures can be formed under the direction of nuclear messenger RNA's. If these bodies could merge with growing mitochondria, newly synthesized mitochondrial enzymes (and even the messenger RNA to synthesize these intra-mitochondrially) would be introduced into the mitochondria without violating known cellular compartmentation. In such a system the contribution made by the mitochondrial DNA to the biosynthesis of the mitochondria could be very modest and it might even become progressively smaller in the course of evolution, if the assumption is retained that the DNA-containing cell organelles are the descendants of autonomous symbionts (38). The implications of this model are numerous but the available space does not allow us to develop it in detail or to point out its flaws.

Outlook

It has lately become common practice to end papers dealing with the possible role of organelle DNA in the biosynthesis of organelles with the optimistic statement that the DNA-containing organelles will be cultured *in vitro* in the near future. From the last section of this paper it is clear that we do not share this optimism since we expect that the biosynthesis of mitochondria, at least, will be found to require the subtle interplay of several different cell components, which may be extremely difficult to reproduce in a sub-cellular system. It is also clear, however, that the rapid advances of the last two years have made the problem of the mechanism of organelle biosynthesis amenable to a concerted biochemical attack from several angles. In the next two years this may be expected to yield clear-cut answers to some of the most complicated and fascinating questions of biochemistry today.

ACKNOWLEDGEMENTS

We are grateful to Professor E. C. Slater for many helpful suggestions and encouragement, to Dr. E. F. J. van Bruggen and Professor M. Gruber for collaborating in the study of the structure of mitochondrial DNA and for providing the electron micrographs; and to Dr. D. Hillenius for advice on taxonomic problems.

REFERENCES

1. Tuppy, H. and Wildner, G., *Biochem. Biophys. Res. Commun.* 20, 733 (1965).
2. Garnjobst, L., Wilson, J. F. and Tatum, E. L., *J. Cell Biol.* 26, 413 (1965).
3. Diacumakos, E. G., Garnjobst, L. and Tatum, E. L., *J. Cell Biol.* 26, 427 (1965).
4. Borst, P. and Ruttenberg, G. J. C. M., *Biochim. Biophys. Acta* 114, 645 (1966).
5. Van Bruggen, E. F. J., Borst, P., Ruttenberg, G. J. C. M., Gruber, M. and Kroon. A. M., *Biochim. Biophys. Acta* 119, 437 (1966).
6. Luck, D. J. L. and Reich, E., *Proc. Natl. Acad. Sci. U.S.* 52, 931 (1964).
7. Wintersberger, E., in Tager, J. M., Papa, S., Quagliariello, E. and Slater, E. C. (Eds.), *Regulation of Metabolic Processes in Mitochondria*, Elsevier, (BBA Library, Vol. 7), Amsterdam, p. 439 (1966).
8. Neubert, D., Helge, H. and Merker, H. J., *Biochem. Z.* 343, 44 (1965).
9. Kroon, A. M., *Protein Synthesis in Mitochondria*, M. D. Thesis, Amsterdam, Drukkerij Noordholland, Hoorn (1966).
10. Gibor, A. and Granick, S., *Science* 145, 890 (1964).
11. Chun, E. H. L., Vaughan, Jr., N. H. and Rich, A., *J. Mol. Biol.* 7, 130 (1963).
12. Leff, J., Mandel, M., Epstein, H. T. and Schiff, J. A., *Biochem. Biophys. Res. Commun.* 13, 126 (1963).
13. Sager, R. and Ishida, M. R., *Proc. Natl. Acad. Sci. U.S.* 50, 725 (1963).
14. Ray, D. S. and Hanawalt, P. C., *J. Mol. Biol.* 9, 812 (1964).
15. Brawerman, G. and Eisenstadt, J. M., *Biochim. Biophys. Acta* 91, 477 (1964).
16. Corneo, G., Moore, C., Sanadi, D. R., Grossman, L. I. and Marmur, J., *Science* 151, 687 (1966).
17. Tewari, K., Jayaraman, J. and Mahler, H., in Mills, R. K., *Some aspects of Yeast Metabolism*, Oxford University Press (In press and personal communication).
18. Suyama, U. and Preer, Jr., J. R., *Genetics* 52, 1051 (1965).
19. Parsons, J. A. and Dickson, R. C., *J. Cell Biol.* 27, 77 A (1965).
20. Edelman, M., Schiff, J. A. and Epstein, H. T., *J. Mol. Biol.* 11, 769 (1965).

21. Ray, D. S. and Hanawalt, P. C., *J. Mol. Biol.* **11**, 760 (1965).
22. Suyama, J. and Bonner, Jr., W. D., *Plant Physiol.* (In press).
23. Kroon, A. M., in Tager, J. M., Papa, S., Quagliariello, E. and Slater, E. C. (Eds.), *Regulation of Metabolic Processes in Mitochondria*, Elsevier, (BBA Library, Vol. 7), Amsterdam, p. 397 (1966).
24. Borst, P. and Ruttenberg, G. J. C. M., in Tager, J. M., Papa, S., Quagliariello, E. and Slater, E. C. (Eds.), *Regulation of Metabolic Processes in Mitochondria*, Elsevier, (BBA Library, Vol. 7), Amsterdam, p. 454 (1966).
25. Borst, P., Kroon, A. M. and Ruttenberg, G. J. C. M., Unpublished observations.
26. Kalf, G. F. and Grece, A., *J. Biol. Chem.* (In press).
27. Schneider, W. C. and Kuff, E. L., *Proc. Natl. Acad. Sci. U.S.* **54**, 1650 (1965).
28. Rabinowitz, M., Sinclair, J., Desalle, L., Haselkorn, R. and Swift, H. H., *Proc. Natl. Acad. Sci. U.S.* **53**, 1126 (1965).
29. Kislev, N., Swift, H. and Bogorad, L., *J. Cell. Biol.* **25**, 327 (1965).
30. Shipp, W. S., Kieras, F. J. and Haselkorn, R., *Proc. Natl. Acad. Sci. U.S.* **54**, 207 (1965).
31. Edelman, M., Cowan, C. A., Epstein, H. T. and Schiff, J. A., *Proc. Natl. Acad. Sci. U.S.* **52**, 1214 (1964).
32. Sueoka, N., *J. Mol. Biol.* **3**, 31 (1961).
33. Hearst, J. E. and Vinograd, J., *Progress in the Chemistry of Organic Natural Products*, Springer Verlag, Berlin (1962).
34. Erikson, R. L. and Szybalski, W., *Virology* **22**, 111 (1964).
35. Schildkraut, C. L., Marmur. J. and Doty, P., *J. Mol. Biol.* **4**, 430 (1962).
36. Thenius, E. and Hofer, H., *Stammesgeschichte der Saügetiere*, Springer Verlag, Berlin (1960).
37. De Duve, C., in Hayashi, T. (Eds.), *Subcellular Particles*, Ronald Press, New York, p. 128 (1959,).
38. Ris, H., in Karlson, P. (Eds.), *Funktionelle und Morphologische Organisation der Zelle*, Springer Verlag, Berlin, p. 3 (1963).
39. Nass, M. M. K. and Nass, S., *J. Cell Biol.* **19**, 593 and 613 (1963).
40. Nass, S. and Nass, M. M. K., *J. Natl. Cancer Inst.* **33**, 777 (1964).
41. Nass, M. M. K., Nass, S. and Afzelius, B. A., *Exptl. Cell Res.* **37**, 516 (1965).
42. Parsons, J. A., *J. Cell Biol.* **25**, 641 (1965).
43. Crawford, L. V., Crawford, E. M., Richardson, J. P. and Slayter, H. S., *J. Mol. Biol.* **14**, 593 (1965).
44. Marmur, J. and Doty, P., *J. Mol. Biol.* **3**, 585 (1961).
45. Marmur, J., Rownd, R. and Schildkraut, C. L., *Progress in Nucleic Acid Res.* **1**, 231 (1963).
46. Rownd, R., *Brit. Med. Bull.* **21**, 187 (1965).
45. Subirana, J. A., *Biochim. Biophys. Acta* **103**, 13 (1965).
48. Hogness, D. S. and Simmons, J. R., *J. Mol. Biol.* **9**, 411 (1964).
49. Vinograd, J., Lebowitz, J., Radloff, R., Watson, R. and Laipis, P., *Proc. Natl. Acad. Sci. U.S.* **53**, 1104 (1965).
50. Jansz, H. S. and Pouwels, P. H., *Biochem. Biophys. Res. Commun.* **18**, 589 (1965).
51. Crawford, L. V., *J. Mol. Biol.* **13**, 362 (1965).
52. Hershey, A. D., Burgi, E. and Ingraham, L., *Proc. Natl. Acad. Sci. U.S.* **49**, 748 (1963).
53. Bode, V. C. and Kaiser, A. D., *J. Mol. Biol.* **14**, 399 (1965).

54. Eigner, J. and Doty, P., *J. Mol. Biol.* **12**, 549 (1965).
55. Kleinschmidt, A. K., Lang, D., Jacherts, D. and Zahn, R. K., *Biochim. Biophys. Acta* **61**, 857 (1962).
56. Cairns, J., *Cold Spring Harbor Symp. on Quant. Biol.* **20**, 43 (1963).
57. Nass, S., Nass, M. M. K. and Hennix, U., *Biochim. Biophys. Acta* **95**, 426 (1965).
58. Schatz, G., Haslbrunner, E. and Tuppy, H., *Monatshefte für Chemie* **95**, 1135 (1964).
59. Kalf, G. F., *Biochemistry* **3**, 1702 (1964).
60. Bahr, G. F. Herbener, G. H. and Glas, U., *Exptl. Cell Res.* **41**, 99 (1966).
61. Allard, C., Mathieu, R., De Lamirande, G. and Cantarow, A., *Cancer Res.* **12**, 407 (1952).
62. Estabrook, R. W. and Holowinsky A., *J. Biophys. Biochem. Cytol.* **9**, 19 (1961).
63. Yoshikawa, H., O'Sullivan, A. and Sueoka, N., *Proc. Natl. Acad. Sci. U.S.* **52**, 973 (1964).
64. Fréderic, J., *Arch. Biol.* (Liège) **69**, 198 (1958).
65. Wildman, S. G., Hongladarom, T. and Honda, S. J., *Science* **138**, 434 (1962).
66. Stocking, C. R. and Gifford, Jr., E. M., *Biochem. Biophys. Res. Commun.* **1**, 159 (1959).
67. Ris, H. and Plaut, W., *J. Cell Biol.* **13**, 383 (1962).
68. Steffensen, O. M. and Sheridan, W. F., *J. Cell Biol.* **25**, 619 (1965).
69. Shephard, D. C., *Biochim. Biophys. Acta* **108**, 635 (1965).
70. Gibor, A. and Isawa, M., *Proc. Natl. Acad. Sci. U.S.* **50**, 1164 (1963).
71. Pollard, C. J., *Arch. Biochem. Biophys.* **105**, 115 (1964).
72. Guttman, H. N. and Wallace, F. G., in Hutner, S. H., *Biochemistry and Physiology of Protoza*, Vol. III, Academic Press, New York, p. 459 (1964).
73. Dubuy, H. G., Mattern, C. F. T. and Riley, F. L., *Science* **147**, 754 (1965).
74. Schildkraut, C. L., Mandel, M., Levisohn, S., Smith–Sonneborn, J. E. and Marmur, J., *Nature* **196**, 795 (1962).
75. Morton, R. K. and Raison, J. K., *Nature* **200**, 429 (1963).
76. Rozijn, T. H. and Tonino, G. J. M., *Biochim. Biophys. Acta* **91**, 105 (1964).
77. Ogur, M., Hickler, S., Lindegren, G. and Lindegren, C. C., *Arch. Biochem. Biophys.* **40**, 175 (1952).
78. Bonner, J., in Bonner, J. and Varner, J. E. (Eds.), *Plant Biochemistry*, Vol. I, Academic Press, New York (1965).
79. Vendrely, J. N., in Chargaff, E. and Davidson, J. N. (Eds.), *The Nucleic Acids*, Academic Press, Vol. II, New York, p. 155 (1955).
80. Avers, C. J., Pepper, C. R. and Rancourt, M. W., *J. Bacteriol.* **90**, 481 (1965).
81. Avers, C. J., Rancourt, M. W. and Lin, F. M., *Proc. Natl. Acad. Sci. U.S.* **54**, 527, (1965).
82. Wilkie, D., *The cytoplasm in heredity*, Methuen & Co., London (1964).
83. Gillespie, D. and Spiegelman, S., *J. Mol. Biol.* **12**, 829 (1965).
84. Wallace, J. and Birnstiel, M. L., *Biochim. Biophys. Acta* **114**, 296 (1966).
85. Hoyer, B. H., McCarthy, B. J. and Bolton, E. T., *Science* **144**, 959 (1964).
86. Humm, D. G. and Humm, J. H., *Proc. Natl. Acad. Sci. U.S.* **55**, 114 (1966).
87. Kroon, A. M., *2nd Meeting Federation of European Biochemical Societies*, Vienna, p. 102 (1965).
88. Cheng, T. Y. and Sueoka, N., *Science* **143**, 1442 (1964).
89. Bienfait, F., Borst, P. and Ruttenberg, G. J. C. M., Unpublished observations.

90. Davidson, N., Widholm, J., Nandi, U. S., Jensen, R., Olivera, B. M. and Wang, J. C., *Proc. Natl. Acad. Sci. U.S.* **53**, 111 (1965).
91. Cheng, T. Y. and Sueoka, N., *Science* **141**, 1194 (1963).
92. Polli, E., Corneo, G., Ginelli, E. and Bianchi, P., *Biochim. Biophys. Acta* **103**, 672 (1965).
93. Chun, E. H. L. and Littlefield, J. W., *J. Mol. Biol.* **7**, 245 (1963).
94. Dawid, I. B., *J. Mol. Biol.* **12**, 581 (1965).
95. Bibring, T., Brachet, J., Gaeta, F. S. and Graziosi, F., *Biochim. Biophys. Acta* **108**, 644 (1965).
96. Chevremont, M., in Harris, R. J. C. (Eds.), *Cell Growth and Cell Division*, Academic Press, New York, p. 323 (1963).
97. Ephrussi, B., *Nucleocytoplasmatic Relations in Microorganisms*, Clarendon Press, Oxford (1953).
98. Sager, R., *Symp. Soc. Genl. Microbiol.* **15**, 324 (1965).
99. Jinks, J. L., *Extrachromosomal Inheritance*, Englewood Cliffs, N. J.; Prentice Hall (1964).
100. Mahler, H. R., Mackler, B., Grandchamp, S. and Slonimski, P. P., *Biochemistry* **3**, 668 (1964).
101. Mackler, B., Douglas, H. C., Will, S., Hawthorne, D. C. and Mahler, H. R., *Biochemistry* **4**, 2016 (1965).
102. Katch, T. and Sanuida, S., *Biochem. Biophys. Res. Commun.* **21**, 373 (1965).
103. Lyman, H., Epstein, H. T. and Schiff, J. A., *Biochim. Biophys. Acta* **50**, 301 (1961)
104. Setlow, R. B., Carrier, W. L. and Bollum, F. J., *Proc. Natl. Acad. Sci. U.S.* **53**, 1111 (1965).
105. Cairns, J., *J. Mol. Biol.* **15**, 372 (1966).
106. Sampson, M., Katch, A., Hotta, Y. and Stern, H., *Proc. Natl. Acad. Sci. U.S.* **50**, 459 (1963).
107. Hotta, Y., Bassel, A. and Stern, H., *J. Cell Biol.* **27**, 451 (1965).
108. Iwamura, T. and Kuwashima, S., *Biochim. Biophys. Acta* **82**, 678 (1964).
109. Iwamura, T. and Muto, N., *Plant and Cell Physiol.* **5**, 359 (1964).
110. Helge, H. and Neubert, D., *Naunyn-Schmiedebergs Arch. Exp. Path. Pharmak.* **251**, 113 (1965).
111. Fletcher, M. S. and Sanadi, D. R., *Biochim. Biophys. Acta* **51**, 356 (1961).
112. Kirk, J. T. O., *Biochem. Biophys. Res. Commun.* **14**, 393 (1964); **16**, 393 (1964).
113. Schweiger, H. G. and Berger, S., *Biochim. Biophys. Acta* **87**, 533 (1964).
114. Wintersberger, E. and Tuppy, H., *Biochem. Z.* **341**, 399 (1965).
115. Sissakian, N. M., Filippovich, J. J., Svetailo, E. N. and Aliyev, K. A., *Biochim. Biophys. Acta* **95**, 474 (1965).
116. Kroon, A. M., *Biochim. Biophys. Acta* **108**, 275 (1965).
117. Kalf, G. F., *Biochemistry* **3**, 1702 (1964).
118. Wintersberger, E., *Biochem. Z.* **341**, 409 (1965).
119. Neubert, D. and Helge, H., *2nd Meeting Federation European Biochemical Societies*, Vienna, p. 84 (1965).
120. Revel, M., Hiatt, H. H. and Revel, J. P., *Science* **146**, 1311 (1964).
121. Honig, G. R. and Rabinowitz, M., *Science* **149**, 1504 (1965).
122. Kroon, A. M., *Biochim. Biophys. Acta* **76**, 165 (1963).
123. App, A. A. and Jagendorf, A. T., *Plant Physiol.* **39**, 772 (1964).

124. Von Der Decken, A., Löw, H. and Sandell, S., in Tager, J. M., Papa, S., Quagliariello, E. and Slater, E. C. (Eds.), *Regulation of Metabolic Processes in Mitochondria*, Elsevier, (BBA Library, Vol. 7), Amsterdam, p. 415 (1966).
125. Roodyn, D. B., in Tager, J. M., Papa, S., Quagliariello, E. and Slater, E. C. (Eds.), *Regulation of Metabolic Processes in Mitochondria*, Elsevier, (BBA Libary, Vol. 7), Amsterdam, p. 383 (1966).
126. Spencer, D. and Wildman, S. G., *Biochemistry* **3**, 954 (1964).
127. Francki, R. J. B., Boardman, N. K. and Wildman, S. G., *Biochemistry* **4**, 865 (1965).
128. Boardman, N. K., Francki, R. J. B. and Wildman, S. G., *Biochemistry* **4**, 872 (1965).
129. Gros, F., Dubert, J. M., Tissieres, A., Bourgeois, S., Michelson, M., Soffer, R. and Legault, L., *Cold Spring Harbor Symp. Quant. Biol.* **28**, 299 (1963).
130. Truman, D. E. S. and Korner, A., *Biochem. J.* **85**, 154 (1962).
131. Kroon, A. M., *Biochim. Biophys. Acta* **72**, 391 (1963).
132. Clark, M. F., *Biochim. Biophys. Acta* **91**, 671 (1964).
133. André, J. and Marinozzi, V., *J. Microscopie* **4**, 615 (1965).
134. Taylor, M. M. and Storck, R., *Proc. Natl. Acad. Sci. U.S.* **52**, 958 (1964).
135. Vasquez, D., *Biochim. Biophys. Acta* **114**, 277, 289 (1966).
136. Weisberger, A. S., Daniel, T. M. and Hoffman, A., *J. Exptl. Med.* **120**, 183 (1964).
137. Huang, M., Biggs, D. R., Clark–Walker, G. D. and Linnane, A. W., *Biochim. Biophys. Acta* **114**, 434 (1966).
138. Truman, D. E. S., *Biochem. J.* **91**, 59 (1964).
139. Bronsert, U. and Neupert, W., in Tager, J. M., Papa, S., Quagliariello, E. and Slater, E. C., (Eds.), *Regulation of Metabolic Processes in Mitochondria*, Elsevier, (BBA Library, Vol. 7), Amsterdam, p. 426 (1966).
140. Lehninger, A. L., *The Mitochondrion*, W. A. Benjamin, New York (1964).
141. Novikoff, A. B., in Brachet, J. and Mirsky, E. (Eds.), *The Cell*, Vol. 2, Academic Press, New York, p. 299 (1960).
142. Luck, D. J. L., *J. Cell Biol.* **16**, 483 (1963).
143. Luck, D. J. L., *J. Cell Biol.* **24**, 445 (1965).
144. Luck, D. J. L., *J. Cell Biol.* **24**, 461 (1965).
145. Yotsuyanagi, Y., *Ultrastructure Res.* **7**, 121, 141 (1962).
146. Wallace, P. G. and Linnane, W. A., *Nature* **201**, 1191 (1964).
147. Polakis, E. S., Bartley, W. and Meek, J. A., *Biochem. J.* **90**, 369 (1964).
148. Ridge, W. M. and Avers, C. J., *J. Cell Biol.* **27**, 84 A (1965).
149. Schatz, G., *Biochem. Biophys. Res. Commun.* **12**, 448 (1963).
150. Bell, P. R. and Mühlethaler, K., *J. Mol. Biol.* **8**, 853 (1964).
151. Schejde, O. A., McCandless, R. G. and Munn, R. Y., *Nature* **203**, 158 (1964).
152. Park, R. B., in Bonner, J. and Varner, J. E. (Eds.), *Plant Biochemistry*, Vol. 1, Academic Press, New York, p. 142 (1965).
153. Green, D. E. and Hechter, C., *Proc. Natl. Acad. Sci. U.S.* **53**, 318 (1965).
154. Fraenkel–Conrat, H. and Williams, R. C., *Proc. Natl. Acad. Sci. U.S.* **41**, 690 (1955).
155. Fraenkel–Conrat, H. and Singer, B., *Virology* **23**, 354 (1964).
156. Gruber, M. and Campagne, R. N., *Proc. Kon. Ned. Ak. Wetenschappen*, Series C **68**, 1 (1965).
157. Adams, J. M. and Capecchi, M. R., *Proc. Natl. Acad. Sci. U.S.* **55**, 147 (1966).
158. Webster, R. E., Engelhardt, D. L. and Zinder, N. D., *Proc. Natl. Acad. Sci. U.S.* **55**, 155 (1966).

159. Borst, P., in Karlson, P. (Eds.), *Funktionelle und Morphologische Organisation der Zelle*, Springer Verlag, Berlin, p. 137 (1963).
160. Penman, S., Becker, Y. and Darnell, J. E., *J. Mol. Biol.* **8**, 541 (1964).
161. Ochoa, S., Weissmann, C., Borst, P., Burdon, R. H. and Billeter, M. A., *Federation Proc.* **23**, 1285 (1964).

DISCUSSION
(Chairman S. G. LALAND)

J. SPONAR: Have you any reason to believe that the strands of mitochondrial DNA are going to separate by heating?

P. BORST: The renaturation follows the kinetics of a second-order reaction.

M. SLUYSER: Is any histone bound to mitochondrial DNA?

P. BORST: We haven't looked. If you would like to investigate this, you are welcome to have some of our DNA.

M. J. EVANS: Have you considered comparing this DNA with the very abundant cytoplasmic DNA present in early amphibian embryos?

P. BORST: This matter is under study in our laboratory at the moment.

I. D. CLARK: Acriflavine can induce akinetoplasty. Does Dr. Borst have any evidence that acriflavine can induce the degradation of mitochondrial DNA?

P. BORST: Acriflavine has been known for a long time to be one of the most efficient mutagenic agents for the induction of the cytoplasmic "petite" mutants in yeast.

T. WILCZOK: You did not mention anything about the hyperchromicity and T_m values of the mitochondrial DNA. Do you have any data? If so, are the GC contents calculated from the buoyant density values and T_m values the same?

P. BORST: The T_m of chick-liver mitochondrial DNA was earlier reported by us to be 90.0° in 0.15 M NaCl, 0.015 M sodium citrate pH 7. The GC content calculated from this T_m and the buoyant density in CsCl are the same. The hyperchromicity of mitochondrial DNA is about 30 per cent but it is difficult to completely remove all U. V. absorbing contaminants, so this may be an underestimation.

D. S. RAY: I would simply like to point out that one of the DNA species which you refer to as a mitochondrial DNA, namely the satellite DNA of *Euglena gracilis* with buoyant density of 1.691, has a molecular weight of about 3×10^6. This species does not have the molecular weight of 11×10^6, which you suggest as a common size for mitochondrial DNA.

P. BORST: I was aware of your work with *Euglena* but it is very difficult to exclude that the DNA which you isolated was not the breakdown product of a much larger DNA.

D. SHUGAR: Under what conditions have you investigated the alkaline denaturation of your mitochondrial DNA? Have you examined in any detail the pH profile for alkaline denaturation? In the light of the facts you have presented, such data may very well provide additional useful information with regard to the homogeneity of your mitochondrial DNA preparations.

P. BORST: We merely added NaOH to 0.1M final concentration; we have not determined alkaline transition profiles.

Mechanisms of Viral RNA Replication

E. M. MARTIN

National Institute for Medical Research, London, N.W.7, England

INTRODUCTION

Our understanding of the nature of virus replication has advanced enormously in the past few years, mainly as a result of intensive studies on the replication of small RNA-containing viruses such as picornaviruses and phages. The pioneer work on the effects of picornavirus infection on the host cell's metabolism revealed a fundamental difference between the mode of synthesis of the RNA component of these viruses and that of the uninfected host cell (1). While mammalian cell RNA synthesis is catalyzed by enzymes using the host cell's DNA as a template and is inhibited by actinomycin D (2), viral RNA synthesis, with a few exceptions, is unaffected by actinomycin or thymidine analogues (2, 3) and occurs independently of the nucleus (4–6). Thus, it appeared evident that neither the host cell's DNA nor newly synthesized DNA is required, and that viral RNA is synthesized by a mechanism which had not been previously encountered in the uninfected cell.

Investigations into the mechanism of viral RNA synthesis entered a new phase in 1963 with the discoveries, firstly, that cells infected with picornaviruses contained a double-stranded form of RNA (7) and secondly that an enzyme could be isolated from infected cells which would catalyze the synthesis of both viral RNA and its double-stranded form (8, 9). Since this time many studies of animal, plant and bacterial virus systems have confirmed these observations (see Levintow, 10), and have given rise to a number of theories concerning replication of viral RNA. In this review I propose to examine these theories and the data on which they are based, and to consider how well they explain the results of more recent investigations.

THE REPRODUCTION OF PICORNAVIRUSES

Picornaviruses are a group of small polyhedral animal viruses which consist of a single molecule of RNA surrounded by a simple protein coat. They include poliovirus, the Columbia SK group (EMC, Mengovirus, etc.), foot-and-mouth disease virus, and the viruses of the common cold group. They grow rapidly in tissue culture cells, where they multiply within the cytoplasm, and they can be readily purified in high yields. Their replication is not affected when the host cell DNA-directed RNA synthesis is blocked by actinomycin D (2, 11, 12), and RNA extracted from the purified virus is infectious. In many respects picornaviruses are similar to the RNA-containing bacteriophages and the small polyhedral plant viruses.

Early events. Infection of the host cell is initiated when the invading virus particle is adsorbed to the cell surface and penetrates the cell. Here it loses its capsid protein, releasing the viral RNA into the cytoplasm. The invading RNA then begins a series of events which ultimately leads to the production of progeny virus particles.

One of the immediate consequences of picornavirus infection is the striking inhibition of cellular RNA synthesis (11, 13–16). This inhibition is paralleled by a decrease in the activity of the nuclear DNA-dependent RNA polymerase (14), and is probably mediated by a protein whose synthesis is directed by viral RNA (17, 18). Inhibition of cellular RNA synthesis is followed by a similar, but less striking inhibition of cellular protein synthesis (11, 13, 15). This is illustrated in Figure 1, which shows the effect of encephalomyocar-

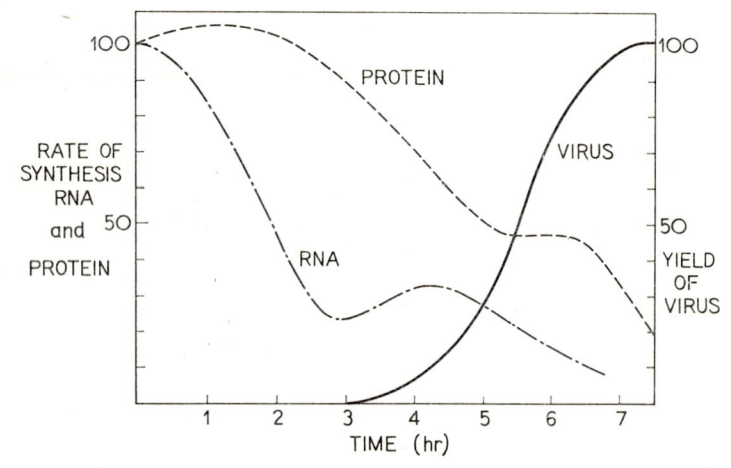

FIGURE 1. Effects of EMC virus infection on rates of incorporation of ^{14}C-orotic acid into RNA and of ^{14}C-leucine into protein of Krebs 2 mouse ascites tumour cells. The results are expressed as a percentage of values for uninfected control cells.

ditis (EMC) virus infection on the rates of protein and RNA synthesis of mouse ascites tumour cells. With this impairment of protein synthetic capacity goes a concomitant decrease in the proportion of ribosomes found in polysomal aggregates of a size characteristic of the normal cell (19).

Synthesis of viral RNA. One of the early proteins found in the cytoplasm of infected cells is an enzyme which catalyzes the RNA-directed synthesis of RNA. This RNA-polymerase was first observed in L cells infected with mengovirus by Baltimore and Franklin (8) and has since been reported in a wide variety of infected animal (20–24), plant (25) and bacterial cells (26–28). It catalyzes the synthesis of viral progeny RNA from nucleoside triphosphate precursors and is responsible for the replication of viral RNA *in vivo*. Its activity is unaffected by actinomycin D or deoxyribonuclease, and therefore is not dependent on a DNA template. The synthesis of the polymerase is under the direction of the viral genome. The enzyme appears to be relatively unstable, as its continued synthesis is required for the synthesis of viral RNA; protein synthesis inhibitors cause a rapid decrease in the polymerase level and in the rate of viral RNA synthesis (29). In all picornavirus systems studied the polymerase is localized in the rapidly sedimenting large-particle fraction of the cytoplasm, and it is within this fraction that the bulk of viral RNA synthesis also takes place (6, 21, 30–33).

The synthesis of viral RNA begins concurrently with the appearance of the viral RNA polymerase (see Figure 2). This synthesis can be readily demonstrated by following the incorporation of labelled RNA precursors

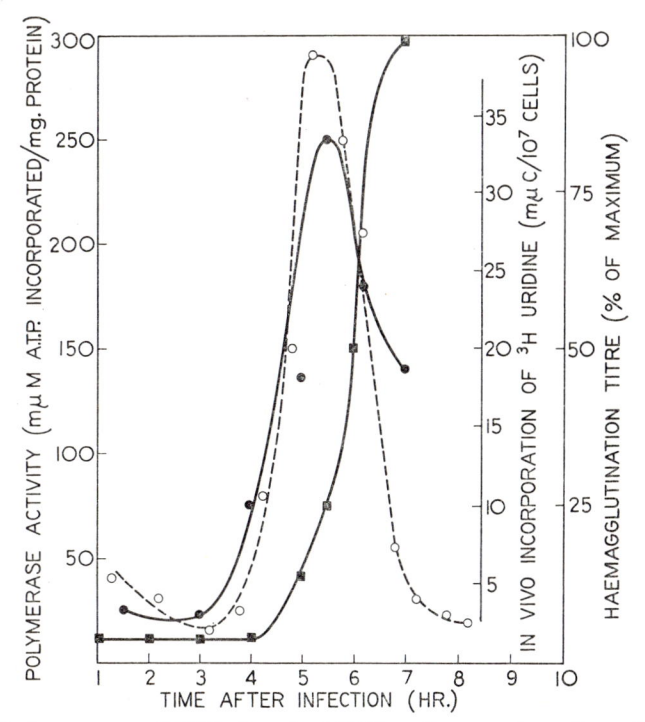

FIGURE 2. Appearance of RNA-dependent RNA polymerase in ascites cells infected with EMC virus in the presence of 5 μg/ml of actinomycin D. ●———● Polymerase activity; ○-----○ Rate of incorporation of ³H-uridine into RNA; ■———■ Virus haemagglutinin titre (see Ref. 22).

into cellular RNA in the presence of actinomycin. Under these conditions more than 90 per cent of the incorporation occurs into virus-specific RNA. It begins after a lag of 1.5 to 3 hours after adsorption of the virus (the "eclipse period") and continues at a rapid rate for the next 2 to 3 hours, finally declining in the last hour (Figure 2). The bulk of the RNA formed during this period has the physical and chemical properties of mature virus RNA. One species of the virus-specific RNA formed in infected cells has properties strikingly different from those of the progeny RNA. It sediments more slowly

($S_{w_{20}}$ = 16 to 18S) in sucrose gradients than mature virus RNA ($S_{w_{20}}$ = 37S), and it is resistant to ribonuclease. It was first observed by Montagnier and Sanders (7) in ascites cells infected with EMC virus and a similar type of RNA has since been reported in a wide variety of animal (34–37), plant (38–40) and bacterial cells (41–44) infected with small RNA-containing viruses. Its chemical and physical properties are those of a double-stranded RNA; its role in the replicative process will be discussed below.

Synthesis of virus capsid protein. The synthesis of viral capsid protein, and of soluble antigens immunologically related to capsid protein, begins

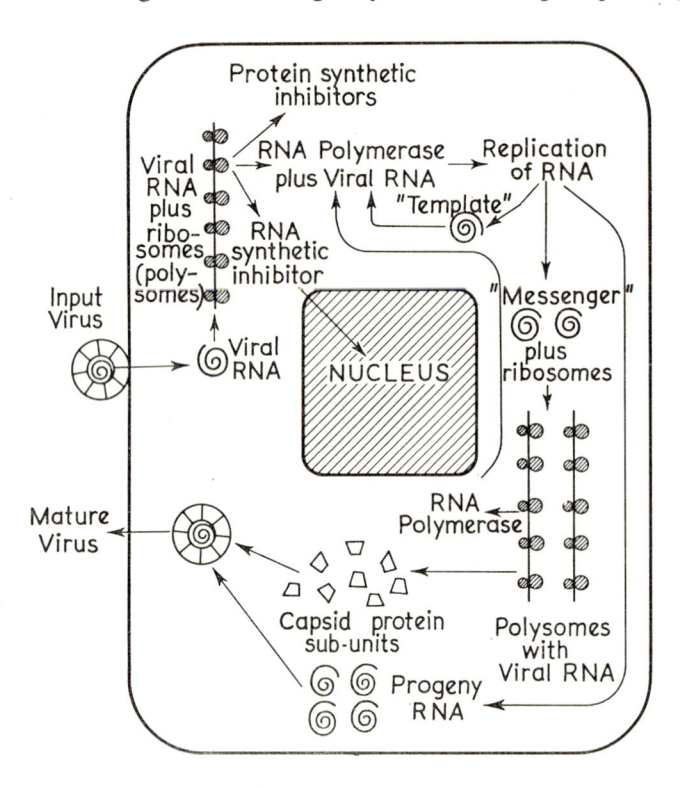

FIGURE 3. Summary of events during replication of a typical picornavirus.

at the same time as viral RNA synthesis; that is, at the end of the eclipse period; thereafter it parallels RNA synthesis either concurrently (45, 46) or a short interval behind (47). It takes place in the cytoplasm of the infected cell, in the large polysomal aggregates which begin to replace the smaller aggregates characteristic of the host cell synthetic apparatus. These virus-specific polysomes are associated with lipid membranous structures (30). RNA polymerase, newly synthesized RNA and mature virus particles are also associated with these structures (6, 21, 30, 34) and can be sedimented together from infected cell homogenates at relatively low centrifugal speeds. Electronmicrographs of poliovirus-infected cells have revealed cytoplasmic structures with properties similar to those found in homogenates (48, 49), and it seems reasonable to suppose that the synthesis of both viral RNA and capsid protein, and the formation of mature particles occurs in close juxtaposition, either within or upon lipoprotein membrane complexes. It is of interest that infection causes a marked stimulation of phospholipid synthesis, presumably to provide material for these lipid structures (50).

The polysomal aggregates characteristic of the infected cell contain RNA of the size and base composition of viral RNA (51, 52). They also contain nascent protein, some of which reacts with antibody directed against dissociated virus capsid protein (19, 52, 53). When isolated and incubated *in vitro* with suitable supplements, the polysomes incorporate amino acids and release soluble protein which is immunologically related to capsid protein subunits (52–54). These observations imply that the synthesis of viral capsid protein is directed by viral RNA, acting as a messenger. When viral RNA is added to ribosomes from uninfected cells, and incubated in the usual cell-free protein synthetic system, it causes a marked stimulation in amino-acid incorporation; however, the products formed in this system have not yet been identified as viral protein (54–57).

Formation of virus particles. After release from the polysomes the virus protein subunits, each of about 25,000 to 30,000 molecular weight, rapidly aggregate to larger units, which either enclose a molecule of viral RNA to form an infectious particle, or form empty capsids (58–60). Mature progeny virus is then released from the cell.

The events taking place during the reproduction of a typical picornavirus are summarized in Figure 3. The virus enters the cell and releases its RNA; this RNA then directs the synthesis of a number of proteins including the metabolic inhibitors and RNA polymerase. The polymerase combines with the input viral RNA, producing copies of it. Some of these "first generation" RNA molecules attach to ribosomes to direct the synthesis of more polymerase, which catalyzes the synthesis of further generations of viral RNA, thus continuing the replication cycle. The polysome aggregates containing viral RNA messenger also produce increasing amounts of capsid protein subunits, and these envelop viral RNA molecules to form the progeny virus.

Two important points arise from this replication scheme. Firstly, it will be seen that viral RNA has three distinct functions: (i) to act as a messenger RNA in the synthesis of "early" proteins and progeny capsid protein; (ii) to act as a template for the synthesis of viral RNA by the polymerase; (iii) to become the RNA of the progeny particles.

Secondly, the act of replication, which is dealt with in more detail below, is a cyclic phenomenon. That is, viral RNA directs the synthesis of RNA polymerase, which makes more viral RNA, which codes for the synthesis of more polymerase, and so on. This cyclic nature of the RNA synthetic process implies that inhibition of either the synthesis of the polymerase or its activity as an RNA replicase will lead to the same result, that is, both polymerase levels and viral RNA synthesis will be depressed.

Viral RNA Polymerase

In vitro synthesis of RNA by polymerase. We can now turn to a more detailed examination of the events taking place during the replication of the viral RNA. A suitable starting point is the polymerase responsible for this replication. This enzyme can be prepared from the cytoplasm of infected mammalian cells as a crude particulate fraction containing polymerase, single- and double-stranded viral RNA, viral protein and some of the intracellular ma-

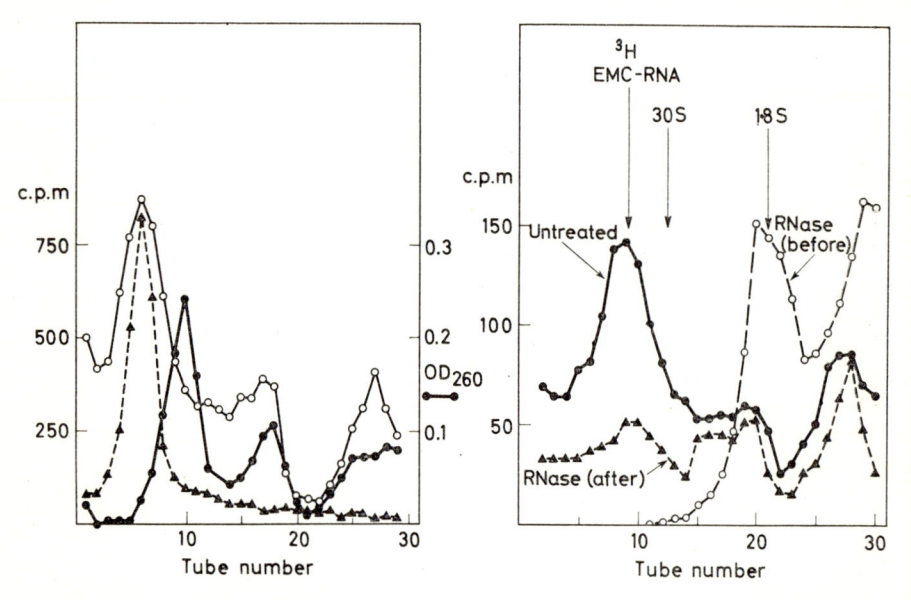

Figure 4. Sucrose gradient analysis of RNA synthesized by the RNA polymerase prepared from ascites cells infected with EMC virus. *Left figure*: ○――――○ Acid-insoluble ^{14}C-radioactivity after incubation of polymerase with ^{14}C-ATP, GTP, CTP and UTP followed by deproteinization with sodium dodecyl sulphate (Ref. 63); ▲----▲ RNA from purified EMC virus grown in the presence of ^{3}H-uridine (acid-insoluble ^{3}H-radioactivity); ●――――● Optical density at 260 mμ (ascites cell ribosomal RNA). *Right figure*: ●――――● Acid-insoluble ^{14}C-radioactivity after incubation of polymerase with ^{14}C-ATP etc.; ○----○ Distribution of radioactivity when RNA extract treated with 1 μg/ml of ribonuclease before sedimentation; ▲----▲ Distribution of radioactivity after each gradient fraction has been treated with ribonuclease. The top of the gradient is to the right of the figures.

ture virus. Attempts at further purification, particularly to separate the polymerase from its template, have not been successful with animal systems, although bacterial enzymes have been highly purified (27, 28, 61). When the crude polymerase preparation is incubated in a suitable medium with all four nucleoside triphosphates (one of which is usually radioactively labelled), radioactivity is incorporated into an acid-insoluble, alkali labile product. When this product is deproteinized and sedimented through a sucrose gradient we get a picture similar to that shown in Figure 4. (I will use examples from our own studies on EMC virus-infected ascites cells for this discussion; however, the results obtained with this system are typical of those obtained with other animal and plant virus systems). The major species of RNA found in the reaction mixture is a 37S component; this is identical in sedimentation properties, ribonuclease sensitivity and buoyant density to the RNA of the mature virus (21, 62, 63).

The second major component is an RNA sedimenting at about 18 to 20S which is resistant to incubation with high levels of ribonuclease (21, 63, 64). A similar component is also found in infected cells pulse-labelled *in vivo*. From its behaviour on heating, its resistance to ribonuclease, its buoyant density and its similarity to reovirus RNA, which is known to be double-stranded, it seems highly probable that the ribonuclease-resistant RNA found either *in vivo* or in polymerase reaction products is a duplex composed of a single strand of viral RNA (the positive strand) hydrogen bonded to its complementary (negative) strand in accordance with the classical Watson–Crick model.

Two other minor components are frequently seen in gradient profiles of the polymerase product. The first of these is in the 4S region; it probably represents incorporation into the terminal position of transfer RNA. Usually it appears only when labelled ATP is used in the reaction mixture, and not labelled GTP. Enzymes capable of terminal addition of adenylic acid residues to pre-existing RNA have been reported as contaminants in viral polymerase preparations (8, 12). The second minor component appears as a peak or shoulder in the region between the single- and double-stranded peaks. Its significance is discussed below.

Dual functions of the polymerase. It was first suggested by Montagnier and Sanders (7) that the double-stranded RNA was a "replicative form", i.e. an obligatory intermediate in the replicative process. This hypothesis, which has been given more concrete form by others (see below), envisages the double-stranded RNA acting as a template for the asymmetric synthesis of single-stranded progeny molecules, similar to the manner in which RNA is synthesized *in vivo* by nuclear DNA-dependent RNA polymerases. Assuming, for the moment, that this is the correct interpretation of the function of double-stranded RNA, this would imply that the synthesis of viral RNA is a two-stage process:

1. The synthesis of a complementary strand of RNA, using the input single-stranded viral RNA as a template, followed by the association of the two strands to form the double-stranded replicative form.

2. The synthesis of single-stranded progeny RNA by a polymerase which utilizes the double-stranded RNA as a template.

It is possible that both these functions could be achieved by a single enzyme; it seems more probable, however, that either two separate enzymes are involved or the polymerase is a polymeric protein with two functionally independent sites. The proportions of single- and double-stranded forms produced by the polymerase show wide variations. Weissmann et al. (64) found that a semi-purified polymerase from E. coli infected with MS2 phage synthesized more than 5 times more positive RNA strands than complementary strands. Different preparations of the EMC viral polymerase showed marked variations in the proportions of these two forms. Gradient analyses of two preparations are shown in Figure 5; the preparation from cells actively

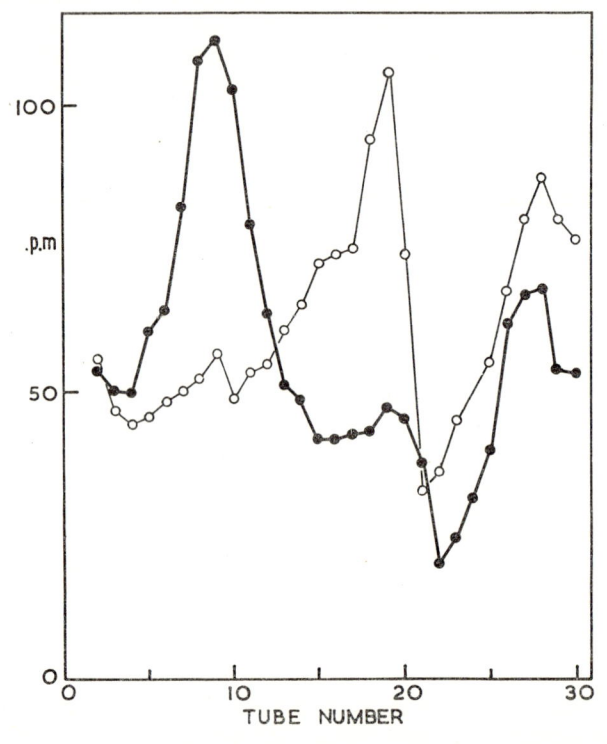

FIGURE 5. Sucrose gradient analysis of reaction products from two RNA polymerase preparations. ●————● From cells in which EMC virus was growing normally; ○————○ From cells in which virus growth was poor. Both enzymes were prepared from cells infected for 5.25 hr.

replicating viral RNA produced far more single-stranded RNA than that from cells which had given a poor yield of virus. It is possible that these profiles are caused by variable amounts of ribonuclease in the RNA extracts, but it is equally possible that they contain variable proportions of enzymes requiring a single-stranded RNA template (polymerase I) and a double-stranded template (polymerase II).

Studies on the effects of protein synthesis inhibitors have also led to the suggestion that more than one polymerase is required for viral RNA replication. Delius and Hofschneider (65) found that, when chloramphenicol was added to *E. coli* 10 to 30 min after infection with M12 phage, viral RNA synthesis was depressed, but continued for far longer than in the absence of the drug; chloramphenicol also increased the ratio of RNA synthetic rate to infectious RNA titre. They concluded that two polymerases were involved in phage RNA replication and that chloramphenicol was preventing the synthesis of the second enzyme (polymerase II); however this conclusion was based on a number of unproven assumptions. A similar experiment, in which the effects of puromycin and cycloheximide on viral RNA synthesis by EMC in ascites cells were followed, has been carried out by the author (66). Both drugs, when added shortly after the start of viral RNA synthesis caused a very rapid fall in RNA synthetic rate, to about 30% of the level prior to addition of the drug. When examined in sucrose gradients the RNA formed in the presence of the inhibitors was mostly of the ribonuclease-resistant 20*S* type. Puromycin, if added directly to the polymerase when assayed *in vitro*, also caused a 70 per cent inhibition of activity; the RNA formed in the presence of the drug was again found to be almost entirely of the double-stranded form. A simple interpretation of these results is that puromycin inhibits the synthetic activity of polymerase II, permitting only the enzyme synthesizing double-stranded RNA on a single-stranded template to remain active.

A more convincing demonstration of the existence of two polymerases is provided by the genetic studies of Lodish and Zinder (67). They isolated a temperature-sensitive mutant of the RNA phage *f*2, which was unable to replicate at 43°, but did so at 34°. At the higher temperature, polymerase activity was absent, no double-stranded RNA was made and labelled parental RNA was not converted to a double-stranded form. If the mutant strain was shifted to the higher temperature 30 min after infection no more double-stranded RNA was made, but the synthesis of single-stranded RNA continued normally, although this yield of progeny was only 10 per cent of normal. The genetic lesion appears to be an inability to make functionally active polymerase I at 43°; there was no impairment in the synthesis or activity of polymerase II.

There is some doubt about the specificity of these polymerases in their requirement for template RNA. Only two groups have been able to purify the enzyme sufficiently to show a dependence upon added RNA. August *et al.* found that polymerase purified 100-fold from *f*2 phage-infected *E. coli* would use viral, ribosomal or transfer RNA as templates (28). It catalyzed the synthesis of RNA of a base composition complementary to that of the template, together with some double-stranded RNA, and it is possible that this preparation contained only polymerase I activity (68). On the other hand, Spiegelman and his colleagues (61), studying a polymerase purified 1000-fold from *E. coli* infected with MS2 and $Q\beta$ phages, found an absolute requirement for the RNA of the infecting virus; no stimulation was obtained with heterologous phage or bacterial RNAs. Spiegelman's preparations catalyzed the synthesis of infectious progeny RNA, i.e. it possessed both polymerase I and II activities (69). It is possible that templage specificity is a property of the whole enzyme complex.

Replication of Viral RNA *in vivo*

At this point, it may be useful to clarify some of the terms commonly used to describe various species of RNA found in virus-infected cells. The term *viral RNA* will be used to refer to the single-stranded RNA which can be extracted from purified virus particles and which is present in extracts of infected cells. It is infectious and its infectivity is destroyed by ribonuclease. The *replicative form* (7) is that RNA which remains after treatment of extracts of infected cells or polymerase reaction products with ribonuclease. It can be seen as a single peak in sucrose gradients of ribonuclease-treated extracts in the 12 to 16S (phage) or 18 to 20S (picornaviruses) regions. The *replicative intermediate* (70) can be operationally defined as any virus-specific RNA synthesized during infection which is *not* viral RNA and can be separated from it by suitable gradient techniques. It is heterogeneous in its sedimentation properties and contains a ribonuclease-resistant component. It probably consists of *template* RNA (double-stranded) to which are attached several partially complete (*nascent*) strands of RNA (see later). It is realized that some of these terms imply a participation as intermediates in the replication of viral RNA. Although such a function has yet to be unequivocally established, these terms will be used in this review, as they appear frequently in the literature.

The fate of the input RNA. In vivo studies on the replication of viral RNA have followed two basic approaches—the host cell is infected with purified virus whose RNA has been suitably labelled with radioactive precursors; the fate of the labelled parental RNA is then followed in the infected cell. Alternatively, infected cells actively synthesizing viral RNA under conditions in which host-cell RNA synthesis is blocked (e.g. with actinomycin or ultraviolet irradiation) can be briefly incubated with suitable radioactive RNA precursors, and the newly-synthesized RNA extracted and examined by physico-chemical methods. These *in vivo* pulse-labelling experiments will be discussed in the next section.

When *E. coli* was infected with labelled R17 phage, Erikson *et al.* (41, 70) found that the parental RNA was rapidly converted into a ribonuclease-resistant double-stranded form. This conversion was inhibited by chloramphenicol, and presumably required the synthesis of protein. All of the single- and double-stranded parental RNA was associated with ribosomes. Similar results were obtained with MS2 phage by Kelly and Sinsheimer (42). Although the parental RNA remained intact within the cell during the course of infection, none of it was incorporated into progeny particles (42, 71). These results are consistent with the suggestion that the input RNA associated with ribosomes to form polysomes, which then synthesize the virus-specific RNA polymerase; this enzyme catalyzes the synthesis of complementary RNA strands which combine with the input RNA to form double-stranded RNA.

The fate of parental RNA in picornavirus infections has been investigated by Tobey (72) and Homma and Graham (73). Both investigators used the mengovirus-L-cell system, and their results were complicated by the fact that more than 80% of the adsorbed virus was not uncoated. Nevertheless it appeared that, while the parental RNA retained its physical integrity throughout the infectious cycle, negligible amounts were incorporated into the progeny particles. Homma and Graham found that much of the parental RNA was localized in the rapidly-sedimenting cytoplasmic particulate fraction very shortly after infection, while Tobey observed that as much as 50% of the input RNA could be found in large polysomal aggregates which were actively synthesizing protein (74). In contrast with the phage situation, neither study showed any evidence that the parental RNA entered into the 20S, ribonuclease-resistant RNA form, although Homma and Graham found that 5% of the pulse-labelled progeny RNA existed in this form. However, Tobey did find a significant proportion of both the parental RNA and the pulse-labelled progeny RNA sedimenting in the 30S region, that is, between the viral RNA and double-stranded RNA (had it been present). All of this RNA was sensitive to ribonuclease.

Thus, the fate of the input picornavirus RNA is similar to that of phage RNA. It is incorporated into polysomes, and acts as a messenger for the synthesis of virus proteins including, presumably, RNA polymerase. It is not incorporated into the progeny virus. The results of Plagemann et al. (75) may have some bearing upon this. They found that single-stranded pulse-labelled RNA which was located in the polysomal aggregates did not become incorporated into mature virus, i.e. viral RNA which was functionally active as a messenger cannot enter the pool from which the progeny receive their complement of viral RNA. The results obtained with phage and picornavirus systems differ in one respect: so far, the parental RNA of the latter has not been found in a double-stranded form. This may perhaps reflect the technical difficulty of detecting it in such systems, where it represents only a small proportion of the total labelled RNA.

In vivo pulse-labelling experiments. When mammalian cells are infected with picornaviruses in the presence of actinomycin (to inhibit host-cell RNA synthesis) and incubated for a short period with a suitable radioactive RNA precursor, the RNA formed during this period can be extracted and examined on a sucrose gradient. A typical profile is shown in Figure 6, taken from ascites cells infected with EMC virus. The major component is 37S progeny viral RNA; the ribonuclease-resistant 20S form is also prominent. A fairly large proportion of the radioactivity sediments in the region between these two RNAs; the amount of this interjacent material varies considerably with the length of the pulse-labelling period and with the time after infection during which the labelling is carried out (6). Similar gradient profiles have been reported with pulse-labelled cells infected with poliovirus (34), ME virus (33, 37), foot-and-mouth disease virus (35) and Semliki Forest virus (36). In all these instances a ribonuclease-resistant component sedimenting in the 18 to 20S region was present. The amount of the interjacent material was very variable, but it was particularly evident in cells infected with foot-and-mouth disease virus.

Basically similar patterns have also been reported for bacterial viruses. In these systems host-cell RNA synthesis has been blocked either by irradiation with ultraviolet or by use of actinomycin-treated protoplasts (41, 76, 77). *E. coli,* pulse-labelled after infection with R17 phage, showed all three types of RNA; the broad 16*S* band containing ribonuclease-resistant RNA predominated over the viral RNA peak (27*S*) with very brief labelling periods,

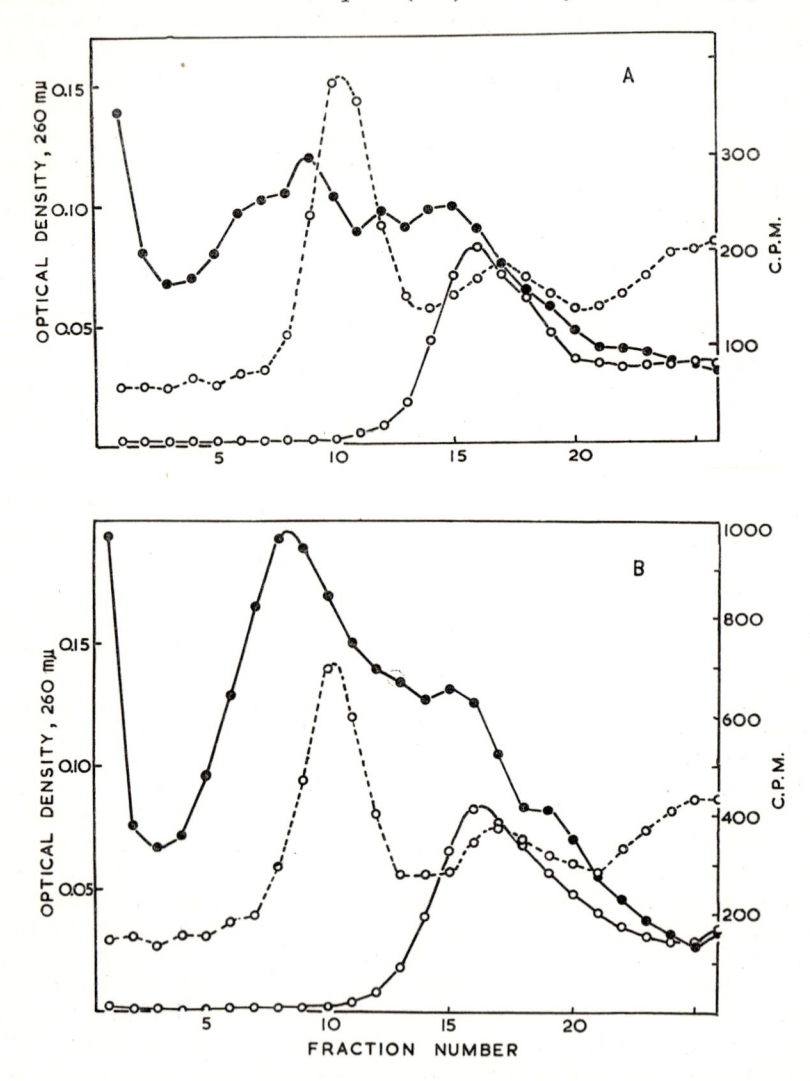

FIGURE 6. Sucrose gradient analysis of RNA extracted from ascites cells infected for 4.5 hr with EMC virus, and labelled for 4 min (top figure) or 8 min (bottom figure) with ^{3}H-uridine. ●——● Acid-insoluble ^{3}H-radioactivity; ○——○ ^{3}H-radioactivity of sample incubated with 1 μg/ml of ribonuclease before gradient analysis; ○----○ optical density at 260 mμ (ascites cell ribosomal RNA). The top of the gradient is to the right. From Ref. 6.

but with longer pulses the situation was reversed (41). In MS2 phage-infected cells, the interjacent RNA was quite prominent, and, in addition, a ribonuclease-resistant peak of about 6S was observed (76).

The nature of the ribonuclease-resistant RNA. The properties of the ribonuclease-resistant replicative form of viral RNA have been the subject of numerous investigations; except for differences in molecular size, all replicative forms which have been studied exhibit similar properties, whether found *in vivo* in animal, plant or bacterial sources, or whether produced *in vitro* by RNA polymerases. Its base composition shows equimolar amounts of adenine and uracil and of guanine and cytosine (however, see below); it has a lower buoyant density than viral RNA; it is resistant to ribonuclease. This resistance is lost after heating; the temperature at which this transition occurs is sharply defined and is dependent upon salt concentration. All of these properties are consistent with those of a double-stranded RNA molecule, the two strands being held together by complementary base pairing as in the Watson–Crick model. X-ray analysis (78) has shown it to possess a structure very similar to that of reovirus RNA, an RNA which is known to be double-stranded (79). When labelled replicative form is heated above its transition temperature the two strands separate. If the heated material is mixed with unlabelled viral RNA and cooled slowly, part of the radioactivity is displaced from the ribonuclease-resistant material, thus indicating the presence of a strand of viral RNA in the duplex (64, 80). A portion of the radioactivity cannot be displaced by excess viral RNA; this residual material can only anneal with homologous viral RNA and cannot be displaced by heterologous RNA (37, 38, 64, 81, 82); it exhibits the base composition which would be expected of the viral complementary strand (83, 84). Therefore, it seems highly probable that the two strands of the replicative form are viral RNA ("positive" strand) and a strand complementary in base composition to viral RNA ("negative" strand). Electron microscopic studies on the purified replicative form show it to consist of molecules of similar rigidity as DNA; the maximum length corresponded to a double-stranded molecule of twice the molecular weight of viral RNA (44). The replicative form found in picornavirus systems is infective (7, 35, 85), but the phage induced form must be separated into its constituent strands by heat denaturation before becoming infective (44).

The role of replicative-form RNA. With one exception (72), a double-stranded replicative form has been observed in all RNA-virus systems in which it has been sought, and it is very tempting to ascribe to it some role as an intermediate in the process of viral RNA replication. The only direct evidence of such an involvement comes from pulse-chase experiments with phage-infected *E. coli*. Fenwick *et al.* (41) pulse-labelled R17-infected cells with ³H-uridine and found that, after very brief pulses, most of the radioactivity appeared in a broad region sedimenting at about 16S on sucrose gradients. When the cells were further incubated with unlabelled medium, there was a marked increase in the proportion of label in the 27S viral RNA, with a corresponding decrease in the amount of ribonuclease-resistant radioactivity. However there was no net decrease in the label in the 16S

region. These results suggest a precursor-product relationship between 16S and 27S RNA but do not necessarily establish this fact.

Both Weissmann et al. (86) and Lodish and Zinder (67) found that when phage-infected bacteria were given pulses of radioactive precursor of less than 20 sec duration most of the radioactive positive strands were in the double-stranded form. During a subsequent chase with unlabelled precursor, Weissmann et al. found that 70 per cent, but no more, of the labelled RNA could be chased from the double to the single-stranded form. Lodish and Zinder repeated this experiment using a temperature-sensitive phage in which polymerase I was inactive at 43°. When the pulse-chase experiment was carried out at the higher temperature, (thus preventing any re-incorporation of labelled positive strands into double-stranded RNA) over 95% of the label was displaced from the double strands.

The picture which emerges from these results is that replication of viral RNA is achieved by the incorporation of nucleoside triphosphates into the positive strand of a double-stranded RNA intermediate, followed by a release of newly synthesized positive strand as a free single-stranded RNA molecule. It may then either act as a template for the synthesis of negative strands by polymerase I and thus be re-incorporated into a double strand, or it may be incorporated into progeny virus or it may act as messenger RNA for the synthesis of protein. The information available from other virus systems on the role of double-stranded RNA is not as extensive. Pulse-chase experiments have not been particularly informative due to the difficulty of rapidly chasing out the labelled precursor. Nevertheless, Hausen (37) has reported that the rate of labelling of the positive strand of the double-stranded form is faster than the rate of formation of new double strands, i.e. label must be leaving the double strands to enter into the pool of single-stranded RNA.

Many of the questions concerning the mechanism of viral RNA synthesis could be resolved by detailed studies of the synthetic reaction catalyzed by purified viral RNA polymerases. Spiegelman and his colleagues have made considerable advances toward this objective. They have prepared a highly purified RNA polymerase from *E. coli* infected with $Q\beta$ phage (61). This enzyme has an absolute requirement for added $Q\beta$ viral RNA and does not respond to any other phage or bacterial RNA; it catalyzes a product which is infectious and which is itself able to act as a template for the further synthesis of infectious RNA—indeed the enzyme appears to be able to synthesize RNA autocatalytically once it has been supplied with an initial small amount of template (69). When examined on sucrose gradients the RNA synthesized by this enzyme showed a pattern typical of many polymerase products, i.e. a major peak of viral RNA with subsidiary peaks in the 15S to 25S region (87). However, Spiegelman interprets this slower sedimenting material as being the products of abortive or incomplete syntheses. Further experiments with this system should prove conclusively whether more than one enzyme is involved in phage viral RNA synthesis and the precise role of any double-stranded intermediates.

Mechanisms of RNA Replication—Conservative Versus Semi-Conservative

If viral RNA is to replicate by the well-established principles of base-pairing (88) it is necessary for a complementary (negative) strand of RNA to be formed during replication. There is overwhelming evidence that this negative strand is formed and that its synthesis is catalyzed by a virus-specific RNA-dependent RNA polymerase. If it is assumed that the negative strand is directly involved in the replicative process, then relatively few feasible hypotheses concerning the mechanism of replication can be proposed. Two of these will be considered in some detail, but first it would be useful to dispose of a proposition common to both theories—that replication is an asymmetric process, only one RNA strand (the positive) being produced in any quantity.

Asymmetry of the replicative process. In all systems examined the progeny RNA formed in infected cells or by viral RNA polymerases has the same base composition as that of RNA prepared from the purified virus. Therefore, in those instances where the base composition of the theoretical complementary strand is sufficiently different from viral RNA (for example, polio, turnip yellow mosaic virus) it can be assumed that no pool of free negative strands accumulates during infection. Tests for ability to anneal with labelled viral RNA have failed to reveal any negative strands in the viral RNA regions of sucrose gradients (81). This implies that the enzymes responsible for the synthesis of viral RNA transcribe only the negative strand; the positive strand, whether it is free or bound in a double-stranded molecule, is either not copied (except during the synthesis of the double-stranded intermediate) or its copy is destroyed immediately after synthesis. This latter possibility is highly unlikely—it is difficult to imagine a hydrolytic process of the required specificity unless the positive strand is specifically protected, say, by capsid protein or ribosomes. In any case, the actual rate of synthesis of the negative strand is very much less than that of the positive strand (37, 64).

Studies on DNA-primed RNA polymerase have shown that, while only one DNA strand is copied *in vivo* (89), either one or both strands can be transcribed by purified enzymes *in vitro*, depending on whether the template is denatured or not (90, 91). This suggests that asymmetry of synthesis is a property of the template or of the enzyme-template complex. It is not difficult to devise mechanisms to account for this asymmetry. Probably the polymerase transcribes the template chain in one direction only (say, from the 5' end to the 3'); it needs only a restriction on which end of the duplex the polymerase can begin the transcription, and asymmetry of synthesis will result. This restriction could be imposed by blocking the forbidden end of the duplex, perhaps by requiring polymerase II to start reading the template before its synthesis by polymerase I is complete. However, if one accepts that the holoenzyme shows an absolute specificity for its template RNA (61), this implies that the polymerase can recognize specific base sequences; if these specific sequences were located at the starting end of the duplex, then the required restriction would be manifest.

The conservative and semi-conservative hypotheses. Only two basically different theories concerning the mechanism of replication of viral RNA have received any measure of support in the literature. These are the asymmetric conservative and the asymmetric semi-conservative hypotheses, and they are illustrated in Figure 7. Both theories assume that viral RNA synthesis is an asymmetric process (see above). The *semi-conservative mechanism* was first proposed by Weissmann et al. (64) and Fenwick et al. (41). According to this hypothesis, the polymerase catalyzes the synthesis of a negative strand, making use of the parental strand as a template [reaction A]. This reaction is common to both theories and results in the synthesis of double-

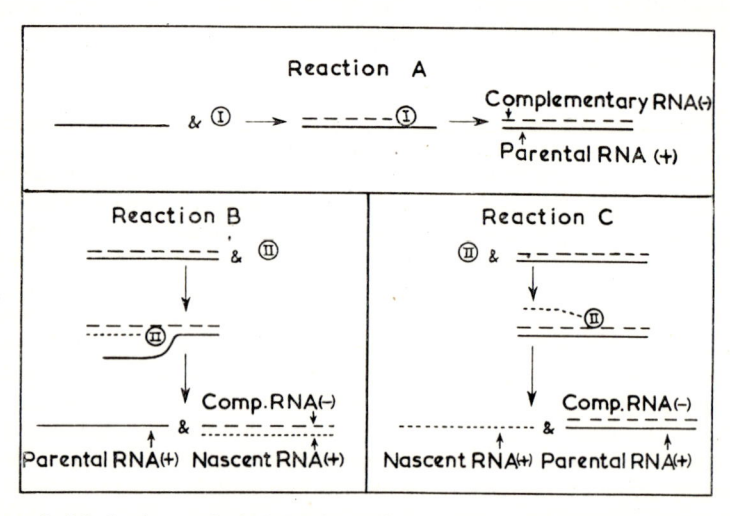

FIGURE 7. Mechanisms of viral RNA replication. Reaction A-synthesis of double-stranded template. Reaction B-semi-conservative synthesis of progeny RNA. Reaction C-conservative synthesis of progeny RNA. ———— Parental RNA; — — — — Complementary (-ve) RNA; – – – – – – Newly synthesized (nascent) RNA; I and II — Viral RNA polymerases.

stranded RNA. Either the same enzyme or a second polymerase then uses the duplex as a template to produce new single strands of viral RNA by displacement of the positive strand of the duplex [Figure 7, reaction B]. An analogous displacement mechanism has been put forward to explain the synthesis of DNA–RNA hybrids when *E. coli* RNA polymerase is provided with a single-stranded DNA template (92, 93). On the other hand, the *conservative mechanism* proposes that the double-stranded template remains intact, and the progeny RNA molecules are laid down on the template without displacement of either strand [Figure 7, reaction C]. This mechanism has well characterized precedents—the synthesis of RNA by the DNA-directed RNA polymerase of *E. coli* takes place by the transcription of the double-stranded DNA in a fully conservative manner (94–97), and the single-stranded DNA phage, ØX174, replicates by the synthesis of progeny molecules from a double-stranded intermediate by a conservative mechanism (98, 99). In many reports on the isolation of polymerases or the detection of double-stranded RNA in infected cells, the authors have concluded that their results were in accord with the semi-concervative hypothesis, and this hypothesis has had the greater number of adherents; however, few have attempted experiments designed to distinguish this particular hypothesis from any alternative one.

Both hypotheses propose that the initial act of replication is the synthesis of a double-stranded intermediate by a polymerase using the single-stranded parental RNA as a template. The important difference between the two mechanisms lies in the fate of the parental strand of the template; in the semi-conservative scheme it is displaced to become available for use as a template, or as a messenger, or to be incorporated into progeny particles; in the conservative mechanism, it is retained in the duplex.

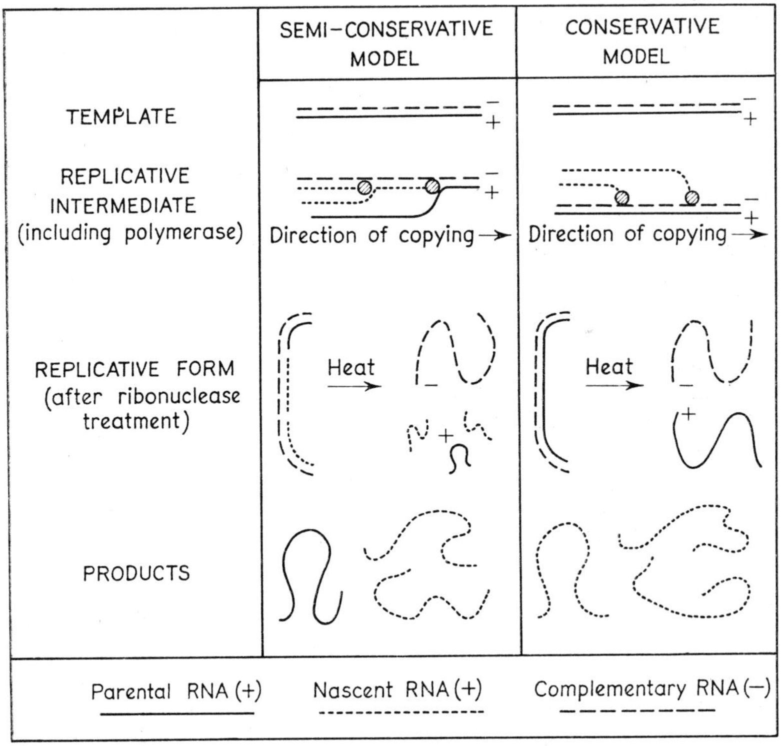

	SEMI-CONSERVATIVE MODEL	CONSERVATIVE MODEL
TEMPLATE		
REPLICATIVE INTERMEDIATE (including polymerase)	Direction of copying →	Direction of copying →
REPLICATIVE FORM (after ribonuclease treatment)	Heat →	Heat →
PRODUCTS		
	Parental RNA (+) Nascent RNA (+)	Complementary RNA (−)

FIGURE 8. Intermediate structures predicted by the two models for viral RNA replication.

An experimental approach which would distinguish between these two mechanisms is illustrated in Figure 8. If the semi-conservative mechanism is operative, parental RNA should appear among the products and the replicative form should consist of an intact complementary chain to which are bound fragments of positive chains. The number and size of these fragments would depend on the number of polymerase molecules which had been attached to the template at the time of isolation; separation of the strands by heating should reveal these fragments. In contrast, the conservative mechanism predicts that only newly-synthetized RNA would appear among the products and that the replicative form would consist of whole positive and negative strands, some of which would be parental RNA.

Experiments along these lines have been reported by Erikson et al. (81), who examined the replicative form prepared from *E. coli* infected with ^{32}P-labelled R17 phage. The ribonuclease-resistant RNA was heated to 95° in the presence of an unlabelled viral RNA marker and rapidly cooled to preserve the separation of the strands. The ^{32}P-labelled parental RNA now sedimented with approximately the same S-value as the unlabelled viral RNA. Further, the distribution on the gradient of material able to specifically anneal with labelled parental RNA (i.e. *negative* strands) was similar to that of the labelled positive strands. They concluded that the replicative form consisted entirely of *intact* positive and negative strands. Brown and Martin (100) have carried out similar experiments with foot-and-mouth disease virus. Heating the pulse-labelled ribonuclease-resistant RNA to 110°, and then rapidly cooling it, resulted in an increase in the average S-value of the label, although it did not sediment as rapidly as unheated viral RNA. However, heating of viral RNA caused a decrease in its sedimentation rate to a value approximating that of the denatured replicative form; no evidence for material of lower S-value than the original replicative form was seen. Both these studies are in complete disagreement with the semi-conservative mechanism. Other evidence would suggest that the positive strands in the replicative forms of a number of other viruses are intact molecules. Replicative forms of EMC (7), polio (85) and foot-and-mouth disease viruses (35) are infectious, while that from M12 phage can be rendered so by heat denaturation (44). It is inconceivable that the complementary strand could produce infectious virus—or at least virus which has properties identical with those characteristic of the positive strand. A single break in infectious RNA is sufficient to render it non-infectious (101, 102). Therefore, one must conclude that the replicative form of these viruses contains an intact positive strand.

It is possible, however, to devise a semi-conservative replicative mechanism which can account for the intactness of the positive strand after ribonuclease treatment of the replicative intermediate. Brown and Martin (100) have made the ingenious suggestion that the negative strand may be a functionally intact circle. In this case the polymerase could work its way round the circle, displacing the newly-synthesized positive strands, producing a continuous thread of progeny viral RNA; the thread would then be chopped into the correct lengths required to fit into the viral capsid. This mechanism was developed to explain the heterogeneity of chain lengths in the progeny RNA pools of cells infected with foot-and-mouth disease virus; however, as far as can be ascertained from examination of sucrose gradient patterns, this particular phenomenon has not been observed in other picornavirus systems. Ribonuclease treatment of this circular replicative intermediate would produce an intact negative strand plus positive strand composed of a circle with one break at the point of attachment of the polymerase; after denaturation, this would produce an intact linear positive strand. This theory poses problems of its own. The break in the positive strand of the replicative form occurs at the site where the polymerase has stopped functioning; this may occur randomly along the length of the RNA chain. As the replicative form of this virus is infectious (35), it seems unlikely that such molecules could act as competent messengers in the synthesis of viral proteins—a necessary first step in the initiation of infection. This theory would also preclude the attachment of more than one polymerase molecule to each complementary strand—any additional molecules would result in additional breaks in the positive strand after treatment with ribonuclease. Despite the demonstration of circular double-stranded intermediates in ØX174-infected cells (103), no report has been published of their existence in cells infected with an RNA virus.

The two hypotheses also differ in their predictions concerning the nature of the single-stranded product. The semi-conservative mechanism would permit the release of the parental RNA into the pool of single-stranded progeny molecules, from which it could be withdrawn for further template production, for association with ribosomes to form polysomes or for incorporation into mature virus. According to the conservative mechanism the parental strand would remain trapped in the replicative form. Studies on the fate of parental RNA are complicated by the very low ratio of infective to total particles. In the case of poliovirus, where only 1 particle in 1000 is infective, much of the input virus is not adsorbed and only a small proportion of the uncoated viral RNA remains intact in the cytoplasm of the infected cell (104). It is possible that only the viral RNA which enters into the replicative form is able to initiate infection—the remainder may be either degraded directly or become trapped in polysome aggregates, to be degraded at the normal rate of messenger breakdown. These considerations throw considerable doubt on the claim by Laduron and Cocito (105) that 30 per cent of coxsackie virus parental RNA could be transferred to the progeny, as no measurement was made of the amount of parental RNA which remained uncoated. However, careful studies have been made by Homma and Graham (73) on mengovirus-infected L-cells; they found that, within the limits of experimental error (about 3%), no parental RNA was transferred to the progeny. A similar conclusion regarding the transfer of phage RNA to progeny was reached by Doi and Spiegelman (71) and Davis and Sinsheimer (106).

In the face of this evidence, the semi-conservative mechanism can only be retained if it is postulated that parental RNA never reaches the pool of single-stranded RNA from which the progeny RNA is derived, and, after release from the replicative form, it can only function either as a polymerase template or a messenger RNA. Such a restriction in function is, in fact, quite possible, since, early in infection, the chance of a particular

RNA molecule being trapped to form a mature particle is slight, as too few capsid subunits have been formed. This would result in its function being restricted to that of a template or messenger. There is some evidence in mengovirus-infected cells (75) that viral RNA which is located in the polysome fraction (i.e. is functioning as a messenger) is never incorporated into mature virus. If this were a general phenomenon there would be a very high probability that most viral RNA molecules present early in infection would be trapped in polysomal aggregates. As these aggregates synthesized more capsid protein, the chances of newly-synthesized viral RNA being incorporated into particles would then become more favourable.

There is some evidence that parental RNA, after entering the replicative form, can be found later in a ribonuclease-sensitive form. Erikson *et al.* (70) found that the proportion of labelled parental R17 phage RNA incorporated into the replicative form reached a maximum of 30 per cent 12 min after infection; thereafter the percentage declined slowly to 17 per cent after 30 min. Weissmann *et al.* (80) found a similar situation in cells infected with labelled MS2 phage. It is difficult to see how these results give positive support to either hypotheses. The semi-conservative mechanism would predict two possible consequences: (i) a rapid equilibrium would be reached in which a portion of the parental RNA would be ribonuclease-resistant (i.e. acting as a template) and a portion sensitive (i.e. messenger), these proportions remaining unchanged throughout infection. This assumes that the RNA is continuously cycled from messenger to template to messenger, etc.; (ii) alternatively, the RNA, after displacement from the replicative form, could be permanently trapped in the polysomal aggregates (as indicated above, incorporation into mature virus is ruled out); it would then become ribonuclease-sensitive. It this were the case, a very rapid fall in ribonuclease resistance, at least as rapid as the earlier increase, could be expected as the parental RNA left the replicative form. As neither of these possibilities is reflected in the actual observations, it seems probable that the slow decline in the proportion of parental RNA which is ribonuclease-resistant represents the gradual degradation of a relatively stable template, caused perhaps by host cell ribonuclease. Either hypothesis could accomodate such a stable template. They differ only in ascribing to the ribonuclease-sensitive parental RNA different origins—as messenger RNA in equilibrium with template RNA (semi-conservative) or as either uncoated virus or RNA uncoated but not taking part in the infectious process (conservative mechanism).

The replicative-intermediate RNA. So far, discussion of the evidence supporting both mechanisms has been confined to studies of the behaviour of parental RNA and the ribonuclease-resistant replicative form. However, any theory must account for all species of RNA actually observed in virus-infected cells. Sucrose gradient analysis of RNA extracted from pulse-labelled infected cells or RNA polymerase reaction mixtures usually shows two or three peaks of radioactivity. One of these is the progeny viral RNA whose S-value depends on its molecular weight—from $27S$ for phages to $45S$ for arboviruses. The second peak is the replicative-intermediate RNA. This sediments at 12 to $14S$ (phage) or 18 to $20S$ (animal viruses) and part of it is resistant to ribonuclease. Between these two species is often found a third component—the interjacent RNA; its significance is discussed in the next section. A typical gradient profile is given in Figure 4, which shows a reaction product of an RNA polymerase from EMC virus-infected ascites cells. If the contribution of progeny RNA to the profile is eliminated one sees a peak of radioactivity at about 20S trailing off into the faster sedimenting region (Figure 9—the interjacent RNA in this preparation is a very minor component). When each fraction is treated with ribonuclease there is again a skew distribution of radioactivity; a greater proportion of the slower sedimenting (20S) material is ribonuclease-resistant, compared with faster sedimenting RNA. If the extract is treated with ribonuclease before gradient analysis, the replicative form so obtained sediments as a sharp peak at 20S containing all of the double-stranded RNA which previously had been spread over the gradient.

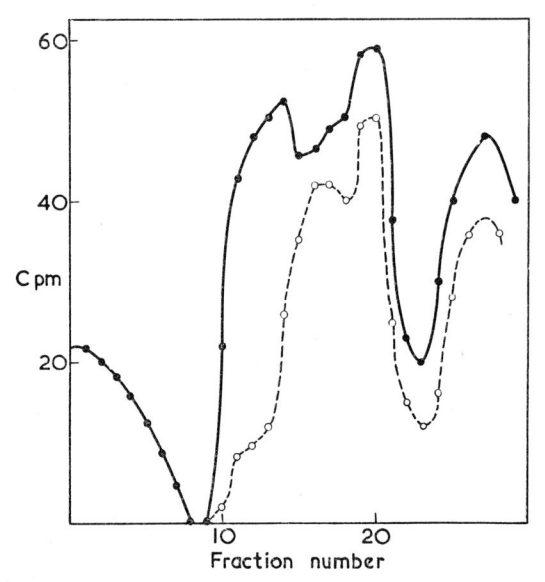

FIGURE 9. Replicative-intermediate RNA. The ^{3}H-labelled EMC-RNA marker has been used to calculate the contribution of single-stranded 37S progeny RNA to the total radioactivity in the gradient illustrated in Fig. 4 (top). This has been subtracted to give the distribution of intermediate forms of viral RNA. ●———● Total ^{14}C-radioactivity; ○- - - -○ ^{14}C-radioactivity resistant to 1 µg/ml of ribonuclease.

Either hypothesis can explain the gradient profiles of the replicative-intermediate RNA. The double-stranded template should have nascent RNA attached to it; the number of nascent molecules would depend upon the number of polymerase molecules simultaneously using the template and the length of each nascent chain would depend on the length of template which had been copied during the period of incubation with precursor. If the nascent RNA is still attached to the template after extraction, the

138   E. M. MARTIN

replicative intermediate will be polydisperse, with a spectrum of RNA species of different sizes and different ratios of template to nascent RNA (i.e. of ribonuclease-resistant to -sensitive RNA). RNA of high S-value will have a greater proportion of ribonuclease-sensitive nascent RNA, and this is precisely the picture which is seen experimentally.

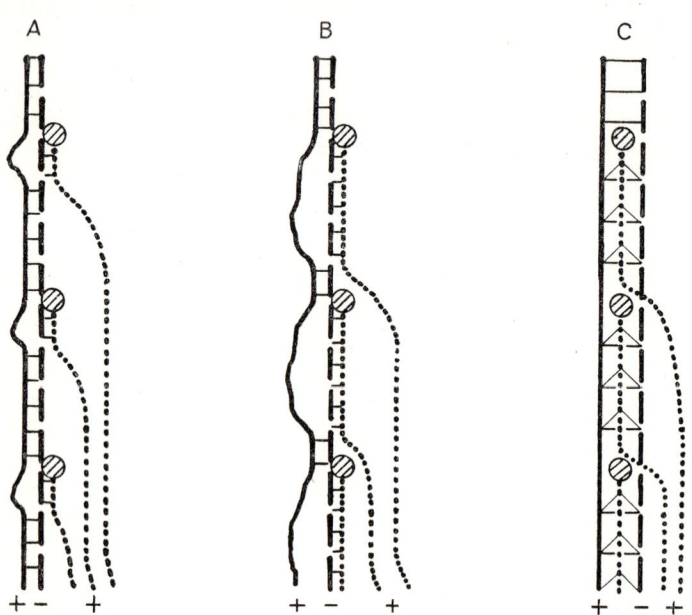

FIGURE 10. Theoretical structures for replicative-intermediate RNA. A and B—Double-stranded templates with displacement of short (A) and long (B) lengths of template positive strand. C—Three-stranded model. ————, Template positive strand; — — —, Template negative strand; · · · · · · · Nascent (positive) RNA.

This interpretation assumes that nascent RNA remains bound to the template after deproteinization. This is in accordance with the semi-conservative mechanism, as the nascent chain would form part of the double-stranded template. However, it is difficult to see how this complex could be stable if the conservative mechanism were to operate simply as illustrated in Figure 8. According to Bremer and Konrad (96) template-nascent RNA hybrids formed by *E. coli* polymerase are stable only in the presence of the enzyme, and deproteinization with detergent causes the dissociation of the complex into free DNA and nascent RNA. The replicative intermediate in many systems is stable to treatment with sodium dodecyl sulphate, although it is possible that phenol may disrupt it (see below). If the conservative mechanism is to be retained it is necessary to postulate that some stable association between template and nascent RNA must occur.

The stability of the replicative form to deproteinization can be explained if it is postulated that the nascent strand of RNA is base-paired with the negative strand of the template in the immediate vicinity of the polymerase molecule, as shown in Figure 10. Presumably, this would result in the displacement of the positive strand of the template from the duplex in the same region. A mechanism similar to this has been proposed for DNA-dependent RNA polymerase by Jehle (97). This model raises certain problems of its own. If the number of polymerase molecules per template is small and the length of the nascent RNA attachment regions is short (Figure 10A), most of the template will consist of intact positive and negative chains base-paired to each other; actual displacement of the parental positive strand and its release from the duplex will not take place. However, if an appreciable length of the template negative strand is complexed to nascent chains (Figure 10B), the likelihood of displacement of the parental chain from the duplex increases, until we are left with the model proposed by the semi-conservative mechanism. Displacement of the parental strand from the template is improbable for the reasons given above, and hence the structure for the replicative intermediate given in Figure 10A is the more likely one.

However, there is no fundamental reason why the positive strand must be displaced, even regionally. It is quite possible that a stable triplex could be formed between the two template strands and the nascent RNA, as shown in Figure 10C; the nascent RNA would be displaced from the triplex during the synthesis of the next nascent RNA molecule. Triple-stranded arrangements of this type have already been suggested as models for the DNA-dependent synthesis of RNA (107, 108), and recently a triplex model for the structure of transfer RNA has been proposed (109). Spacial considerations have shown that such structures are quite feasible (107, 108); indeed, triplexes of the type poly (A+2U) are relatively simple to prepare *in vitro* (110, 111). The triplex model has certain advantages when compared with any of the duplex models. The scheme illustrated in Figure 10A has two obvious faults. It must invoke a mechanism which would limit the length of the displaced template-positive strand, perhaps to only a few nucleotides. Also, it is difficult to see why the lengths of the template-positive strand which have been displaced are not sensitive to ribonuclease. These objections do not apply to the triplex model (Figure 10C), and for this reason it seems a more attractive hypothesis.

Few studies on the isolated replicative intermediate have been reported. Bishop *et al.* (84) have measured the caesium sulphate buoyant density and base composition of replicative intermediate RNA extracted from cells infected with poliovirus in the presence of ^{32}P. When the RNA was extracted with phenol and dodecyl sulphate at 60°, the replicative intermediate exhibited the same buoyant density and base composition as ribonuclease-treated RNA. The base composition corresponded to that of double-stranded RNA containing one positive and one negative strand. However, if only detergent were used at room temperature, the base composition was similar to that expected of a triple-stranded structure—two strands of positive RNA and one of negative. Treatment with ribonuclease did not alter this base ratio. A simple interpretation of these results is that the nascent RNA has remained associated with its template after detergent extraction, but not after treatment with hot phenol. This is in agreement with the conservative mechanism, although one would have expected the replicative intermediate to contain a whole spectrum of nascent RNA chain lengths, with RNA of base ratios equal to those of the first half of the RNA chain predominating. Also, on the basis of the model shown in Figure 10A, one would not have expected all three chains of the triplex to be resistant to ribonuclease. However, these results can be accommodated by the triplex model (Figure 10C). This predicts that the ribonuclease-treated replicative intermediate would have the base composition of one negative and two positive strands, as observed by Bishop *et al*. If relatively few polymerase molecules used each template the base composition would not be expected to alter appreciably after removal of the nascent RNA "tails" with ribonuclease—the change in ratio of positive to negative strands may be from 2.5/1.0 to 2.0/1.0. If it is assumed that the triplex is unstable at 60°, and reforms into a duplex plus free nascent RNA chains, then the two-stranded base composition of replicative forms can be accounted for (see below). Nevertheless, unless a large number of polymerase molecules use the template simultaneously, the triplex model would require much of the nascent RNA to be resistant to ribonuclease. If this is the case, then much of the ribonuclease-sensitive radioactivity seen in gradient profiles would remain unexplained. Perhaps re-arrangement of the *in vivo* replicative intermediate to form stable triplexes may have taken place during its preparation by Bishop *et al*.

Base compositions of the ribonuclease-resistant replicative forms of tobacco mosaic virus (40, 83) turnip yellow mosaic virus (40) and some phages (43, 77) have been reported; all were extracted with phenol and all showed base ratios characteristic of double-stranded RNA. Presumably, the nascent RNA was lost during phenol extraction or ribonuclease treatment. Zimmerman *et al*. (112) measured the base composition of the 16S replicative intermediate from polio-infected cells and found it more closely approximated viral RNA, rather than double-stranded RNA. This would suggest agreement with Bishop *et al*.; however, hot phenol was used in extracting the RNA and this, presumably, would dissociate any nascent RNA from the templates. Ammann *et al*. (44) have isolated the replicative intermediate from M12 phage-infected *E. coli*, without the use of ribonuclease. It possessed all the properties commonly associated with double-stranded RNA, and no evidence was obtained for significant amounts of a third strand. However, phenol and high temperatures had been used during its purification, and this may have caused dissociation of the nascent RNA. Obviously, the problem of whether the replicative intermediate contains nascent RNA chains, and if so, in what form they are associated with the template, is a complex one, which will be clarified only when amounts of intermediate RNA sufficient for physico-chemical characterization have been isolated.

Other replicative mechanisms. Models for the replication of viral RNA, other than the two schemes which have been discussed here, are theoretically possible. Hausen (37) has considered some of these. It is possible that the double-stranded RNA, once formed, could replicate in the same manner as DNA, with the positive strands being stripped off for packaging into progeny and the negative strands being destroyed. Such a symmetric conservative synthesis of both chains is at variance with all data on the rates of synthesis of the two types of strand; also, the specific destruction of the negative strand is improbable. Alternatively, after a period of symmetric, conservative synthesis of the double-stranded RNA, single positive strands could be read off these templates by either of the asymmetric mechanisms. This three-stage synthetic scheme is precisely analogous to the mode of replication of ØX174 phage DNA (99). In this scheme the major replicative step occurs by increasing the number of double-stranded templates, so that these form an appreciable proportion of the newly synthesized RNA. In all the RNA virus systems investigated, double-stranded RNA is a very minor fraction of the total virus-specific RNA—no more than 0.4 to 4 per cent (34, 37, 40, 43, 44, 73). The major replicative step is, therefore, the transcription of single strands from the double-stranded template; replication of the template is not necessary nor is there any evidence that it occurs. Of course, it is possible that double-stranded RNA is not involved in replication, but is merely a by-product. The polymerase could use single complementary strands as templates for the synthesis of positive strands. Kinetic considerations (37), the unequal numbers of positive and negative strands, and the failure to detect free negative strands argue against this mechanism. Finally, models could be constructed which omit the need to synthesize complementary strands (71), but, as no examples of the non-complementary synthesis of polynucleotides of defined sequence have yet appeared, it seems fruitless to discuss them at this stage, particularly as more attractive schemes exist.

Interjacent RNA

Sucrose gradient profiles of RNA extracted from viral polymerase reaction products or, more prominently, from cells pulse-labelled with radioactive RNA precursors, frequently show a peak of radioactivity in the region between the replicative form and the progeny viral RNA. This interjacent peak is clearly seen in extracts of pulse-labelled cells infected with MS2 phage (76), EMC virus (6), foot-and-mouth disease virus (35), ME virus (33) and Semliki forest virus (36); it is also seen in cells infected with labelled R17 (70) or MS-2 (76) phages, and in the products synthesized by RNA polymerases from poliovirus (21) and EMC virus-infected cells (Figure 6). It is unlikely that all instances of interjacent RNA peaks can be ascribed to a single cause; nevertheless, this material is such a persistent feature in sucrose gradients of intracellular viral RNA that it demands explanation. In particular, it is important to know whether it is concerned in the process of viral RNA replication, and if so, whether our theories must be modified accordingly.

20S RNA of MS2 phage. The most extensive investigation into the nature of interjacent RNA has been carried out by Sinsheimer's group (76, 113), who found a ribonuclease-sensitive component in MS2-infected *E. coli* sedimenting at 20*S*, between viral RNA (27*S*) and the replicative intermediate (15*S*). It was apparent both in cells infected with RNA-labelled input phage and in infected cells pulse-labelleed with ³H-uracil. An additional ribonuclease-resistant component at 6S was also seen. The buoyant density of 20S component (1.888 g/cm³) was close to, but not identical with, that of the phage viral RNA (1.893), and higher than that of the replicative intermediate (1.866). When sedimented through gradients of low ionic strength, its S-value was reduced (to 12.5S), but not to the same extent as that of viral RNA (12.0S). These observations are consistent with the conclusion that 20S RNA is single-stranded but is less compact than the 27S RNA derived from mature phage. As the viral RNA has too high an S-value (i.e. is too

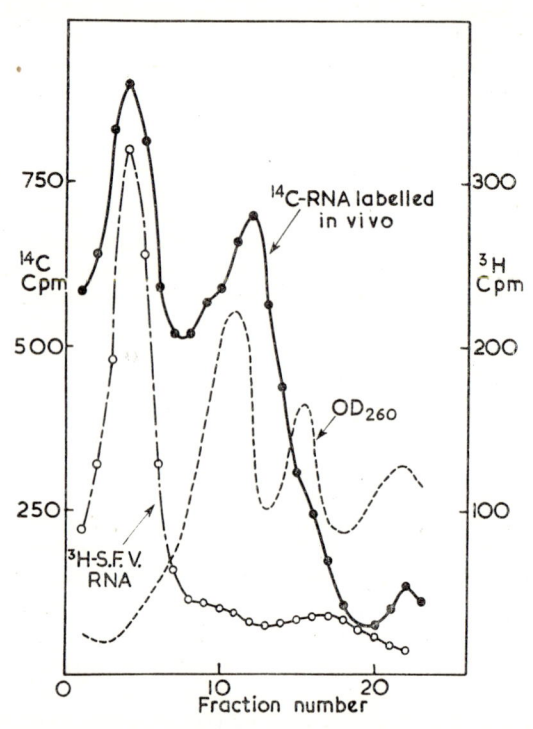

FIGURE 11. Sucrose gradient analysis of a phenol-sodium dodecyl sulphate extract of chick fibroblasts incubated with actinomycin D and ¹⁴C-adenosine from 4 to 6 hr after infection with Semliki forest virus. ●————● Acid-insoluble ¹⁴C-radioactivity; ○—·—·—○ ³H-radioactivity of RNA extracted from purified Semliki forest virus, grown in presence of ³H-uridine; ———— optical density at 260 mμ (chick ribosomal RNA). The top of the gradient is to the right.

compact) for a random-coiled molecule of its estimated molecular weight (1×10^6), Kelly et al. (76) concluded that the interjacent RNA was the random coiled form of viral RNA.

26S RNA of Semliki forest virus. Further information about interjacent RNA has come from studies by Sonnabend and Mécs on the replication of Semliki forest virus (SFV) (114, 115). SFV is an arbovirus containing lipid and RNA, and its replication, like that of picornaviruses, is not affected by actinomycin. When RNA was extracted from chick fibroblasts infected with SFV in the presence of actinomycin and ^{3}H-adenosine, the sucrose gradient profiles exhibited two major peaks as shown in Figure 11. Peak I (45S) sedimented in an identical fashion to RNA extracted from purified virus and was probably progeny viral RNA. The second peak (II) sedimented at 26S and was partially resistant to ribonuclease. The proportions of the two major peaks varied considerably—early in infection (2 to 3 hr p.i.) or after only brief labelling periods (15 min) the 26S form predominated. Later on (4 to 6 hr p.i.) or when the labelling period was extended, a greater proportion of the label appeared in the 45S RNA. For this reason, and because of its ribonuclease resistance, the 26S form was at first considered to be the replicative intermediate (36).

During the course of these investigations an enzyme was isolated from SFV-infected chick cells which catalyzed the incorporation of labelled nucleoside triphosphates into an acid-insoluble product. It had all the properties usually associated with a virus-induced RNA-dependent RNA polymerase—it was not affected by deoxyribonuclease or actinomycin, it required all four nucleoside triphosphates, magnesium and an energy generating system, and its activity increased during the period of rapid viral RNA synthesis (115). When the enzyme reaction product was extracted with detergent or phenol and examined on sucrose gradients, it was found to consist of a single species of RNA sedimenting at 20S (Figure 12). It was resistant to ribonuclease and this resistance was abolished by heating ($T_m = 103°$ in 0.15 M sodium chloride); its buoyant density (1.630) was less than that of 45S viral RNA (1.680). The polymerase reaction product therefore exhibits all the properties typical of double-stranded RNA. However, despite exhaustive investigation, it has not been possible to obtain synthesis of 45S viral RNA with this enzyme preparation. This observation provides strong evidence to support the idea that two enzymes are involved in the replication of viral RNA—a polymerase synthesizing double-stranded RNA from a single-stranded template, such as the one isolated in these studies, and a second enzyme which uses the double-stranded RNA as a template for the synthesis of progeny viral RNA.

When polymerase reaction products and *in vivo* pulse-labelled RNA from infected cells were compared on the same gradient, it was found that the S-value of the enzyme product (20S) was significantly less than that of the 26S peak of the RNA labelled *in vivo* (Figure 12); however, its S-value was identical to that of the *in vivo* RNA which had been treated with ribonuclease before gradient analysis. This suggested that the 26S material was a mixture of replicative intermediate and a species of RNA similar to the interjacent RNA described below. Partial separation of these two types of RNA was achieved by sedimentation of the labelled RNA through gradients containing 0.5 M potassium chloride, when the interjacent RNA moved sufficiently faster than the replicative intermediate to show a definite separation of radioactivity. In these high-salt gradients, the 45S RNA increased its S-value to a greater extent than did the 26S RNA. Better separation was obtained in caesium sulphate gradients—when material from the 20 to 30S region of a sucrose gradient was re-run in caesium sulphate, two sharp bands were seen—the major peak at 1.680 g/cm³ which was ribonuclease-sensitive and was presumably the interjacent RNA, and a second ribonuclease-resistant peak (1.625 g/cm³) which corresponded to the density of the polymerase reaction

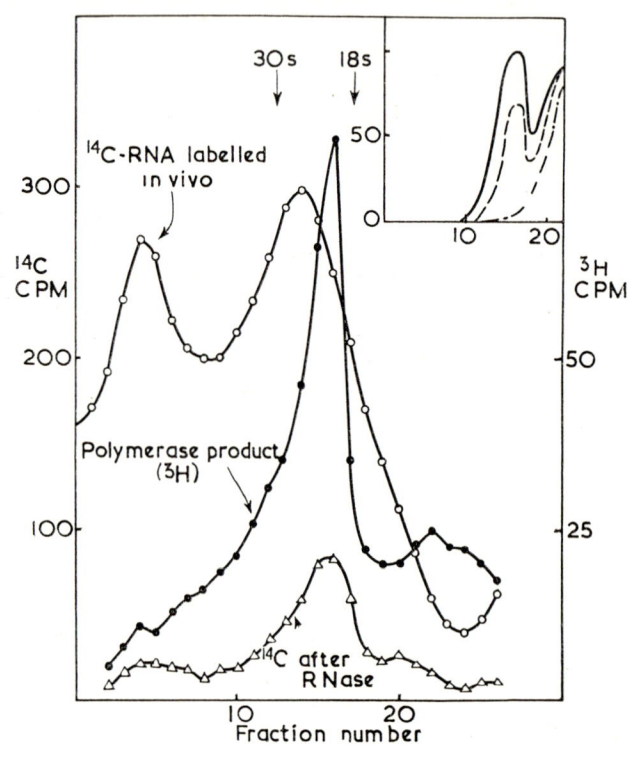

FIGURE 12. Sucrose gradient analysis of a reaction product obtained by incubation of RNA polymerase from Semliki forest virus-infected cells with ³H-GTP, and extracting the RNA with sodium dodecyl sulphate. ●———● Acid-insoluble ³H-radioactivity; ○———○ phenol-SDS extract of Semliki forest virus-infected cells incubated with actinomycin and ¹⁴C-uridine for 4 to 6 hr after infection — acid insoluble ¹⁴C-radioactivity; △———△ ¹⁴C-radioactivity after incubation of each fraction with 1 μg/ml of ribonuclease. *Insert*: Distribution of acid-insoluble radioactivity after incubation of a mixture of polymerase reaction product and *in vivo* labelled RNA with 0.5 μg/ml of ribonuclease for 20 min at 30°. ——— ³H-radioactivity (reaction product); ———— ¹⁴C-radioactivity (*in vivo* labelled RNA). The top of the gradient is to the right.

product (see Fig. 13). Although preliminary experiments suggested that the interjacent RNA was resistant to very low levels of ribonuclease (116), kinetic studies on the digestion of 26S and 45S RNAs showed no significant differences in the initial rates of breakdown of the two RNA species, although both showed some resistance to low levels of ribonuclease. However, there was a marked difference in the amount of nuclease-resistant core in the two preparations, e.g. after incubation with 10 µg of ribonuclease for 60 min at 37°, 9.0 per cent of the 26S RNA remained acid-insoluble, while only 2.0% of the 45S RNA remained. Assays for infective RNA were carried out on pulse-labelled RNA separated on a sucrose gradient. The results (Figure 14) show that, within the limits of the sensitivity of this method, the 26S RNA is not infective.

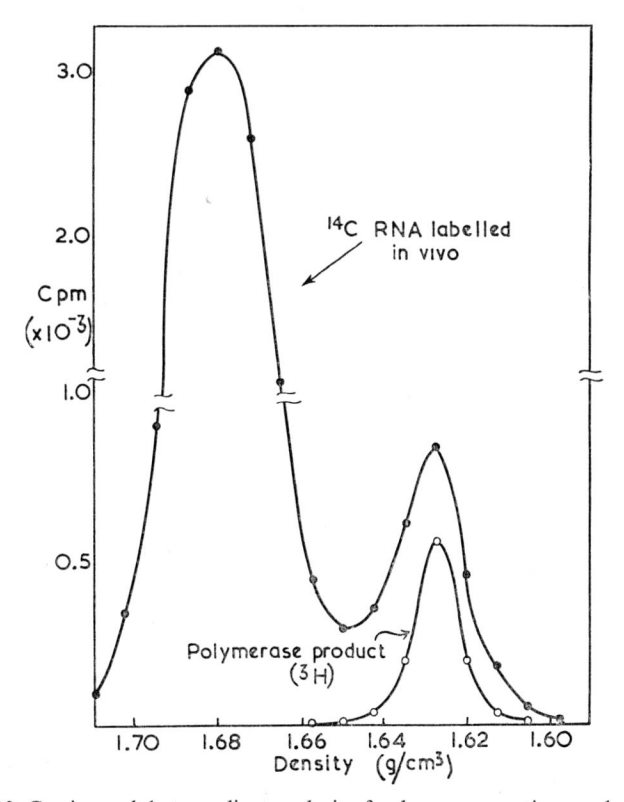

FIGURE 13. Caesium sulphate gradient analysis of polymerase reaction product and *in vivo* labelled RNA. A phenol-S.D.S extract of cells infected with Semliki forset virus and incubated with actinomycin and ^{14}C-adenosine for 5 to 6 hr post infection, was sedimented on a sucrose gradient. Fractions containing RNA sedimenting in the 18S to 30S region were pooled, mixed with a polymerase reaction product (for details, see Figure 12), tris-HCl (pH 7.7) and 60% w/w caesium sulphate solution, and centrifuged for 65 hr at 31,000 r.p.m. (15°C). ●————● *In vivo* pulse-labelled ^{14}C-radioactivity; ○————○ ^{3}H-radioactivity (polymerase reaction product).

From the results obtained to date Sonnabend et al. (116) have reached preliminary conclusions about the nature of the 26S RNA which are similar to those of Kelly et al. (76)—i.e. that it is a single-stranded viral RNA differing from the 45S RNA of the mature virus only in its secondary structure, which may be less compact. It is possible, as suggested by Roberts et al. (33), that interjacent RNA is an artefact produced either by nuclease digestion of viral RNA or by association of viral RNA with ribosomal RNA. It does not seem probable that ribonuclease breakdown would produce a single species of RNA with such well-defined and reproducible sedimentation properties; in any case, such an explanation would not account for the variation in proportions of 45S and 26S RNAs seen in cells infected or pulsed for different periods. No association of viral RNA and any other species of RNA has been seen on any gradients.

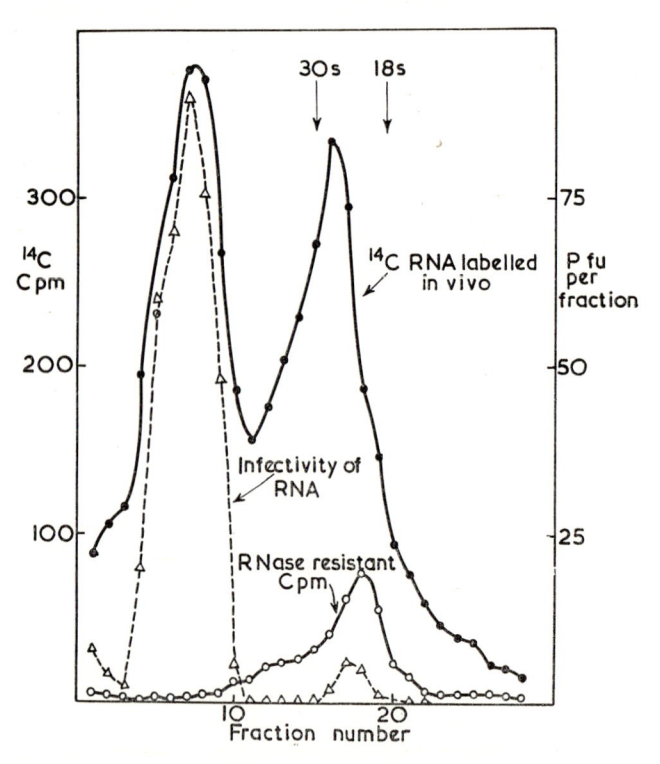

FIGURE 14. Distribution of infective RNA on sucrose gradient of *in vivo* labelled RNA. ●————● ^{14}C-radioactivity of RNA extract from cells labelled with ^{14}C-uridine from 0 to 6 hr after infection with Semliki forest virus; ○————○ ribonuclease-resistant ^{14}C-radioactivity; △----△ titre of infective RNA.

The function of interjacent RNA. It is possible that interjacent RNA is an essential intermediate in the process of viral RNA replication; the kinetics of appearance of labelled precursors into interjacent RNA are not inconsistent with such a role. However, its rate of synthesis often differs considerably from that of the double-stranded replicative form of RNA, and the amounts of label in the interjacent region are very variable— far more is seen in pulse-labelled RNA than in polymerase reaction products. These differing rates of synthesis suggest mutually exclusive functions. On the evidence to date it seems more likely that double-stranded RNA rather than interjacent RNA is the true intermediate in the replicative process. Further information about interjacent RNA may alter this picture entirely.

A more probable explanation for the function of interjacent RNA is that suggested by Kelly et al. (76). They observed during the first few minutes of synchronized infection that the labelled parental RNA in cell lysates went through periodic changes in ribonuclease resistance—becoming resistant, then sensitive, then resistant again. When the RNA was deproteinized and examined on gradients, it was found to be predominantly 27S (viral RNA) during the ribonuclease-sensitive phases and 20S (interjacent RNA) during the resistant phases. Movement into and out of the double-stranded template was precluded, as both types of RNA were ribonuclease-sensitive after extraction. Entry of the parental RNA into the 20S form was blocked by chloramphenicol. They suggested that the periodic changes could be due to the entry of the parental RNA into a ribonuclease-resistant protein-synthesizing complex (polysomes), and that in such a complex it became structurally altered to the 20S form, whose configuration was more like that of a random coil. In this form it could function as a messenger RNA. This hypothesis—that interjacent RNA is the messenger form of viral RNA—is consistent with our own observations on the 26S RNA of SFV infected cells, particularly with the time-course of its appearance during the infection cycle.

It should be relatively simple to test this hypothesis experimentally. Viral and interjacent RNAs should have the same base composition and should be interconvertible under controlled conditions *in vitro*. Virus-specific polysomes should contain interjacent rather than viral RNA; Godson and Sinsheimer have already reported that polysomes from cells infected with ^{32}P-labelled MS2 phage contain labelled RNA which is mainly 20S and 16S (113). In cell-free protein-synthetic systems 20S RNA should be more effective as a messenger than viral RNA, although it is possible that the transition from one to the other is brought about by ribosomes. It may be possible to influence the type of RNA synthesized by polymerases *in vitro* by adding either ribosomes, to encourage interjacent RNA formation, or viral capsid protein, to stimulate viral RNA formation. A full understanding of the role of interjacent RNA will have to await the results of such experiments.

Dependence of RNA Replication on Protein Synthesis

Viral RNA synthesis can be inhibited *in vivo* by general inhibitors of protein synthesis, such as puromycin, cycloheximide or high concentrations of *p*-fluorophenylalanine (66, 117–121). Inhibition is complete when these drugs are added before the commencement of viral RNA polymerase synthesis. Obviously if protein synthesis is blocked before the formation of any virus-specific RNA polymerase, synthesis of viral RNA will not occur. However, when the inhibitors are added after some polymerase has already been formed, i.e. during the period of rapid viral RNA synthesis, the incorporation of precursors into viral RNA ceases immediately and the rise in infectious RNA titre is blocked; thus, continuing synthesis of protein is required for the replication of viral RNA. Baltimore and Franklin (121) and Scharff et al. (119) have suggested that this dependence of RNA synthesis on continuing protein synthesis is due to the very high lability of viral RNA polymerase, which, it is suggested, is rapidly inactivated with a half-life of about 1 hr.

It is difficult to see how this explanation can account for all of the reported observations concerning the action of puromycin on viral RNA synthesis. The striking feature of puromycin action is the rapidity with which it brings viral RNA synthesis to a halt. When added to mengovirus-infected cells shortly after the beginning of viral RNA synthesis, a profound inhibition of viral RNA synthetic rate is seen 30 minutes later (121); when added to cells 3.5 hr after infection with EMC virus, 2×10^{-4} M puromycin reduced the rate of incorporation of precursors into protein and viral RNA to 2% and 30% respectively within 10 min (66); similar results were obtained with 5 μg/ml of cycloheximide. There is an equally rapid recovery of viral RNA synthesis when puromycin is removed (120, 121).

The rapidity with which these inhibitors act and the acute sensitivity of viral RNA synthesis to their action would suggest that they may act by inhibiting viral polymerase action, rather than blocking the synthesis of an enzyme which is rapidly turning over. Baltimore and Franklin (122) found that the mengovirus RNA polymerase was not sensitive to the addition of puromycin *in vitro*, although the author has observed an inhibition of EMC virus polymerase activity when puromycin was added to the assay mixture (66). Despite this observation, it does not seem very likely that puromycin blocks polymerase action directly; that is, interferes with either the active site of the enzyme or the formation of an enzyme-template complex. The fact that a number of protein synthetic inhibitors can exert the same inhibitory effect suggests that these drugs are interfering with an RNA synthetic system which is coupled to a protein synthetic system in such a way that the rate of the latter governs the rate of the former. To be more specific, it is entirely feasible that newly-synthesized viral RNA must be removed from the template, either by ribosomes (if the RNA is to become a messenger) or capsid proteins (if the RNA is to become incorporated into progeny), in order to prevent clogging up of the template with nascent RNA. Release of viral RNA would then be dependent either on its attachment to ribosomes to form polysomes which actively synthesize protein, or on the availability of capsid proteins, which could remove the viral RNA in the form of mature virus.

The mandatory coupling of messenger RNA synthesis to protein synthetic activity in bacteria has been suggested by Stent (123), and there is much evidence in favour of this theory. In the *in vitro* synthesis of RNA by the purified DNA-dependent polymerase of *E. coli*, Bremer and Konrad (96) found that most of the newly-synthesized RNA remained bound to the template at the end of the reaction. Removal of the product with ribonuclease has been shown to stimulate the rate and extent of polynucleotide formation (124). A similar stimulation can be achieved by the addition of ribosomes to the reaction mixture (125); in this system an association of nascent RNA and polysomal aggregates was observed. A cell-free protein synthetic system which was dependent upon DNA and a concomitant RNA synthetic capacity has been reported (126), and polysomes apparently containing DNA strands were seen in the electron microscope. The direction of synthesis of messenger RNA, from the 5' end to the 3' (127) is the same as the direction of reading of the message (128, 129). All these observations are in accord with the idea that movement of messenger RNA through polysomes can facilitate its transcription from a DNA template by removing newly-synthesized RNA as it is formed.

There is ample supporting evidence that a similar situation may exist in the case of viral RNA replication. The close association in picornavirus-infected cells of viral RNA polymerase, polysomes of a characteristic size, viral antigen and mature virus in lipoprotein complexes (6, 29, 30, 33) suggests a highly integrated system. Under these conditions it is quite conceivable that viral RNA synthesis is directly coupled to its utilization as a messenger or as progeny. This hypothesis would provide a simple explanation for the action of protein synthesis inhibitors. It may also explain why different proportions of double- and single-stranded RNAs are formed in the reaction products of different viral RNA polymerase (e.g. Figures 5 and 12). The polymerase catalyzing the synthesis of the double-stranded template (Figure 7, reaction A) should function in the absence of protein synthesis, as its product must necessarily remain bound to its template. However, in transcribing the double-stranded template (Figure 7, reaction B and C) removal of the products is essential, and, if the means to do so were lacking in the enzyme preparation, the synthesis of progeny RNA would be curtailed.

Also, if a mechanism existed for the removal of nascent RNA, such as is proposed here, it would remove one of the objections to the model of RNA replication shown in Figure 10A—it suggests a means whereby the length of the displaced template-positive strand could be limited. One objection to this hypothesis is that puromycin, which is active in blocking viral RNA synthesis, is thought to act by releasing the peptide chain from its ribosome (130), and thus cause the ribosomes to move along the messenger RNA at an increased rate (131). If this were so, one would expect the rate of removal of nascent RNA from its template to be stimulated by puromycin rather than the reverse. However, as cycloheximide, a drug which does not affect release of nascent polypeptide chains (132), also causes inhibition of viral RNA synthesis, it seems reasonable to assume that puromycin does not stimulate messenger RNA movement, at least in this instance.

Levintow et al. (117) found that low levels of p-fluorophenylalanine permitted polioviral RNA and protein synthesis, but virus maturation was prevented; increasing quantities of the drug produced a parallel decrease in the rates of viral RNA and protein synthesis, as would be expected if the two processes were coupled. When the inhibition was reversed with phenylalanine, mature virus was formed—however, none of the viral RNA which had been made in the presence of the low levels of inhibitor was incorporated into particles. Hence, it seems likely that the integration of the protein and RNA synthetic activities may extend to the maturation of the virus particle, and that maturation occurs only when both viral RNA and protein synthetic capacities are unimpaired.

CONCLUSIONS AND SUMMARY

Much of the relevant data concerning the replication of viral RNA in animal and bacterial cells infected with small RNA-containing viruses have been reviewed. The most prominent features of this process are the synthesis of specific polymerases which catalyze the synthesis of viral RNA, and the appearance in infected cells of a ribonuclease-resistant double-stranded RNA. All evidence from a variety of systems indicates that the ribonuclease-resistant RNA is a duplex composed of one strand each of viral (positive) and complementary (negative) RNAs, and that this double-stranded RNA functions as a template for the synthesis of progeny viral RNA. The major problem yet to be resolved is precisely how this asymmetric synthesis is achieved.

Two main schemes have been proposed—the *semi-conservative* mechanism, which postulates that the nascent RNA displaces the positive strand of the duplex, releasing it to become functional progeny RNA; and the *conservative* mechanism, which proposes that the newly-synthesized RNA is laid down on the double-stranded template without displacement of either chain, i.e. the duplex is stable. Both schemes assume that it is the negative strand only which is transcribed, and that this strand is originally synthesized by a polymerase which uses the invading parental RNA as a template.

Few of the results of experiments designed to test these hypotheses have favoured the semi-conservative model. This theory predicts that the parental RNA, once it has entered the double-stranded duplex, should be partially displaced from it, and is thus rendered partially sensitive to ribonuclease. However, ribonuclease treatment of the replicative intermediate containing labelled parental RNA, followed by heat denaturation, shows that most of the parental RNA is intact. The infectivity of ribonuclease-resistant RNA would also imply a similar conclusion.

In contrast, results of nearly all these studies can be readily explained by the conservative mechanism, which predicts that the positive strand of the template duplex would remain intact after ribonuclease treatment, as it envisages a stable double-stranded template. Certain difficulties do arise when considering the stability of the template-nascent RNA complex to deproteinization. However, these difficulties can be resolved by minor modifications of the model—if one postulates either that displacement of the template positive strand occurs in the region where the polymerase is actually transcribing the negative strand, or that the nascent RNA forms a stable triplex with its double-stranded template, then the conservative model can account for most of the observations made on various systems by different techniques. It is certainly more in accord with the experimental data than the semi-conservative mechanism.

In many systems a third type of RNA can be seen in sucrose gradients of extracts from infected cells. This interjacent RNA is single-stranded and has many properties in common with the RNA of the mature virus particle. It differs mainly in its sedimentation properties and these differences can be explained if it is postulated that viral RNA is a more compact form of interjacent RNA. The function of this RNA has yet to be clarified, but preliminary studies on bacterial and animal systems suggest that it could be the form in which viral RNA functions as a messenger in protein synthesis.

Finally, there appears to be a close correlation between the ability of a system to synthesize viral RNA and its capacity to synthesize protein. The rapidity with which RNA synthesis ceases after the addition of protein-synthesis inhibitors to infected cells suggests that the connexion between the two processes is closer than can be accounted for by a requirement for the synthesis of a very labile polymerase. The results of inhibitor studies are more readily explained by the hypothesis that continued synthesis of viral RNA depends on the uninterrupted removal from the template of newly-formed RNA, either by ribosomes or viral capsid protein. Hence, whether the viral RNA is destined to become a messenger or to be incorporated into progeny, an unimpaired protein synthetic apparatus is essential for RNA synthesis. It is also possible that all the events of virus replication, from the synthesis of the template RNA to the maturation of the virus particle, are interdependent, and that each synthetic process must be functionally active and coupled in an integrated system before viral replication can proceed to completion.

REFERENCES

1. Cohen, S. S., *Ann. Review Biochem.* **32**, 83 (1963).
2. Reich, E., Franklin, R. M., Shatkin, A. J. and Tatum, E. L., *Proc. Natl. Acad. Sci. U.S.* **48**, 1238 (1962).
3. Salzman, N. P., *Virology* **10**, 150 (1960).
4. Franklin, R. M. and Rosner, J. *Biochim. Biophys. Acta* **55**, 240 (1962).
5. Holland, J. J. and Bassett, D. W., *Virology* **23**, 164 (1964).
6. Dalgarno, L. and Martin, E. M., *Virology* **26**, 450 (1965).
7. Montagnier, L. and Sanders, F. K., *Nature* **199**, 664 (1963).
8. Baltimore, D. and Franklin, R. M., *J. Biol. Chem.* **238**, 3395 (1963).
9. Weissmann, C. and Borst, P., *Science* **142**, 1188 (1963).
10. Levintow, L., *Annual Review of Biochem.* **34**, 487 (1965).
11. Hausen, P. and Verwoerd, D. W., *Virology* **21**, 617 (1963).
12. Horton, E., Liu S-L., Martin, E. M. and Work, T. S., *J. Molecular Biol.* **15**, 62 (1966).
13. Martin, E. M., Malec, J., Sved, S. and Work, T. S., *Biochem. J.* **80**, 585 (1961).
14. Baltimore, D. and Franklin, R. M., *Proc. Natl. Acad. Sci. U.S.* **48**, 1383 (1962).
15. Baltimore, D., Franklin, R. M. and Callender, J., *Biochim. Biophys. Acta* **76**, 425 (1963).
16. Holland, J. J., *Proc. Natl. Acad. Sci. U.S.* **49**, 23 (1963).
17. Balandin, I. G. and Franklin, R. M., *Biochem. Biophys. Research Communs.* **15**, 27 (1964).

18. Verwoerd, D. W. and Hausen, P., *Virology* **21**, 628 (1963).
19. Penman, S., Scherrer, K., Becker, Y. and Darnell, J. E., *Proc. Natl. Acad. Sci. U.S.* **49**, 654 (1963).
20. Cline, M. J., Eason, R. and Smellie, R. M. S., *J. Biol. Chem.* **238**, 1788 (1963).
21. Baltimore, D., *Proc. Natl. Acad. Sci. U.S.* **51**, 450 (1964).
22. Horton, E., Liu S.-L., Dalgarno, L., Martin, E. M. and Work, T. S., *Nature* **204**, 247 (1964).
23. Wilson, R. G. and Bader, J. P., *Biochim. Biophys. Acta* **103**, 549 (1965).
24. Ho, P. P. K. and Walters, C. P., *Biochemistry* **5**, 231 (1966).
25. Astier-Manifacier, S. and Cornuet, P., *Biochem. Biophys. Research Communs.* **18**, 283 (1965).
26. Weissmann, C., Simon, L. and Ochoa, S., *Proc. Natl. Acad. Sci. U.S.* **49**, 407 (1963).
27. Haruna, I., Nozu, K., Ohtaka, Y. and Spiegelman, S., *Proc. Natl. Acad. Sci. U.S.* **50**, 905 (1963).
28. August, J. T., Shapiro, L. and Eoyang, L., *J. Molecular Biol.* **11**, 257 (1965).
29. Eggers, H. J., Baltimore, D. and Tamm, I., *Virology* **21**, 281 (1963).
30. Penman, S., Becker, Y. and Darnell, J. E., *J. Molecular Biol.* **8**, 541 (1964).
31. Bellett, A. J. D., Harris, R. G. and Sanders, F. K., *J. Gen. Microbiol.* **38**, 299 (1965).
32. Eason, R. and Smellie, R. M. S., *J. Biol. Chem.* **240**, 2580 (1965).
33. Roberts, W. K., Newman, J. F. E. and Rueckert, R. R., *J. Molecular Biol.* **15**, 92 (1966).
34. Baltimore, D., Becker, Y. and Darnell, J. E., *Science* **143**, 1034 (1964).
35. Brown, F. and Cartwright, B., *Nature* **204**, 855 (1964).
36. Sonnabend, J. A., Friedman, R. M., Dalgarno, L. and Martin, E. M., *Biochem. Biophys. Research Communs.* **17**, 455 (1964).
37. Hausen, P., *Virology* **25**, 523 (1965).
38. Shipp. W. and Haselkorn, R., *Proc. Natl. Acad. Sci. U.S.* **52**, 401 (1964).
39. Mandel, H. G., Matthews, R. E. F., Matus, A. and Ralph, R. K., *Biochem. Biophys. Research Communs.* **16**, 604 (1964).
40. Ralph, R. K., Matthews, R. E. F., Matus, A. I. and Mandel, H. G., *J. Molecular Biol.* **11**, 202 (1965).
41. Fenwick, M. L., Erikson, R. L. and Franklin, R. M., *Science* **146**, 527 (1964).
42. Kelly, R. B. and Sinsheimer, R. L., *J. Molecular Biol.* **8**, 602 (1964).
43. Kaerner, H. C. and Hoffmann-Berling, H., *Nature* **202**, 1012 (1964).
44. Ammann, J., Delius, H. and Hofschneider, P. H., *J. Molecular Biol.* **10**, 557 (1964).
45. Scharff, M. D. and Levintow, L., *Virology* **19**, 491 (1963).
46. Krug, R. M. and Franklin, R. M., *Virology* **22**, 48 (1964).
47. Martin, E. M. and Work, T. S., *Biochem. J.* **83**, 574 (1962).
48. Horne, R. W. and Nagington, J., *J. Molecular Biol.* **1**, 333 (1959).
49. Dales, S., Eggers, H. J., Tamm, I. and Palade, G. E., *Virology* **26**, 379 (1965).
50. Penman, S., *Virology* **25**, 148 (1965).
51. Attardi, G. and Smith, J., *Cold Spring Harbor Symp. Quant. Biol.* **27**, 271 (1962).
52. Summers, D. F. and Levintow, L., *Virology* **27**, 44 (1965).
53. Scharff, M. D., Shatkin, A. J. and Levintow, L., *Proc. Natl. Acad. Sci. U.S.* **50**, 686 (1963).
54. Summers, D. F., McElvain, N. F., Thorén, M. M. and Levintow, L., *Biochem. Biophys. Research Communs.* **15**, 290 (1964).

55. Kerr, I. M., Martin, E. M., Hamilton, M. G. and Work, T. S., *Cold Spring Harbor Symp. Quant. Biol.* **27**, 259 (1962).
56. Darnell, J. E., *Cold Spring Harbor Symp. Quant. Biol.* **27**, 149 (1962).
57. Baltimore, D., Eggers, H. J. and Tamm, I., *Biochim. Biophys. Acta* **76**, 644 (1963).
58. Watanabe, Y., Watanabe, K. and Hinuma, Y., *Biochim. Biophys. Acta* **61**, 976 (1962)
59. Scharff, M.D., Maizel, J. V., Jr. and Levintow, L., *Proc. Natl. Acad. Sci. U.S.* **51**, 329 (1964).
60. Kerr, I. M., Martin, E. M., Hamilton, M. G. and Work, T. S., *Biochem. J.* **94**, 337 (1965).
61. Haruna, I. and Spiegelman, S., *Proc. Natl. Acad. Sci. U.S.* **54**, 579 (1965).
62. Weissmann, C., *Proc. Natl. Acad. Sci. U.S.* **54**, 202 (1965).
63. Dalgarno, L., Martin, E. M., Liu S.-L. and Work, T. S., *J. Molecular Biol.* **15**, 77 (1966).
64. Weissmann, C., Borst, P., Burdon, R. H., Billeter, M. A. and Ochoa, S., *Proc. Natl. Acad. Sci. U.S.* **51**, 890 (1964).
65. Delius, H. and Hofschneider, P. H., *J. Molecular Biol.* **10**, 554 (1964).
66. Martin, E. M. Unpublished experiments.
67. Lodish, H. and Zinder, N., *Science* **152**, 372 (1966).
68. Shapiro, L. and August, J. T., *J. Molecular Biol.* **11**, 272 (1965).
69. Spiegelman, S. Haruna, I., Holland, I. B., Beaudreau, G. and Mills, D., *Proc. Natl. Acad. Sci. U.S.* **54**, 919 (1965).
70. Erikson, R. L., Fenwick, M. L. and Franklin, R. M., *J. Molecular Biol.* **10**, 519 (1964).
71. Doi, R. H. and Spiegelman, S., *Proc. Natl. Acad. Sci. U.S.* **49**, 353 (1963).
72. Tobey, R. A., *Virology* **23**, 10 (1964).
73. Homma, M. and Graham, A. F., *J. Bacteriology* **89**, 64 (1965).
74. Tobey, R. A., *Virology* **23**, 23 (1964).
75. Plagemann, P. G. W., Hoshino, M. and Swim, H. E., *Federation Proc.* **24**, 597 (1965).
76. Kelly, R. B., Gould, J. L. and Sinsheimer, R. L., *J. Molecular Biol.* **11**, 562 (1965).
77. Nonoyama, M. and Ikeda, Y., *J. Molecular Biol.* **9**, 763 (1964).
78. Langridge, R., Billeter, M. A., Borst, P., Burdon, R. H. and Weissmann, C., *Proc. Natl. Acad. Sci. U.S.* **52**, 114 (1964).
79. Langridge, R. and Gomatos, P. J., *Science* **141**, 694 (1963).
80. Weissmann, C., Borst, P., Burdon, R. H., Billeter, M. A. and Ochoa, S., *Proc. Natl. Acad. Sci. U.S.* **51**, 682 (1964).
81. Erikson, R. L., Fenwick, M. L. and Franklin, R. M., *J. Molecular Biol.* **13**, 399 (1965)
82. Watanabe, Y., *Biochim. Biophys. Acta* **95**, 515 (1965).
83. Weissmann, C., Billeter, M. A., Schneider, M. C., Knight, C. A. and Ochoa, S., *Proc. Natl. Acad. Sci. U.S.* **53**, 653 (1965).
84. Bishop, J. M., Summers, D. F. and Levintow, L., *Proc. Natl. Acad. Sci. U.S.* **54**, 1273 (1965).
85. Pons, M., *Virology* **24**, 467 (1964).
86. Weissmann, C., Billeter, M. A., Vinuela, E. and Libonati, M., Conference on Plant Viruses, Wageningen, Holland (July 1965) (In press).
87. Haruna, I. and Spiegelman, S., *Science* **150**, 884 (1965).
88. Watson, J. D. and Crick, F. H. C., *Nature* **171**, 964 (1953).
89. Tocchini-Valentini, G. P., Stodolsky, M., Aurisicchio, A., Sarnat, M., Graziosi, F., Weiss, S. B. and Geiduschek, E. P., *Proc. Natl. Acad. Sci. U.S.* **50**, 935 (1963).

90. Hayashi, M., Hayashi, M. N. and Spiegelman, S., *Proc. Natl. Acad. Sci. U.S.* **50**, 664 (1963).
91. Hayashi, M., Hayashi, M. N. and Spiegelman, S., *Proc. Natl. Acad. Sci. U.S.* **51**, 351 (1964).
92. Sinsheimer, R. L. and Lawrence, M., *J. Molecular Biol.* **8**, 289 (1964).
93. Chamberlin, M. and Berg, P., *J. Molecular Biol.* **8**, 297 (1964).
94. Geiduschek, E. P., Nakamoto, T. and Weiss, S. B., *Proc. Natl. Acad. Sci. U.S.* **47**, 1405 (1961).
95. Chamberlin, M., Baldwin, R. L. and Berg, P., *J. Molecular Biol.* **7**, 334 (1963).
96. Bremer, H. and Konrad, M. W., *Proc. Natl. Acad. Sci. U.S.* **51**, 801 (1964).
97. Jehle, H., *Proc. Natl. Acad. Sci. U.S.* **53**, 1451 (1965).
98. Sinsheimer. R. L., Starman, B., Nagler, C. and Guthrie, S., *J. Molecular Biol.* **4**, 142 (1962).
99. Denhardt, D. T. and Sinsheimer, R. L., *J. Molecular Biol.* **12**, 647 (1965).
100. Brown, F. and Martin, S. J., *Nature* **208**, 861 (1965).
101. Geirer, A., *Nature* **179**, 1297 (1957).
102. Ginoza, W., *Nature* **181**, 958 (1958).
103. Fiers, W. and Sinsheimer, R. L., *J. Molecular Biol.* **5**, 408 (1962).
104. Joklik, W. K. and Darnell, J. E., *Virology* **13**, 439 (1961).
105. Laduron, P. and Cocito, C., *Biochem. Biophys. Research Communs.* **13**, 32 (1963).
106. Davis, J. E. and Sinsheimer, R. L., *J. Molecular Biol.* **6**, 203 (1963).
107. Stent, G. S., *Advances in Virus Research* **5**, 95 (1958).
108. Zubay, G., *Proc. Natl. Acad. Sci. U.S.* **48**, 456 (1962).
109. Guschlbauer, W., *Nature* **209**, 258 (1966).
110. Massoulié, J., Guschlbauer, W., Klotz, L. and Fresco, J., *C.R. Acad. Sci.*, Paris **260**, 1285 (1965).
111. Stevens, C. L. and Felsenfeld, G., *Biopolymers* **2**, 293 (1964).
112. Zimmerman, E. F., Heeter, M. and Darnell, J. E., *Virology* **19**, 400 (1963).
113. Godson, G. N. and Sinsheimer, R. L., Abstracts Biophysical Soc., Xth Meeting, Boston, p. 6 (Feb. 1966).
114. Sonnabend, J. A., Mécs, E. and Martin, E. M., Nature (In press).
115. Martin, E. M. and Sonnabend, J. A., Virology (In press).
116. Sonnabend, J. A., Mécs, E. and Martin, E. M., Abstracts, Federation of European Biochemical Soc., IIIrd Meeting, Warsaw (April 1966).
117. Levintow, L., Thorén, M. M., Darnell, J. E. and Hooper, J. L., *Virology* **16**, 220 (1962)
118. Wecker, E., Hummeler, K. and Goetz, O., *Virology* **17**, 110 (1962).
119. Scharff, M. D., Thorén, M. M., McElvain, N. F. and Levintow, L., *Biochem. Biophys. Research Communs.* **10**, 127 (1963).
120. Wecker, E., *Nature* **197**, 1277 (1963).
121. Baltimore, D. and Franklin, R. M., *Biochim. Biophys. Acta* **76**, 431 (1963).
122. Baltimore, D. and Franklin, R. M., *Biochem. Biophys. Research Communs.* **9**, 388 (1962).
123. Stent, G., *Proc. Royal Soc. B*, **164**, 181 (1966).
124. Krakow, J. S., *J. Biol. Chem.* **241**, 1830 (1966).
125. Shin, D. H. and Moldave, K., *Biochem. Biophys. Research Communs.* **22**, 232 (1966).
126. Byrne, R., Levin, J. G., Bladen, H. A. and Nirenberg, M. W., *Proc. Natl. Acad. Sci. U.S.* **52**, 140 (1964).

127. Bremer, H., Konrad, M. W., Gaines, K. and Stent, G. S., *J. Molecular Biol.* **13**, 540 (1965).
128. Thach, R. E., Cecere, M. A., Sundararajan, T. A. and Doty, P., *Proc. Natl. Acad. Sci. U.S.* **54**, 1167 (1965).
129. Salas, M., Smith, M. A., Stanley, W. M., Jr., Wahba, A. J. and Ochoa, S., *J. Biol. Chem.* **240**, 3988 (1965).
130. Nathans, D., *Proc. Natl. Acad. Sci. U.S.* **51**, 585 (1964).
131. Williamson, A. R. and Schweet, R., *J. Molecular Biol.* **11**, 358 (1965).
132. Ennis, H. L. and Lubin, M., *Science,* **146**, 1474 (1964).

DISCUSSION

(Chairman: S. G. LALAND)

W. FULLER: 1. In your three-stranded model for the conservative replication of viral RNA, how was the third RNA strand attached to the conservative duplex?

2. I would like to make a brief comment on the point made by Dr. Borst regarding the use of poly-AUU as a model compound for studying the resistance of three-stranded RNA complexes in enzyme attack. The way in which the three polynucleotide chains are linked together in poly-AUU has not been determined and it may be that its behaviour could be quite different from that of the three-stranded RNA.

E. M. MARTIN: 1. The three-stranded model shown in Fig. 10C was based on the suggestions of Stent and of Zubay that triple-stranded structures involving hydrogen bonding between one base and two complementary bases were structurally quite feasible. I should stress that the model is a hypothetical one. There is, as yet, only meagre evidence for the existence of triple-stranded intermediates. As soon as convincing evidence for their existence is obtained, then perhaps we shall present to you X-ray crystallographers the problem of finding out how the third strand is attached.

2. Certainly the presence of G–C pairs in the viral RNA should increase the stability of a triple-stranded structure with respect to that of synthetic poly-AUU.

A. B. STONE: Presumably the low infectivity of the replicative form of RNA viruses is due to their inability to act as messengers for the first and essential step of virus production: the synthesis of RNA-replicase. Is it possible that one could pre-infect the cells with a different strain of the virus, so that they already contain a supply of replicase enzyme when they are superinfected with replicative form?

E. M. MARTIN: To my knowledge no one has performed such an experiment, although it should not be very difficult technically. The result would certainly be interesting.

S. OCHOA: In connection with Dr. Martin's statements relative to the role of double-stranded RNA in viral RNA replication, I should like to recall that Borst and Weissmann (*Proc. Natl. Acad. Sci. U.S.,* **54**, 982 (1965)) found with the MS2 RNA synthetase system that very little RNase-resistant material was formed when the incubation was conducted at low temperatures unless the reaction mixture was extracted with phenol or treated with certain detergents. It appears then that the "plus" and "minus" strands of the replicating intermediate are not held together by hydrogen bonds to form a conventional double-stranded helix. This can subsequently arise through phenolization or other treatments.

P. BORST: I would like to comment that all of the objections that Dr. Martin has brought forward against the semi-conservative mechanism have been dealt with in the papers of the Weissmann–Ochoa group. On the other hand, Weissmann and I have never been able to obtain any evidence for the crux of the conservative replication hypothesis, to wit, the resistance to ribonuclease of the third strand in the postulated triple-stranded model. Do you have any evidence to show that it is or might be resistant?

E. M. MARTIN: The work of Erikson et al. (81) and of Brown and Martin (100) on the intactness of the parental strand after ribonuclease treatment of the replicative intermediate appeared after the publication of the major papers of your group (9, 26, 62, 64, 78, 80), and have not been discussed by them. These observations, together with the resistance of the replicative-form infectivity to ribonuclease, constitute the major objections to the semi-conservative hypothesis.

I agree that no evidence is at present available for the existence of a triple-stranded structure. My own attempts with reaction products from SFV and EMC polymerases have not been successful. The point I wish to emphasize is that the semi-conservative hypothesis has failed to explain important observations and we must look for some alternative. The models I have suggested may be equally unsatisfactory, but it is an attempt to broaden our concepts about replicative mechanisms. The main difficulty with the conservative hypothesis is its inability to explain the stability of nascent RNA to deproteinization. It is possible that the ribonuclease-sensitive RNA apparently associated with the replicative intermediate may be interjacent RNA. If this were so, it would be unnecessary to postulate a triple-stranded intermediate.

ADDENDUM (added in proof): The role of double-stranded RNA in virus replication has been queried in two recent papers. Plagemann & Swim (Bact. Rev. 30, 288) found that, when RNA polymerase from mengovirus-infected cells was incubated, first with ^{3}H-GTP, then with cold precursor, label could be chased from double-stranded into single-stranded RNA. When pulse-chase experiments were performed in vivo, chasing caused no loss of label from duplex RNA. They propose that in vivo the polymerase template is a single negative strand of RNA and that the product is complexed with a factor to prevent formation of double strands; these would be formed only in the absence of the factor, e.g. in vitro. Haruna & Spiegelman (Proc. Natl. Acad. Sci. U.S. 55, 1256) found no evidence for negative-strand formation during synthesis of Qβ phage RNA by purified replicase, thus ruling out mechanisms involving complementarity. However, Weissmann & Felix (Proc. Natl. Acad. Sci. U.S. 55, 1264) did find appreciable amounts of negative strands in similar experiments with the same replicase, although not all of these strands occurred in the double-stranded form.

Studies on Temperature Conditional Mutants of *E. coli* with a Modified Aminoacyl RNA Synthetase

M. Yaniv and F. Gros

Institut de Biologie Physico-Chimique, 13, rue Pierre Curie, Paris, France

Introduction

In the course of the last five years a very clear and complete picture has emerged concerning the general mechanisms involved in the formation and the genetic role of nucleic acids and proteins, but much remains to be learnt about the detailed structure and functioning of the machinery involved in macromolecular synthesis.

This machinery can be schematically considered as being composed of specific templates, recognition enzymes and particulate reading units, the ribosomes. Little is known about the factors involved in the recognition of specific polynucleotide sequences by the "transcription" or the "activating" enzymes (more particularly, as far as the nature of their active sites and their general conformation or architecture is concerned). The role of ribosomal proteins in the functioning of ribosomes has just begun to be explored. The same is true about the detailed operations of the DNA replicating systems and their mode of initiation. Although it is obvious that our knowledge regarding these problems may be advanced by means of modern biochemical techniques, investigations in this field would nonetheless be greatly facilitated if mutants were available in which some of the fundamental steps in the synthesis of macromolecules were altered. One might expect, in this way, to better interpret normal behaviour, e.g. the origin and control of some of the key elements in macromolecular syntheses, if it were possible to analyze the altered functions of the corresponding elements in the mutated cells.

In addition it would be very interesting in itself to locate genetically the determinants controlling the nucleic acid and protein forming machinery in order to determine whether the principal functions involved in the synthesis of these macromolecules depend, as in many viruses, on groups of genes topologically assembled in specific regions of the chromosome.

For a long time it was implicitly assumed that the essential mechanisms of macromolecular syntheses would escape genetic analysis because the relevant mutants would be lethal. Such an attitude has been completely revised during the last three years, as a result of the work of phage biologists, who have shown that the genetics of virus replication or maturation could be studied using conditional mutants in which specific steps in these general processes were blocked at a supraoptimal temperature range (1).

It has become possible to extrapolate such a methodology to the analysis of macromolecular biosynthetic processes in bacterial cells.

Recent and striking illustrations of the importance of conditional mutants in delineating fundamental steps in the functioning of the cell machinery come from the use of various suppressor strains to indicate the mechanisms of polypeptide chain propagation, or of streptomycin conditional dependent strains whose existence sheds a new light on the specificity of messenger RNA recognition by ribosomes (2).

We should like to discuss other categories of conditional mutants which seem to open up new approaches to the study of nucleic acid and protein synthesis. These are organisms unable to perform the synthesis of one specific type of macromolecule beyond a certain temperature at which this biosynthetic process would proceed normally in the wild type.

A general table (Table I) illustrates the main types of temperature sensitive mutants thus far investigated.

TABLE I

Macromolecule whose synthesis is blocked over the critical temperature	Mechanism of the mutation	References
DNA chromosomal	Inhibition of initiation of a new cycle of replication	(3)
DNA episomic	Inhibition of episomic replication chromosomal replication continues	(4)
Cellular division	Inhibition of cell division Macromolecular syntheses not inhibited	(5)
Proteins	Thermosensitive phenylalanyl-s-RNA synthetase in a relaxed strain of E. Coli	(6,7)
Synthesis of false proteins?	Altered ribosomes in a suppressor strain of E. Coli	(8)
Proteins RNA (group "A")	Thermosensitive valyl-s-RNA synthetase (stringent strain)	(6,9)
RNA and to a moderate degree proteins (mesausred by aa radioactive incorporation) (group"B")	Thermosensitive alanyl-s-RNA synthetase	
DNA, RNA, Proteins (Lethal)	Energy production	(10)

Five major classes of mutants can be distinguished depending upon the type of macromolecule they are unable to synthesize above the subcritical temperature:

(a) Those which cannot initiate a new round of DNA replication, but can only complete a cycle of DNA synthesis. In some cases the mutation concerns the replication of chromosomal DNA; whereas it specifically affects in some instances that of episomal DNA's without causing any defect at the chromosomal level (4, 3, 11).

(b) A possibly related class of temperature sensitive organisms must also be considered, in which the process of cell bipartition appears to be inhibited at 42°C, leading to filamentous forms (5, 11).

(c) A third class includes organisms which cannot synthesize proteins as judged by a defect in their capacity to incorporate radioactive amino acids. The great majority of these mutants harbor an alteration at the level of one specific activating enzyme, which displays abnormal instability at the temperature causing growth inhibition. As indicated in Table II, several groups of amino acid activating enzyme mutants are presently known. When the strain in which the mutation has occurred has a stringent "RC" locus, RNA synthesis is also blocked above the critical temperature range (6, 9, 12). In contrast, when the "RC locus is relaxed" the strain can manufacture RNA without synthesizing proteins (6)* (such a case is known in a mutant with a modified phenylalanyl-RNA synthetase).

Among the temperature conditional mutants in which the synthesis of proteins is the sensitive step, another type of organism, recently described by Apirion (8), is worth mentioning, since it would harbor a defect in the ribosome structure. This type of mutant has been selected on the basis of its suppressor activity for a mutation in tryptophane synthetase at 30°C and its inability to form colonies at 42°C. The alteration is manifested by a loss of biological activity of ribosomes at temperatures close to 45°C when tested *in vitro* with poly U as messenger RNA. *In vivo* incorporation of radioactive amino acids continues linearly at 42°C, suggesting that biologically altered proteins could be formed.

(d) A fourth group has a phenotype characterized by the cessation of RNA synthesis long before the cessation of DNA or protein synthesis when the organisms considered are placed at the critical temperature. Two of these mutants have been shown by us to contain a modified alanine-activating enzyme.

(e) Finally a type of mutant described by Cousin and Jacob (10) ceases to manufacture DNA, RNA and proteins when transferred at 40°C. Although

---

* In passing, other mutations of the amino acid activating systems have been found which are not conditional, like those just described, but "gratuitous", for they lead to no change in the growth rate of the corresponding strain under a great variety of conditions but rather to a specific drug resistance: an example of this case is given by an *E. coli* strain isolated by Fangman and Neidhardt (13,14) as a mutant resistant to an amino acid analogue, the compound parafluorophenylalanine. In such a strain the phenylalanyl RNA synthetase, contrary to the wild type enzyme, does not activate the analogue but keeps using phenylalanine as a substrate.

TABLE II. Comparative Activities of Aminoacyl sRNA Synthetases in Strain CR-34 (Wild Type) and Mutants T-2 and T-140

ACTIVITIES OF AMINOACYL S RNA SYNTHETASE IN WILD TYPE AND TEMPERATURE SENSITIVE MUTANTS

amino acid	CR 34		T 2		T 140	
	30	41	30	41	20	40
alanine			136	100	**1.9**	**1.2**
arginine	15	10	12	10	28	24
aspartic acid	57	72	26	19	44	42
cysteine					33	15
glutamic acid					33	41
glutamine	79	69	17	16	51	46
glycine	70	89	107	127	19.4	19
histidine	45	48	43	46	104	101
isoleucine	40	27	24	20	21	24
leucine	63	31	52	36	201	214
lysine	82	87	57	76	89	85
methionine	60	41	32	22	84	89
phenylalanine	41	27	20	24	29	22
proline	14	13	6	6	40	40
serine	52	29	20	10	77	74
threonine	86	52	64	39	117	111
tryptophane						
tyrosine	26	17	25	8	10	7
valine	112	76	**1.6**	**0**	191	183

The values are given in $\mu\mu$moles amino acid charged on sRNA. Incubation mixture contains: 90 μg sRNA (prepared from ML 308) in the case of CR-34 and T-2, or purchased from general biochemicals, (T-140); 4 μg protein (CR-34, T-2) or 16 μg protein (T-140) from partially purified extracts (9); ATP, 0.1 μmoles; 0.0125 μmoles of each of 18 non radioactive amino acids plus 0.0125 μmoles of each of the ^{14}C amino acid tested. Total volume, 0.1 ml. Incubation, 15 minutes.

the mechanism of this mutation is not yet clear, it appears to be based upon an alteration of one general energy yielding process. It is interesting to note from this classification that only in the cases of mutants with altered amino acid activating enzymes has the biochemical mechanism of the mutation been specified. No mutants have been found, so far, in which the "DNA" or "RNA" polymerase systems display temperature sensitivity. A place is also opened for mutants with thermally unstable transfer factors or sRNA's.

We shall restrict the following discussion to a survey of the metabolic properties of the class of mutants unable (as judged by incorporation studies) to synthesize both protein and RNA, or RNA only, above a critical temperature range.

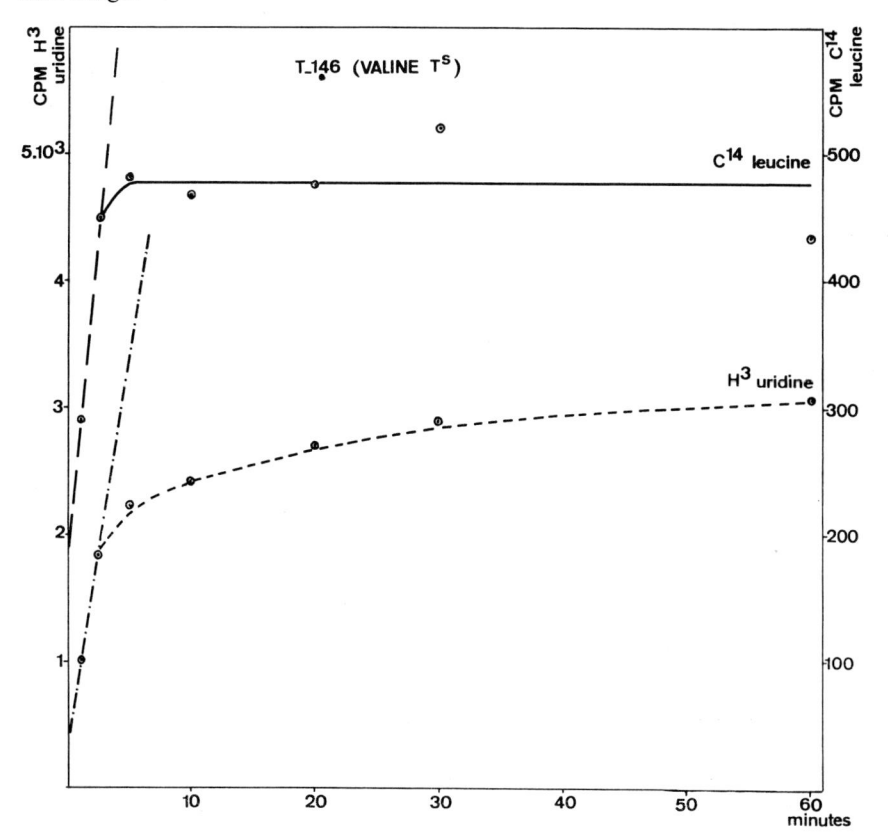

FIGURE 1. ^{14}C leucine and ^{3}H uridine incorporation into 5% TCA precipitable material in mutant strain T-146. Bacteria grown in 63 medium supplemented with thymine, 40 μg per ml; B$_1$, 1 μg per ml; threonine, leucine and methionine, 50 μg per ml of each; 0.4% glucose and containing 10% of broth. At time zero the radioisotopes: 0.5 μc per ml ^{14}C leucine (43 μc/μm) and ^{3}H uridine, 1 μc per ml (200 μc/μm) were added and the culture transferred to 40°C. At the intervals indicated 0.5 ml samples were precipitated with 10% cold TCA, filtered on millipores and washed with 5% cold TCA. The millipores were counted with a scintillation counter.

General Properties of Mutants With Heat Sensitive Aminoacyl RNA Synthetases

A. MACROMOLECULAR SYNTHESES

As shown in Figure 1, several representatives (which we shall refer to as group A) are characterized by a very rapid loss in their capacity to incorporate *both* protein and RNA precursors at the temperature known to inhibit growth. As will be shown later, the great majority of these organisms, whether obtained after the mutagenic action of nitrosoguanidine, or ethyl–methane sulfonate, harbor an altered valyl-RNA synthetase. The reason

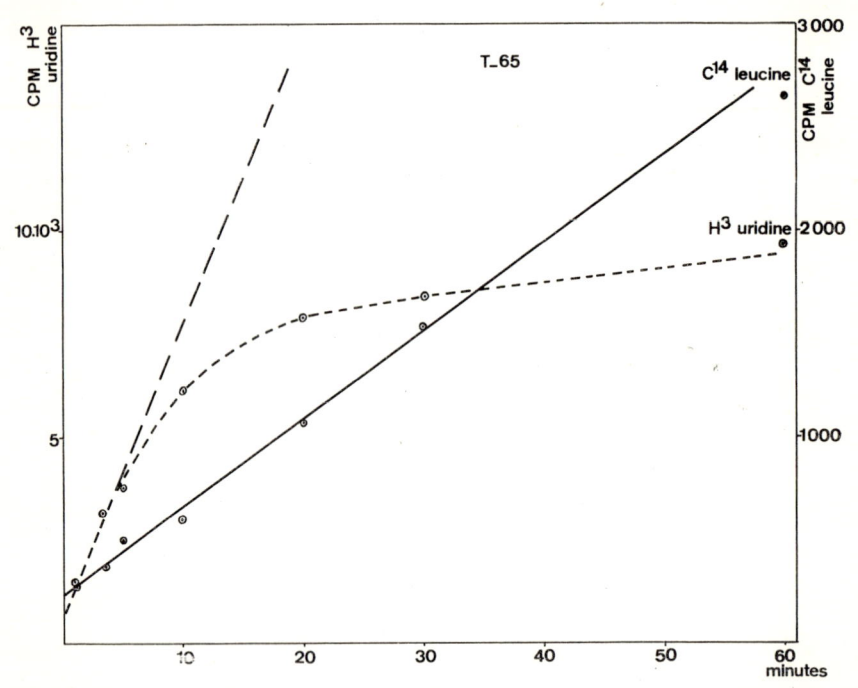

FIGURE 2. ^{14}C leucine and ^{3}H uridine incorporation into 5% TCA precipitable material in mutant strain T-65. Same procedure as described in Figure 1.

for the apparently very high frequency of this mutation is unknown. It is very likely that some sort of selective pressure establishes itself in the course of mutant isolation, which causes the preferential appearance of organisms with a modification in the enzyme activating valine.

Another category of mutants (which we shall designate by group B) displays a different phenotypic behavior: when cultivated at 40°C, the bacteria incorporate ^{14}C leucine linearly into proteins over a 2 hour period, but stop manufacturing RNA within the first 10 minutes (Figure 2). Two of these organisms have been shown to possess an altered alanyl RNA synthetase as demonstrated by the *in vitro* studies which will be described later. Accordingly when ^{14}C-L-alanine is used as a protein precursor, instead of ^{14}C-L-leucine, it is found with the mutant T-65 that this amino acid is incorporated at a decreasing rate (Figure 3). It is therefore likely that the continuous incorporation of radioactive leucine reflects the synthesis of proteins whose content of alanine is especially low, or of abnormal polypeptide derivatives (see Addendum, p. 177). We have not studied the nature of

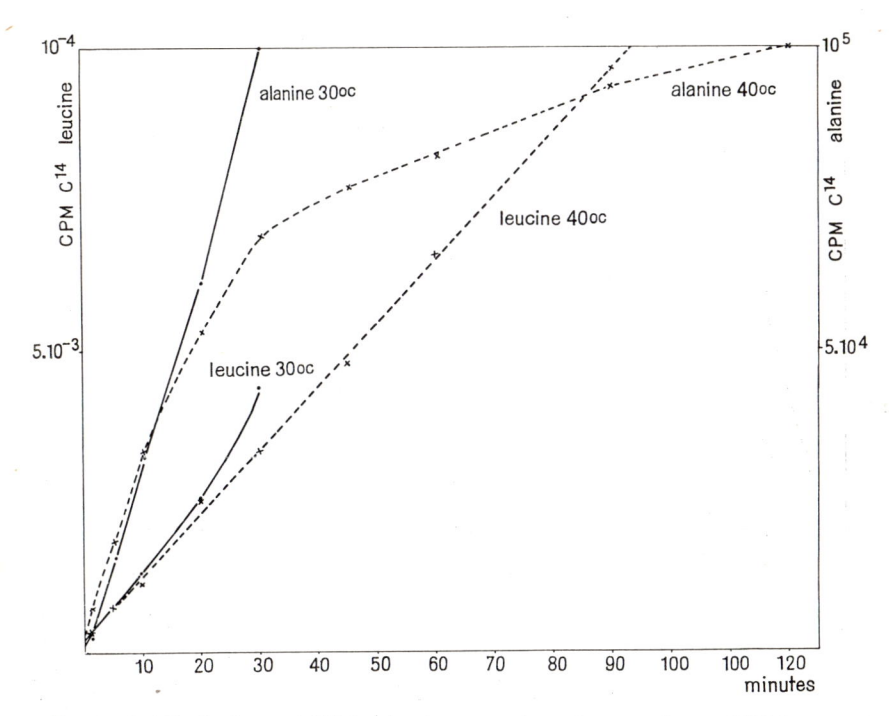

FIGURE 3. ^{14}C alanine and ^{14}C leucine incorporation after transfer to 40°C in mutant T-65. Bacteria grown in 63 medium supplemented with thymine, B$_1$, threonine, leucine, methionine and glucose. At time zero ^{14}C-L-alanine 0.5 μc per ml (34 μc/μm) or ^{14}C-L-leucine 0.5 μc per ml (43 μc/μm) were added, samples incubated at 30°C or 40°C and aliquots taken and precipitated with 10% cold TCA.

the proteins formed under these conditions, but it is worth noting that transfer at 40°C of the strain T-65 (which is lactose positive), does not allow the increase of a β-galactosidase activity following IPTG induction even if the induction period has taken place at 30°C and the carbon source eliminated so as to prevent catabolite repression. The strain T-65 also does not support the growth of bacteriophage T-4 at 40°C.

We have seen that, in all the mutants described, RNA synthesis comes to a rapid halt after transfer at 40°C. It must be noted that the organisms studied are of "stringent" type, i.e. cannot sustain RNA synthesis in the absence of one essential amino acid. As already stressed by Neidhardt (7) this situation supports the view that the discharging of one single sRNA acceptor site suffices to block RNA synthesis.

As expected, and as shown in Figure 4, the addition of chloramphenicol, by protecting the preexisting aminoacyl-sRNA from being utilized for

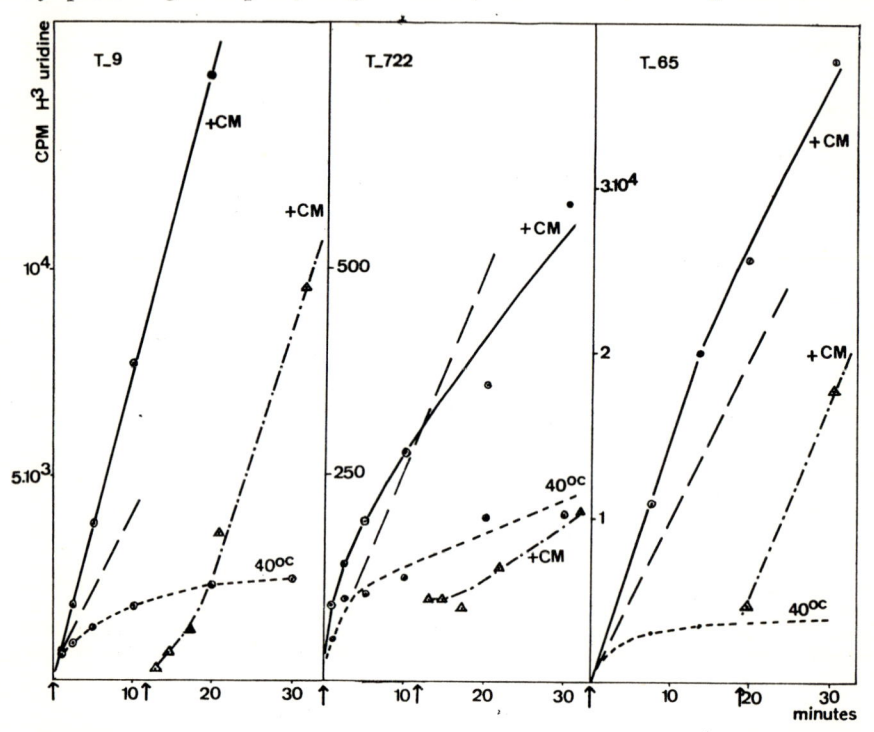

FIGURE 4. RNA synthesis in mutants with a heat sensitive valyl or alanyl sRNA synthetase—effect of chloramphenicol. Bacteria grown in 63 medium supplemented with thymine, B_1, threonine, leucine, methionine, glucose. ^{3}H uridine, 1 μc per ml (200 μc/μm) was added and the culture transferred to 40°C. Incorporation of uridine was followed in samples that were supplemented with 50 μg per ml chloramphenicol (CM) at time zero as well as in samples in which CM was added after 12 minutes (T-9, T-722) or 19 minutes (T-65) incubation at 40°C.

protein synthesis, restores RNA synthesis to its maximum rate. The stimulating effect is lost more or less rapidly after transfer at 40°C, depending presumably upon the degree of unstability of the valyl RNA synthetase in the mutant considered. Chloramphenicol also stimulates RNA synthesis in mutants of group B preincubated at 40°C for a period sufficiently long to inhibit this process. It is clear that a partial discharge of the alanyl accepting sites of sRNA inhibits RNA synthesis at least 10 minutes before inhibiting incorporation of alanine into proteins, suggesting that the appearance of a small fraction of uncharged sRNA molecules represses total RNA synthesis to a marked extent.

Before abandoning these considerations on macromolecular synthesis, it is worth pointing out that the loss in the capacity to synthesize proteins, in the mutants of group A, is a rapidly reversible process. If a mutant of group

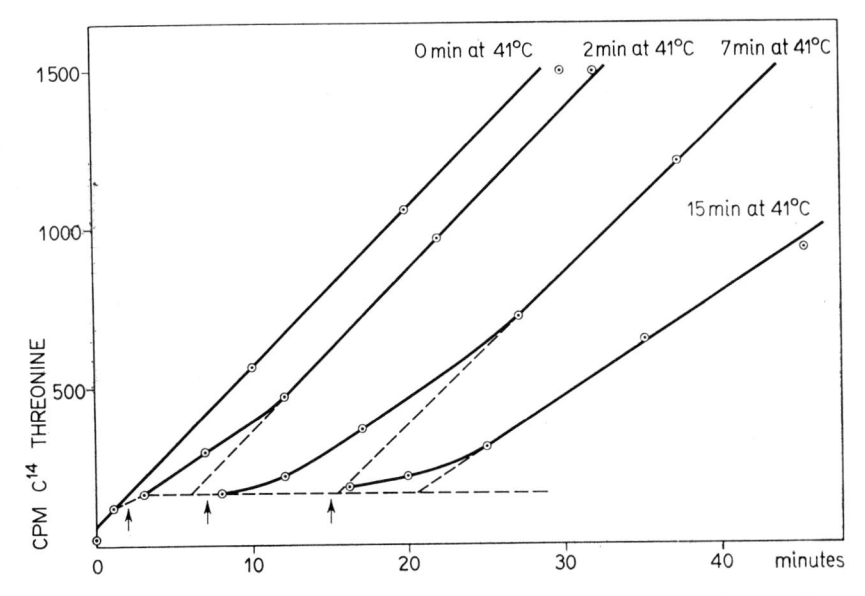

FIGURE 5. Effect of preincubation at 41°C on the capacity to synthetize proteins at 30°C in mutant T-2. At zero time, a culture of T-2 was transferred to 41°C; 0.2 μc per ml ^{14}C threonine (10 μc/μm) was added and, at time 0, 2, 7, 15 minutes, aliquots were transferred to 30°C and the incorporation of ^{14}C threonine into TCA precipitable material was followed.

A such as T-2 is placed at 40°C and transferred back to 30°C, following different heating periods, there is a rapid resumption of protein synthesis, suggesting a direct renaturation of the heat inactivated RNA synthetase (Figure 5).

B. SELECTIVE ALTERATIONS IN AMINOACYL RNA SYNTHETASES OF HEAT SENSITIVE MUTANTS

Preliminary studies in our laboratory have shown that the failure of mutants belonging to group A to synthesize proteins above 37°C is due in most cases to a specific alteration in the enzyme attaching valine to its cognate sRNA's (9). More recently, we have shown that the linear synthesis of proteins observed in mutants of group B and the rapid cessation of alanine incorporation, has its biochemical basis in an alteration of the alanyl–RNA synthetase enzyme.

The specificity of each mutation considered is illustrated in Table II in which the aminoacyl RNA synthetases activities of mutants within groups A and B have been compared to those of the wild type strain at two different temperatures.

In purified extracts from T-2 mutants belonging to group A, valine is the only amino acid whose esterification to sRNA is found considerably reduced at 30°C and is practically negligible at 40°C. As far as mutant T-140 from group B is concerned, we observe the same specificity of alteration for the alanyl RNA synthetase.

TABLE III. Strain T-918 or T-140 F⁻, were crossed at 30° with Hfr 808 (thermoresistant), and the following thermoresistant recombinants were selected: **917**: thy, B_1, tl, met; **1403**: thy, B_1, tl, met.

COMPARATIVE AMINOACYL-RNA SYNTHETASE ACTIVITIES IN THERMO-SENSITIVE MUTANTS AND THEIR THERMORESISTANT RECOMBINANTS					
Mutants T^s (μμ aa incorporated)			Recombinants T^r (μμ aa incorporated)		
Strain	20°C	40°C	Strain	20°C	40°C
T. 918 (valine)	27.3	1.4	917	68.2	80.0
T. 140 (alanine)	1.9	1.2	1403	74	32.5
T. 65 (alanine)	7.6	1.8			

The question arises whether growth thermosensitivity of the conditional mutants previously described is directly related to their possession of a particular altered aminoacyl RNA synthetase. That this is well the case is shown by studying the properties of this enzyme among thermoresistant recombinants obtained from the thermosensitive strain. It is found that the recombinants which are thermoresistant contain valyl or alanyl RNA synthetases with "wild type" characteristics (Table III).

C. ACTIVITIES OF VALYL sRNA SYNTHETASES IN DIFFERENT THERMOSENSITIVE MUTANTS FROM GROUP A

In view of the extremely high frequency of thermosensitive mutants harboring an altered valyl RNA synthetase, one might ask to what extent the organisms so obtained result from mutations at independent loci of the gene responsible for this particular enzyme.

Table IV allows us to compare, in this respect, the different levels of valyl RNA synthetase activities, among ten independently isolated clones of thermosensitive mutants, whether obtained after nitrosoguanidine or ethylmethane sulfonate treatments. Particularly interesting is the fact that purified extracts derived from each of these mutants (grown at 25°C) harbor markedly dif-

TABLE IV

VALYL-s-RNA SYNTHETASES ACTIVITIES FROM DIFFERENT THERMO-SENSITIVE MUTANTS OF GROUP A					
mutants	activity· 20°	activity· 40°	origin	mutagen	genetic character
T 2	11.7	5.9	CR 34	NG	thy B$_1$ tl
T 7	1.3	0.46	CR 34	NG	thy B$_1$ tl ser gly
T 9	45.2	1.1	CR 341	NG	thy B$_1$ tl met arg
T 16	2.7	0.46	CR 34	NG	thy B$_1$ tl arg tyr
T 722	2.18	0.40	CR 34	NG	thy B$_1$ tl
T 141	83.4	3.2	CR 341	NG	thy B$_1$ tl met
T 146	22	0.7	CR 341	NG	thy B$_1$ tl met x
T 536	20	4.2	CR 341	EMS	thy B$_1$ tl met
T 537	10	0.5	CR 341	EMS	thy B$_1$ tl met
CR 341	152	172			thy B$_1$ tl met

activity· μμ moles aminoacids incorporated per 100 γ s-RNA

ferent levels of enzyme activity, even when the test is performed at 20°C. The mutant activity can range between less than 1% of the wild type activity at the same temperature to more than 50%. These variations are suggestive of mutations causing different degrees of conformational instability among the valine specific enzymes considered, such that, in some cases, the enzyme although being functional *in vivo* at a subcritical temperature does not retain its activity *in vitro* at the same temperature.

D. KINETICS OF INACTIVATION OF THE MODIFIED VALYL s-RNA SYNTHETASE IN MUTANT T-9.

That the valyl RNA synthetases of thermosensitive mutants can undergo very rapid conformational changes leading to inactivation at temperatures close to those inhibiting growth, is illustrated by studying the stability of one of the mutant enzymes *in vitro*. A purified extract derived from mutant T-9 was kept at different temperatures ranging from 20 to 40°C in the absence of its substrate and of the effectors for the reaction. At various stages throughout the incubation, samples were taken and assayed at 20°C for their capacity to attach valine to sRNA. A comparative study was performed with the wild type enzyme (CR-34). Figure 6 clearly shows that in extracts from strain T-9 the valyl RNA synthetase activity decays very rapidly at temperatures above 30°C. At 40°C, the first part of the decay curve reveals a half life of 20″, whereas under the same conditions the wild type extract retains its original activity.

Interestingly all decay curves in the case of T-9 extracts are biphasic, suggesting some sort of an equilibrium between active enzyme molecules and inactive products, perhaps subunits.

The purification of the "modified" enzymes, with a view to comparing heir general properties with the enzyme from the wild type, is underway.

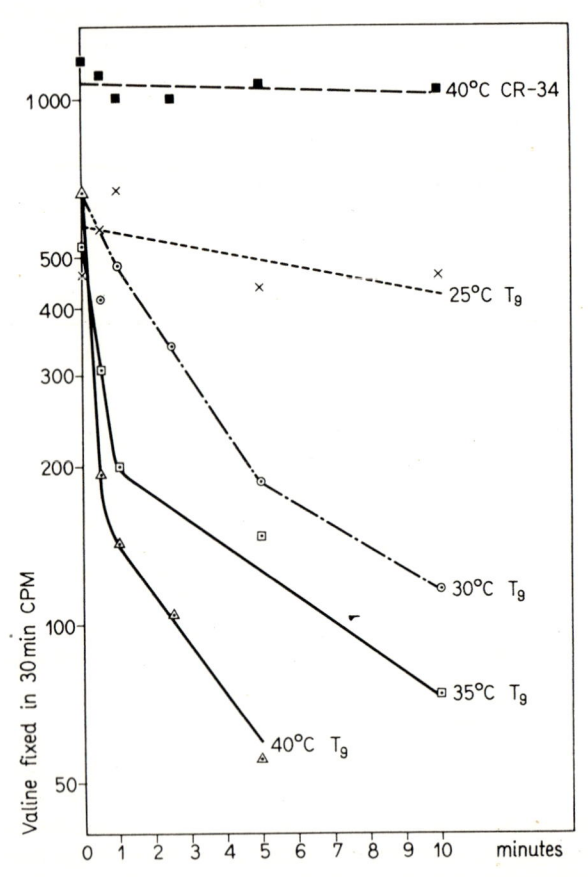

FIGURE 6. Thermal inactivation of valine sRNA synthetase of T-9. Partially purified extracts of CR-34 and T-9 were incubated at the temperatures indicated, in Tris buffer pH 7.8, 10^{-2} M, containing Mg acetate 8.10^{-3} M, NH_4Cl 6.10^{-2} M and mercaptoethanol 6.10^{-3} M. At intervals, samples containing 8 μg of protein were mixed with 0.05 ml of mixture containing: 45 μg sRNA, Tris-HCl pH 7.8, 2 μmoles; NH_4Cl, 36 μmoles; Mg acetate 0.25 μmoles; ATP 0.1 μmole; PEP 0.2 μmoles; pyruvate kinase 0.8 μg; mercaptoethanol 0.3 μmoles; ^{14}C valine 12 millimicromoles (22 μc/μm) plus a mixture of non-radioactive amino acids, 6 millimicromoles of each. Incubation, 30 minutes at 20°C.

E. REVERSION OF THERMOSENSITIVE STRAINS

Valine is known to exert a bacteriostatic effect on normal strains of *E. coli* K12. Thermosensitive mutants of group A (presumably because they harbor a defect in their capacity to activate valine) are resistant to its inhibitory effect to a degree which depends upon the strain and the temperature of cultivation. This phenomenon was profited from to obtain different classes of revertants, depending upon whether the selection occurred in the presence or absence of valine. When a search was made for simple thermoresistant revertants in the absence of valine, a fraction of the organisms obtained apparently recovered a normal valyl RNA synthetase activity since they exhibited the same growth characteristics as the wild type. Other revertants of this group are probably partial revertants since they do not grow faster at 40°C than at 30°C. All these revertants are valine sensitive.

If one selects for thermoresistant valine resistant revertants, several classes can also be obtained whose growth properties at different temperatures are represented in Table V. One, which consists of true revertants, includes organisms which can grow at 40°C with a smaller generation time than at 30°C, like the wild type organism. In a second more interesting class, we encounter revertants which we call "conditional", for they grow at 40°C on solid medium but not on liquid medium, unless valine itself be present.

TABLE V

GROWTH CHARACTERISTICS OF THERMORESISTANT, VALINE RESISTANT REVERTANTS

types of strains	broth (solid) 30°C	broth (solid) 40°C	broth (liquid) 30°C	broth (liquid) 40°C	$40+10^{-2}$M valine	synthetic (solid) 30°C	synthetic (solid) 40°C
1	+	+	+	+		+	+
2	+	+	+	−	+	+	+
3	+	−	+	−		−	+

One likely explanation for such a behavior is that these particular organisms have recovered an activating enzyme, the conformation of which is unstable at 40°C but can be stabilized by valine. As these organisms are also valine resistant (and consequently derepressed for valine biosynthesis) they excrete valine, and establish a sufficiently high amino acid concentration gradient on solid but not on liquid medium.

The properties of purified valyl RNA synthetases of such conditional revertants are under investigation.

A last group is temperature sensitive in broth but thermoresistant on synthetic medium, suggesting that the valyl RNA synthetase of such organisms is inhibited by some factor present in broth.

Derepression of Valine Biosynthetic Enzymes in Thermosensitive Mutants of Group A

Each individual amino acid is not only required as a building block for protein synthesis, or as a catalyst in RNA synthesis, but it can also control, by feedback and repression, the enzymes involved in its own synthesis from the general metabolic precursors. That amino acids do not act directly as corepressors for their own biosynthetic pathway has in fact been demonstrated by Schlesinger and Magasanik (15) who found that α-methyl histidine, which is an analogue of histidine, blocks specifically the activation of this amino acid and behaves as an inducer for the histidine biosynthetic enzymes, showing that activation is the prerequisite for corepression. This conclusion has been further supported by Eidlic and Neidhardt (6) who observed that thermosensitive mutants with modified valyl RNA synthetases are derepressed for certain enzymes involved in the formation of branched amino acids (16).

We have confirmed Eidlic and Neidhardt's observation, and Figure 7 illustrates this point by showing the extents of derepression of threonine deaminase among several mutants of group A cultivated at different temperatures. Interestingly, the extents and rates of derepression as a function of the temperatures of cultivation vary markedly from one mutant to another illustrating, once again, the different degrees of stability of the modified enzymes as previously mentioned. Before discussing the possible relationship between activation and corepression in more general terms, we must note that, to the extent that the mutants just described harbor different indices of derepression at a given temperature, they also harbor different degrees of resistance to valine. There exists in other words a direct, if not proportional, relationship between derepression of the biosynthetic pathway for branched amino acids and the resistance to valine (Figure 8). This result supports very well the proposal by Adelberg (17) that valine resistance is due to derepression of isoleucine synthesis. In view of the current general interest in the regulatory mechanisms, it was worth analyzing further the finding that corepression by amino acids requires them to be activated. To obtain more information on this problem, we have attempted to investigate if it is possible to dissociate somehow the effects caused by alterations in the valine activating systems on corepression, from those on protein synthesis. The rationale for these studies was to determine if repression would not involve a different pathway of amino acid activation (perhaps a different kind of sRNA) than protein synthesis itself. These considerations underline our interest for isolating thermoresistant revertants of group A that would have recovered normal growth characteristics at 40°C without recovering repressibility by valine. Unfortunately all the organisms hithertofore isolated, which fulfil this criterion, appear to be double mutants as shown by transduction tests (18) and their derepression probably lies on mutations in the regulator or operator gene. That a modification of the activating system can nonetheless cause a certain uncoupling between the corepressor activity of amino acids and their use for protein synthesis is shown by the experiment of Figure 9. Using strain T-725, both the level of derepression of threonine deaminase and the generation times have been plotted versus the temperatures of cultivation. Quite a high extent of derepression is produced by raising the temperature before any appreciable retardation on the growth rate is observable. More specifically there is a range of temperature within which

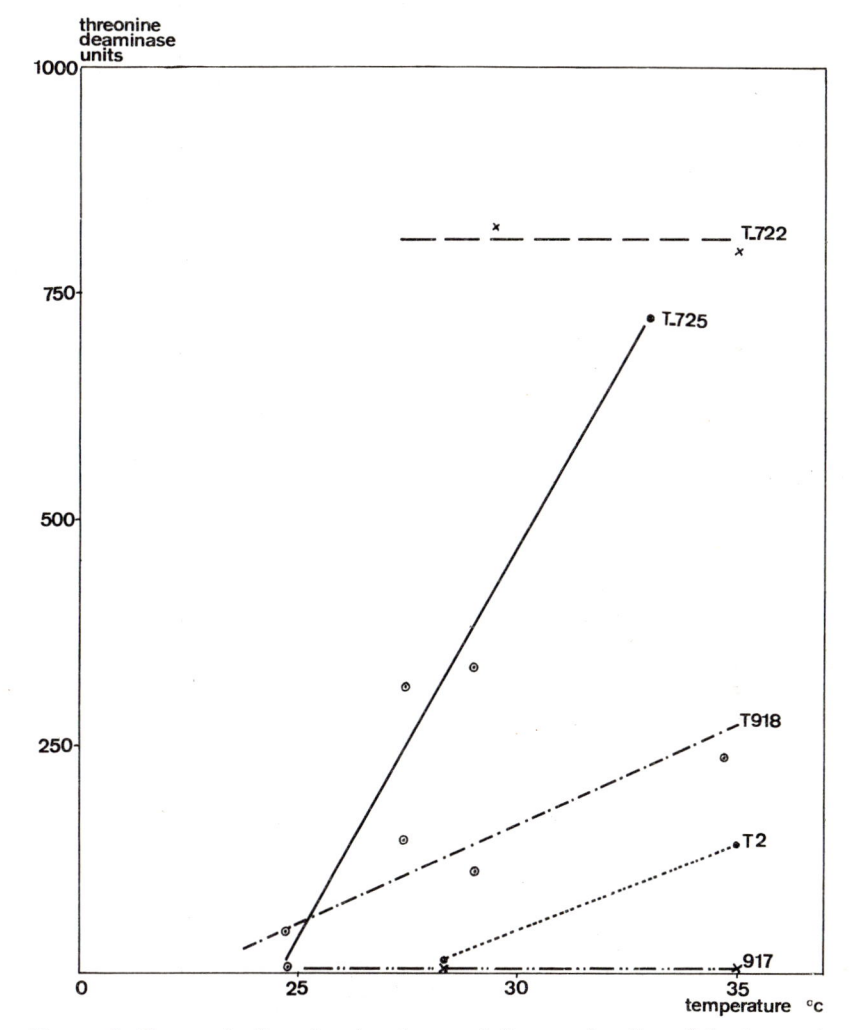

FIGURE 7. Changes in threonine deaminase activity as a function of the temperature of cultivation. Details described in Figure 9.

the rate at which valine is transferred from sRNA to proteins remains constant, whereas derepression of threonine deaminase keeps increasing.

These results are reminiscent of an earlier experiment by Gorini (20) in which various amounts of a given amino acid were added to a culture

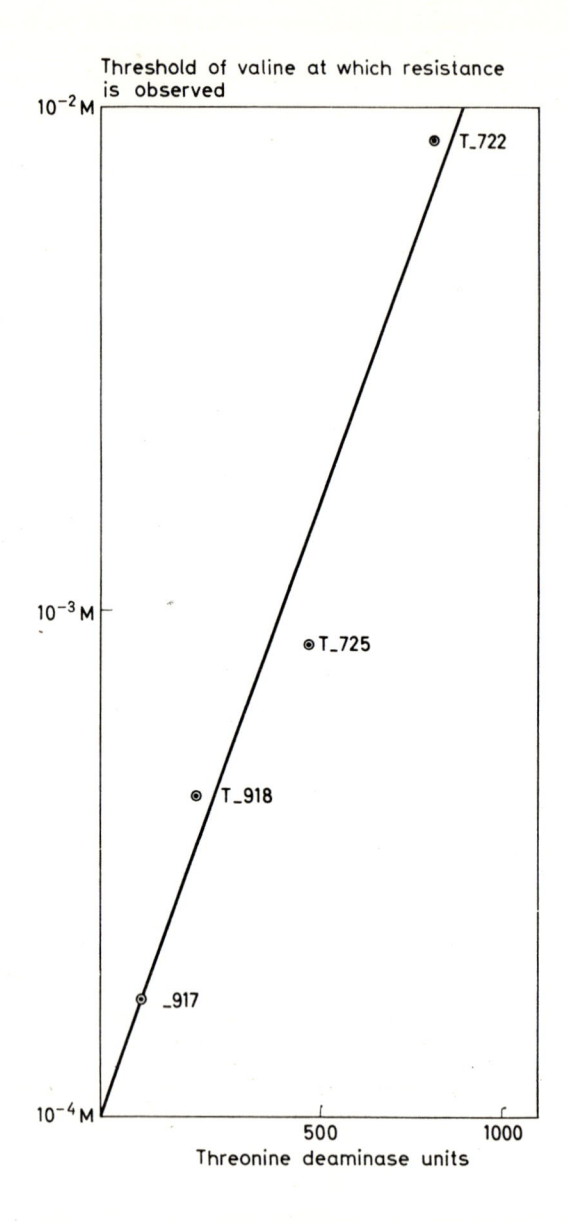

FIGURE 8. The level of valine resistance for different mutants as function of their threonine deaminase activity. Enzyme activity assayed in extracts of bacteria, grown at 30°C (we define here by valine resistance the ability to form colonies on plates of synthetic medium 63, supplemented with thymine, 40 μg per ml; B_1, 0.5 μg per ml; threonine, leucine and methionine, 50 μg per ml of each; glucose, 0.4%, plus variable concentrations of valine).

of a prototrophic organism maintained in steady state conditions of growth. It was found that the amino acid could be added up to a certain range of concentration *without* causing repression or changing the growth rate (presumably because of a feedback inhibition of endogenous syntheses). In other

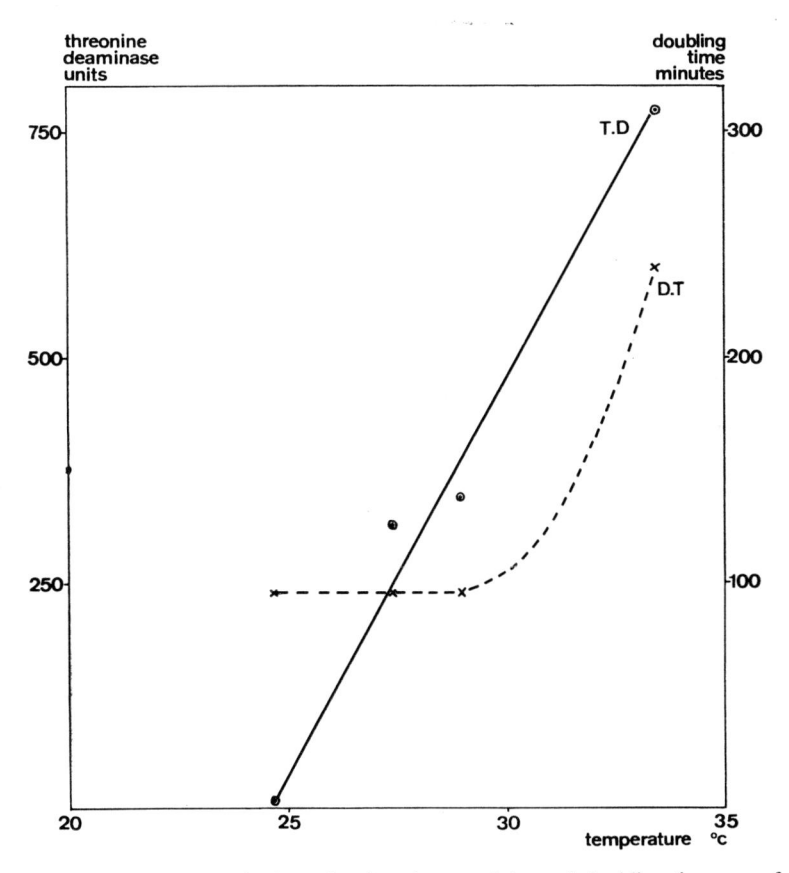

FIGURE 9. T-725 changes in threonine deaminase activity and doubling times as a function of the temperature of cultivation. The bacteria were grown in broth at 24.7°C. Samples were transferred at the given temperatures. Doubling times were determined by optical density measurements. After half a generation of growth at the temperature considered, bacteria were harvested and their threonine deaminase activity was assayed according to Changeux (19), and we have plotted increases in specific enzymatic activity per generation.

words, it was only after the protein forming machinery had been saturated by the amino acid in use that repression began to operate. It appears therefore as if the rate constant (K_2) of the key reaction relating amino acid activation and corepression is smaller than the constant which characterizes the channelling of activated amino acids into proteins (K_1).

As shown in Figure 10, several alternatives can account for the participation of amino acid activating systems in corepression. According to equation 1, aminoacyl sRNA could directly react with the apo-repressor to cause repression, but both repression and protein synthesis would require the prior esterification of specific amino acids to sRNA. As suggested by

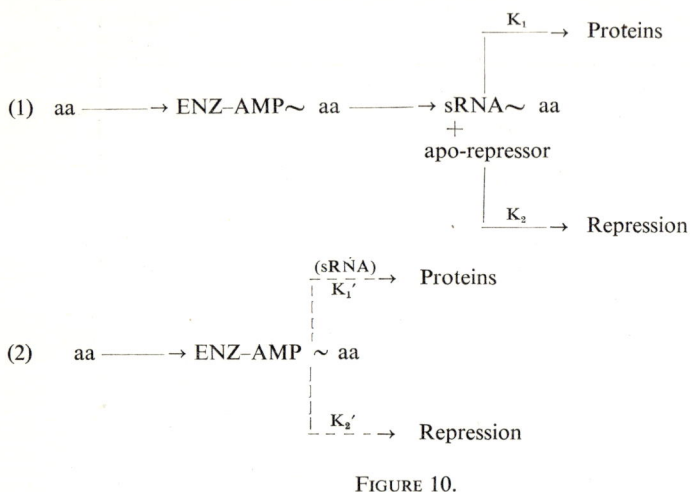

(1) $\quad aa \longrightarrow ENZ\text{–}AMP\sim aa \longrightarrow sRNA\sim aa \quad \begin{array}{c} \xrightarrow{K_1} \text{Proteins} \\ + \\ \text{apo-repressor} \\ \xrightarrow{K_2} \text{Repression} \end{array}$

(2) $\quad aa \longrightarrow ENZ\text{–}AMP \sim aa \quad \begin{array}{c} \xrightarrow{\overline{(sRNA)}\;K_1'} \text{Proteins} \\ \xrightarrow{K_2'} \text{Repression} \end{array}$

FIGURE 10.

equation 2, many more possibilities would be opened among which the following ones can be considered:

(a) there is a distinct category of acceptor sRNA's for repression and for protein synthesis;

(b) the ternary complex reacts with an amino acid acceptor X differing from sRNA;

(c) the ternary complex (ENZ–AMP$\sim$aa) is the corepressor. It need *not* react with another acceptor besides the aporepressor itself (see Addendum, p. 177).

APPROXIMATE GENETIC LOCATIONS OF VALYL AND ALANYL RNA SYNTHETASE MUTATIONS

By means of suitable crosses and the use of the interrupted mating procedure, the thermoresistant alleles corresponding to two thermosensitive mutants of group A and B have been grossly mapped (21, 9). Figure 11 gives the approximate locations of the sites whose mutations lead to production of heat sensitive valyl RNA synthetase and alanyl RNA synthetase respectively. It is clear that these sites are far apart on the map, indicating that the genes for activating enzymes are certainly not all clustered, but the question as to whether these genes belong to *independent* clusters is of course entirely open.

DISCUSSION

The first conclusion to be drawn from the present series of experiments, as well as from those previously reported by Eidlic and Neidhardt (6), is that a single activating enzyme seems to be involved in the esterification of the various acceptor-sRNA's that correspond to each particular amino acid. This is shown by the high frequency of single point mutations, causing an absolute loss in the capacity to attach a specific amino acid to the bulk sRNA *in vitro*. Whether each activating enzyme is accompanied in the cell by "satellite components", endowed with restricted specificities of recog-

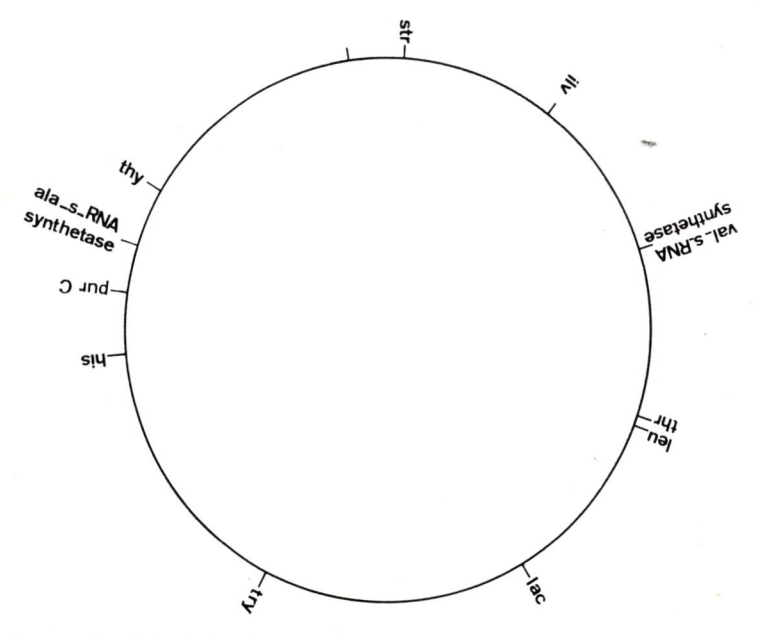

FIGURE 11. Genetic locations of aminoacyl RNA synthetase mutational sites.

nition for degenerate sRNA's, is difficult to establish, but their contribution to the overall esterification process must be quantitatively small. The existence of such satellite RNA synthetases for phenylalanine in the case of Neurospora cells (22) and for leucine with *E. coli* cells has in fact been reported.

From a more general standpoint, it is clear that the use of temperature sensitive conditional mutants should open up new approaches to many interesting problems related to the biosynthesis of macromolecules, its regulation and genetic determinism.

As illustrated by the present work, albeit in a very limited number of cases, the mapping of the determinants corresponding to the conditional mutation involved should ultimately enable us to specify the genetic determinism of the main elements in the protein and nucleic acid forming machinery. It would be of great interest for instance to investigate if the genes controlling the properties of amino acid activating enzymes are topologically grouped in one or several discrete regions of the chromosome as are apparently the genes for ribosomal and sRNA (23). Is there a topological relationship between the genes for RNA synthetases and the "regulator genes" involved in the biosynthesis of the amino acids which they activate?

It is equally obvious that in the near future more information should be forthcoming on the relationship between "activation" and "corepression". The question dealing with the possible existence of special sRNA's for corepression (distinct from those "adapting" the protein subunits) is open, as is the search for any specific amino acid acceptor, distinct from sRNA, which would be recognized by activating enzymes and mediate corepression.

Finally a comparison between "modified" and normal activating enzymes should give precious indications regarding the nature of the two active sites of these molecules and about their tertiary and quaternary structure in general.

ACKNOWLEDGMENTS

We are indebted to Dr. Masamichi Kohiyama and Dr. Denise Cousin for the generous gift of several mutant strains and we wish to thank Dr. François Jacob for many valuable discussions.

This work was supported by grants from the Fonds de Développement de la Recherche Scientifique et Technique, the Commissariat à l'Energie Atomique, the Centre National de la Recherche Scientifique, the Ligue Nationale Française contre le Cancer and the Fondation pour la Recherche Médicale Française.

REFERENCES

1. Epstein et al., *Cold Spring Harbor Sym.* **28**, 375 (1963).
2. Gorini, L. and Kataja, E., *PNAS* **51**, 883 (1964).
3. Kohiyama, M., Lamfrom, H., Brenner, S. and Jacob, F., *Comp. rend. Acad. Sci.* **257**, 1979 (1963).
4. Jacob, F., Brenner, S. and Cuzin, F., *Cold Spring Harbor Sym.* **28**, 329 (1963).
5. Van de Putte, P., Van Dillewijn, J. and Rörsch, A., *Mutation Research* **1**, 121 (1964).
6. Eidlic, L. and Neidhardt, F. C., *PNAS* **53**, 539 (1965); *J. Bact.* **89**, 706 (1965).
7. Neidhardt, F. C., *Progress in Nucleic Acid Research and Molecular Biology*, Edited by J. N. Davidson and W. E. Cohn **3**, pp. 145–179 (1964).
8. Apirion, D., *J. Mol. Biol.* (In press).
9. Yaniv, M., Jacob, F. and Gros, F., *Bull. Soc. Chim. Biol.* **47**, 1609 (1965).
10. Cousin, D. and Jacob, F., Personal communication.
11. Kohiyama, M., Cousin, D., Ryter, A. and Jacob, F., *Annales de l'Institut Pasteur* **110**, 465 (1966).
12. Yaniv, M., Kohiyama, M., Jacob, F. and Gros, F., *Comp. rend. Acad. Sci.* **260**, 6734 (1965).
13. Fangman, W. L. and Neidhardt, F. C., *J.B.C.* **239**, 1844 (1964).

14. Fangman, W. L., Nass, G. and Neidhardt, F. C., *J. Mol. Biol.* **13**, 202 (1965).
15. Schlesinger, S. and Magasanik, B., *J. Mol. Biol.* **9**, 670 (1964).
16. Freundlich, M., Burns, R. O. and Umbarger, H. E., *PNAS* **48**, 1804 (1962).
17. Ramakrishnan, T. and Adelberg, E. A., *J. Bact.* **87**, 566 (1964); **89**, 654 (1965).
18. Yaniv, M., Unpublished results.
19. Changeux, J. P., *Bull. Soc. Chim. Biol.* **46**, 927 (1964).
20. Gorini, L., *Bull. Soc. Chim. Biol.* **40**, 1939 (1958).
21. Lazar, D., Unpublished results.
22. Barnett, W. E. and Elper, J. L., *PNAS* **55**, 184 (1966).
23. Dubnau, D., Smith, I. and Marmur, J., *PNAS* **54**, 724 (1965).

DISCUSSION

(Chairman: E. Fredericq)

P. Borst: You mentioned that in the strain with the temperature-sensitive alanine RNA synthetase, a shift to 40° immediately stopped the induction of β-galactosidase, although alanine incorporation went on for another 20 minutes at normal rates. Does this mean that limited unloading of sRNA is sufficient to block induction, while protein synthesis only stops when sRNA is largely or completely unloaded, or do you have another explanation for this effect?

M. Yaniv: As we noted, in the mutant T-65 harboring a modified alanyl-sRNA synthetase, RNA synthesis at 40° drops to a low level, at least 15 minutes before the incorporation of ^{14}C-L-alanine begins to diminish. We assume that a partial decrease in the ratio of uncharged to charged alanyl-sRNA is enough to decrease RNA synthesis, before having a marked effect on protein synthesis. Ribosome formation is almost totally inhibited and there is a partial accumulation of neosomes sedimenting from 18S to 28S.

It is likely, in view of the extensive incorporation of leucine into proteins, that messenger RNA synthesis continues at 40°C, but we have not been able to induce β-galactosidase formation (even in the absence of a carbon source), nor to get phage T_4 production after infection at 40°C. The possibility exists, however, that false proteins are synthesized in this case and we are trying to examine the existence after β-galactosidase induction of a material that could be serologically related to this enzyme.

P. Borst: You mentioned the finding of multiple sRNA synthetase in Neurospora. I wonder whether one of the two enzymes found might be located in the mitochondria. Do you know whether any one has studied the sub-cellular distribution of these enzymes in this case?

M. Yaniv: I don't know if anyone has tested whether the second phenylalanine synthetase in Neurospora is a mitochondrial enzyme. Such a hypothesis could account for the low concentration of the second enzyme.

Addendum (added in proof): There are two additional points to which we would like to draw attention:

An alternative to the proposal advanced at the top of page 163 is that leucine incorporation actually represents a specific amino group addition in the N-terminal position of ribosomal proteins as described by Kaji (*J. Biol.*

Chem. **241**, 3294 (1966)). This would not be observed with the valine mutants, possibly because of a better protection of the polysome structure in the absence of the required aminoacyl sRNA. The leakiness of the alanine mutants would allow reading and destruction of bound mRNA and so expose ribosome to NH_2-group addition.

Recently Ames *et al.* (*Cold Spring Harbour Symposium Quantitative Biology*, 1966) have described in *Salmonella* mutations which cause simultaneously a change in the level of the histidine acceptor sRNA and a marked derepression of the histidine enzyme system. This finding strongly suggests that the same categories of sRNA molecules function as amino acid adaptors and as corepressors, as illustrated in equation (1) on page 174.

The Mechanism of Peptide Bond Formation in Protein Synthesis

R. E. Monro, B. E. H. Maden* and R. R. Traut[†]

Medical Research Council Laboratory of Molecular Biology, Hills Road, Cambridge, England

Introduction

Translation of the genetic message involves the organized interaction of sRNA and mRNA on ribosomes, and requires in addition two or more supernatant proteins, GTP, divalent cations and monovalent cations. Polypeptide chain growth proceeds by a cyclical process culminating each time in the formation of a new peptide bond. Two covalent reactions and several weak interactions are known to participate. Current research on the mechanism of protein synthesis centres on the fuller characterization of these reactions and interactions and on the roles of the various components. In the first part of this paper we briefly survey the state of knowledge in this field. We then discuss models for protein synthesis and define a terminology for the several steps which are distinguished. The second part of the paper is concerned with the demonstration by the use of puromycin that the peptide bond-forming step of protein synthesis occurs on the 50S ribosomal subunit. An enzyme for this reaction is postulated and its nature discussed.

PART I — GENERAL

Components of Protein Synthesis

Some relevant properties of the components of protein synthesis will now be summarized. Emphasis will be laid on the *E. coli* system.

mRNA. The sequence of amino acids in a given protein is determined by the sequence of bases in the nucleic acid of the corresponding gene. The two sequences are colinear (1, 2). mRNA carries the information from gene to ribosome, where it provides the template for protein synthesis (3, 4, 5). The 5′ end of the nucleic acid template corresponds to the α-NH$_2$ end of the protein (6, 7, 8, 9). The general nature of the genetic code is now well known (10). It is:

---

* Medical Research Council Scholar for Training in Research Methods at: Department of Radiotherapeutics, University of Cambridge, Hills Road, Cambridge, England.

† Present address: Laboratoire de Biophysique, Université de Genève, 24 quai de l'École de Médecine, Genève, Switzerland.

(i) Triplet: i.e. each amino acid is specified by a group of three nucleotides, called the codon (11, 12);
(ii) Non-overlapping: any given base is "read" in only one codon (10, 13);
(iii) Degenerate: i.e. a given amino acid can be specified by more than one codon (11, 14, 15).

sRNA plays a central role in protein synthesis. One or more sRNA molecules are specific for each amino acid, with which they are enzymatically esterified to form activated intermediates for peptide bond formation (16, 17, 18). According to the adaptor hypothesis (19, 20, 21) selection of the correct amino acid during translation is accomplished by interaction between a codon on mRNA and a complementary group of bases (the anticodon) on sRNA. Furthermore, sRNA undergoes specific interactions with other components of the protein synthesizing system and may also undergo conformational changes. sRNA might be thought of as a species of multifunctional coenzymes. It is analogous to coenzyme A in fatty acid synthesis, but has important additional functions.

Ribosomes. Since the process of translation is organized on ribosomes, any successful model of protein synthesis must envisage how interactions of the various components are integrated on these cell organelles. This demands some knowledge of the structure of the ribosome and of the interactions which it undergoes with other components.

Ribosomes from all sources are ribonucleoprotein particles, which consist of two subunits of unequal size (22). In *E. coli* these subunits have sedimentation coefficients of $50S$ and $30S$ (23). Each subunit has a composition of 63% RNA and 37% protein. The molecular weight of the $50S$ subunit is 1.8×10^6 and of the $30S$ subunit 0.8×10^6. Each $50S$ subunit contains two molecules of RNA, with sedimentation constants of $23S$ (24) and $5S$ (25). Each $30S$ subunit contains only one molecule of RNA with a sedimentation constant of $16S$ (24). These three species of RNA differ not only in sedimentation constant but also in base sequences (26, 27).

The ribosomal proteins are a heterogeneous group of proteins. End group analysis (28, 29) and electrophoretic studies (28, 30, 31, 32) reveal the presence of some 35 or more different proteins, of which 15 are from the $30S$ subunit and 20 from the $50S$. We cannot be sure at present whether this many proteins are present in a given subunit or whether they arise from heterogeneity of the ribosome population. Nor can we know to what extent the observed species of protein differ from one another until finger print studies or sequence determinations have been carried out on the resolved components.

At magnesium concentrations approximating intracellular conditions, the two subunits combine 1:1 to form a $70S$ ribosome (23). This is the active unit, and only the $70S$ ribosome carries out the integrated process of protein synthesis (33). Moreover only the complete $70S$ ribosome catalyzes the supernatant-linked splitting of GTP to GDP and P_i (see below).

The universal occurrence of two separable subunits in ribosomes suggests that this phenomenon is an essential feature of protein synthesis which any model of the process should seek to explain. At present the need for two separable subunits is not understood, but it may be that relative movement or separation occur as an obligatory part of their function in protein synthesis.

Regardless of whether or not the ribosomal subunits move relative to one another in protein synthesis, it is clear that they have distinct functions. Specification of aminoacyl sRNA takes place on the 30S subunit. This is demonstrated by the observation that mRNA associates specifically with this subunit (34) and the codon-directed binding of aminoacyl sRNA can also take place with isolated 30S subunits (35). The 30S subunit thus has the capacity to stabilize the weak interaction between triplet codon and anticodon, probably through interactions with parts of the molecules in the neighbourhood of the triplets.

Isolated 50S subunits do not bind mRNA or codon-directed aminoacyl sRNA. They do, however, enhance the codon-directed binding of aminoacyl sRNA to 30S subunits (36, 37). This effect could arise through stabilization of the weak 30S complex by interaction of the 50S subunit with another part of the sRNA molecule. It is also possible that a new site on the 30S subunit is opened up as a result of interaction with the 50S subunit (36, 37).

The enhancement of specific binding to the 30S subunit by the 50S subunit is not surprising since there is independent evidence for sRNA binding sites on the 50S subunit. A nascent protein–sRNA complex is firmly associated with this subunit during protein synthesis (38). There is also a single, weak, non-specific binding site for sRNA on the 50S subunit (39). There is some evidence that the sites involved in these two types of binding are identical (40) but further clarification is necessary. In any case, during protein synthesis two sRNA molecules must be capable of contiguous alignment on the 50S subunit, since it is on this subunit that peptide bond formation takes place. Evidence for this is the subject of the second part of this paper.

When ribosomes engaged in protein synthesis are isolated, at least two sRNA molecules are found firmly bound per active 70S ribosome (41, 42, 43). It is possible that at some stages in the cycle of peptide growth there are fewer or more sRNA molecules bound, but it is clear that there are at least two binding sites for sRNA on the ribosome.

The diagram in Figure 1 summarizes the above information. A part of the sRNA molecule bearing the anticodon interacts with the 30S subunit while another part of the sRNA, including the adenosine end and the amino acid or peptide, interacts with the 50S subunit. Two molecules can be bound in protein synthesis, one of which bears the growing peptide and the other an amino acid. The functioning of this complex in protein synthesis will be further discussed in a later section.

Supernatant proteins. During protein synthesis all the required components must at one stage or another be associated on the ribosome to form the complex of mRNA, sRNA, ribosome and supernatant factors. In the assembled active complex both ribosomal and supernatant proteins are present. The functions of a single representative of either group are unknown. The ribosomal proteins clearly contribute to the structure of the ribosome, but they could also have enzymic functions. It is therefore premature to refer to ribosomal proteins as purely structural and the supernatant proteins as enzymic. It is nevertheless possible to distinguish the two types of proteins. The ribosomal proteins are in stable association with RNA from which they can only be separated by drastic treatment with high salt concentrations or detergents. On the other hand, ribosomes isolated by mild procedures are made more active for protein synthesis by addition of post-ribosomal supernatant protein, and washing under mild conditions is sufficient to make them almost completely dependent on the addition of supernatant factors. Ribosomal proteins have not been found free in the post-ribosomal supernatant fraction (32); supernatant proteins are found predominantly there. We think, then, that the supernatant proteins are reversibly associated with the active complex and may come on and off either during each cycle of peptide bond formation or possibly only once during synthesis of a complete polypeptide.

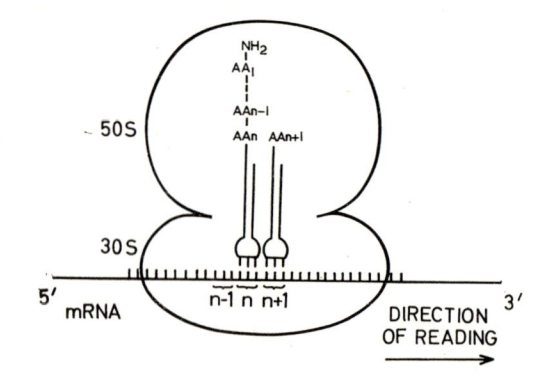

FIGURE 1. Schematic representation of interactions between ribosome, sRNA and mRNA in protein synthesis (modified from Watson (44)).

Investigation of supernatant factors in the *E. coli* system has been intensively developed in Lipmann's laboratory. Demonstration of the involvement of a supernatant fraction and its partial purification (45) was followed by its resolution into two complementary fractions (46). These were further purified and termed the T and G factors (47). The active components in the T and G fractions are presumably protein since they are non-dialysable, heat labile, sensitive to sulphydryl inhibitors and chromatograph like proteins (in contra-distinction to nucleic acids). The G factor is thought to have a function which involves GTP since the GTP hydrolysis is catalyzed by the combination of G factor and ribosomes (see below). Apart from this clue, little is known about the functions of the T and G proteins. Very recently evidence for a third supernatant factor has been reported (48).

GTP. The ester bond in aminoacyl sRNA possesses sufficient free energy to drive peptide bond formation strongly in the forward direction (for further discussion see next section on basic mechanism). However, it has long been thought that an additional source of energy is needed in the form of GTP (49). The nature of the GTP reaction in protein synthesis has been difficult to investigate in view of the presence of very active GTP degrading enzymes present in most cell-free preparations. The use of specific labelling techniques showed that the reaction does not involve pyrophosphate elimination or GMP incorporation (50, 51). At about the same time the ribosome-supernatant-linked hydrolysis of GTP to GDP and P_i was discovered (51). This reaction was thought to arise through uncoupling of the normal GTP reaction in protein synthesis. Subsequent studies provided further circumstantial evidence that the release of inorganic phosphate from GTP is closely related to protein synthesis (52, 53). Finally, by further purification of the system, evidence was obtained for stoichiometry between release of inorganic phosphate and amino acid incorporation (47).

Recent studies with a new GTP analogue (54) lend support to the evidence that amino acid incorporation is coupled to hydrolysis of the terminal phosphate of GTP. The structure of this analogue, guanylyl methylenediphosphate (GMP–PCP, Figure 2), was designed to test the nature of the GTP

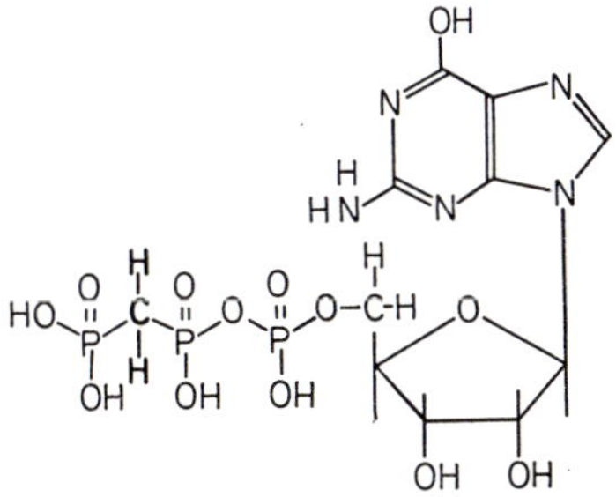

FIGURE 2. Structure of guanylyl methylenediphosphonate (GMP–PCP), shown in protonated form.

reaction. GMP–PCP has a methylene group in place of oxygen between the β and γ phosphorus atoms, thus preventing enzymic cleavage at this position without major alteration of configuration (cf. corresponding ATP analogue studied by Myers (55, 56)). When assayed in the cell-free system GMP–PCP inhibits protein synthesis by specific interference with the GTP reaction. The inhibitor presumably enters the GTP site and blocks the terminal phosphate reaction. This specific inhibitor provides a useful tool for investigation of the GTP reaction, and is referred to again in a later section of this paper.

A BASIC MODEL OF POLYPEPTIDE CHAIN GROWTH

Peptide bond formation is thought to occur by the mechanism shown in Equation 1 (45):

$$\begin{array}{c} \text{sRNA}_n \qquad\qquad \text{sRNA}_{n+1} \\ | \qquad\qquad\qquad | \\ \text{---NH.CH.CO---NH.CH.CO} + \text{HNH.CH.CO} \quad \rightarrow \\ |\qquad\qquad\quad |\qquad\qquad\qquad | \\ R_{n-1}\qquad\quad R_n \qquad\qquad\quad R_{n+1} \end{array} \qquad (1)$$

$$\begin{array}{c} \text{sRNA}_{n+1} \\ | \\ \text{---NH.CH.CO---NH.CH.CO---NH.CH.CO} + \text{sRNA}_n \\ |\qquad\qquad\quad |\qquad\qquad\qquad | \\ R_{n-1}\qquad\quad R_n \qquad\qquad\quad R_{n+1} \end{array}$$

The peptide grows stepwise from the N-terminal to the C-terminal end. The peptide is attached to sRNA through an ester linkage to the terminal adenosine. Peptide bond formation takes place by transfer of the peptidyl group to an incoming molecule of aminoacyl sRNA, with resultant attachment of the augmented peptide to the new molecule of sRNA and elimination of the sRNA to which the peptide was previously attached.

Support for this mechanism comes from (a) kinetic evidence that peptide chain grows from N-to C-terminus (57) in a stepwise process (58); (b) isolation of polypeptide bound to a molecule with characteristics similar to sRNA (38); (c) demonstration that peptides are attached by ester linkage to ribose in the terminal adenosine of sRNA (59, 60); (d) demonstration that the peptidyl sRNA-ribosome complex reacts rapidly with puromycin to form a new peptide bond without the occurrence of appreciable amounts of intermediates in which the peptide is unattached to sRNA (38, 66; see also later section of this paper).

Equation 1 can be elaborated in view of the general nature of the genetic code, outlined above. It is clear that successive groups of three nucleotides (codons) on the mRNA (template) must be read and that the position on the template must be accurately marked so as to maintain reading in the correct frame. Figure 1 incorporates these features. The incoming aminoacyl-sRNA is specified by interaction with a nucleotide triplet on the template. The position for reading is marked by continued interaction of peptidyl-sRNA with the previous codon.

On the basis of this model the following steps (66) are deduced to take place during the formation of each successive peptide bond (numbering does not imply temporal order):

1. *Specification* of the correct aminoacyl sRNA takes place through interaction between codon and anticodon. This interaction is necessary for entry to the peptidyl transfer site. Thus in the presence of a codon only the correct aminoacyl sRNA is incorporated, whereas in the absence of a codon nothing is incorporated. The latter result is demonstrated by the observation that short oligonucleotides direct synthesis of short oligopeptides which contain only the specified amino acids (67). Some mechanism must exist for exclusion of unspecified aminoacyl sRNA. Interaction of aminoacyl sRNA with its codon may effect entry to the peptidyl transfer site solely through orienting the sRNA molecule with reference to the ribo-

some. It may be, however, that aminoacyl sRNA or even the ribosome has to undergo a conformational change in order to allow entry to the peptidyl transfer site, and that codon-anticodon interaction is necessary for such a change.

2. *Peptide bond formation* takes place through transfer of peptidyl group from sRNA to aminoacyl sRNA. It is reasonable to suppose that this reaction is catalyzed by an enzyme, and in the discussion of terminology (below) the name *peptidyl transferase* is proposed. Characterization of the peptidyl transfer reaction forms the subject matter of Part II in this paper.

3. *Attachment and release of sRNA.* The process of chain growth involves (a) firm binding of peptidyl sRNA in association with codon and (b) release of sRNA after transfer of peptide. Although other mechanisms are possible, this is the simplest and most attractive way in which the reading frame could be correctly maintained, i.e., there is continued interaction of codon and anticodon from the time that a molecule of aminoacyl sRNA is specified to the time that the next molecule has been specified and the peptide transferred to it.

4. *Movement.* The template must move relative to the growing point of the peptide by a distance of three nucleotides for each amino acid which is added.

These points outline several characteristics of the translation mechanism. The resultant basic model can be elaborated in various ways, but it should be emphasized that if this analysis is correct, any valid model of protein synthesis will have to fit into the above framework.

Equilibrium. According to Equation 1, the net change in free energy corresponds to the formation of a peptide bond and the hydrolysis of aminoacyl sRNA. Even though it is the bond in peptidyl sRNA which is severed, the free energy change is equivalent to that of the ester bond between amino acid and sRNA. It is important to realize this in calculating the theoretical equilibrium of the reaction since the free energy of hydrolysis of peptidyl sRNA is lower than that of aminoacyl sRNA.

We should expect the equilibrium of the reaction (Equation 1) to be emphatically towards peptide bond formation for the following reasons:

(i) The free energy of hydrolysis of aminoacyl sRNA is of the same order of magnitude as that for a pyrophosphate bond (61, 62, 63) and is much greater than that of a peptide bond (64).

(ii) A high concentration ratio of charged to uncharged sRNA is favoured in the cell because the aminoacylation of sRNA is coupled to the hydrolysis of two pyrophosphate bonds (pyrophosphate elimination from ATP followed by pyrophosphate hydrolysis (65)).

Although the free energy changes in Equation 1 strongly favour peptide bond formation, there is evidence, as noted above, that the reaction is coupled to hydrolysis of yet another pyrophosphate bond, that of GTP. This is not surprising since the simple reaction shown in equation 1 is controlled by a complex mechanism involving several steps, and is linked to the movement of mRNA. The excess expenditure of free energy in peptide bond formation results from the involvement of a complex mechanism, and in a certain sense it is the price paid for fidelity of information transfer in the translation process.

MODELS TO DESCRIBE RIBOSOMAL ORGANIZATION OF PEPTIDE ASSEMBLY

Although there is no direct evidence, it will be assumed on theoretical grounds that a peptide chain remains associated with a single ribosome throughout its synthesis. Peptide bond formation involves transfer of the growing peptide from sRNA in one site (A) to sRNA in a second site (B). Several models can be distinguished for events which then follow.

(a) *Translocation models*: This type of model was first proposed by Watson (44). After transfer of peptide to site B, the freed sRNA is eliminated and the peptidyl sRNA is shifted back into site A. Template movement is coupled to peptidyl sRNA movement. Site B is thus regenerated and the appropriate codon is in position for specification of the next sRNA. This model has the advantage of explaining in a defined way the unidirectional relative movement of ribosome and mRNA. In an elaboration of this scheme, we suggested that GTP and a supernatant enzyme might be involved in the movement of sRNA between the two sites and proposed the name *sRNA translocase*, for this enzyme, pending elucidation of a more defined reaction (66). GTP could alternatively be involved directly in mRNA movement or in the binding (42) or release of sRNA.

The translocation model can be readily extended to include more sites, as for instance in the three-site model proposed by Wettstein and Noll (43). A characteristic feature of all models of this type is that each site has a unique function and there is asymmetry of the ribosome. This model therefore predicts the occurrence of unique sites for interaction with sRNA and with specific inhibitors.

(b) *Double site model.* A two-site model which does not involve translocation has been put forward by Schweet and coworkers (68). In this model the two sites are equal and no translocation of sRNA takes place. Thus after transfer of peptide from site A to B the next aminoacyl sRNA enters site A and the peptide is transferred back to site A as the next peptide bond is formed. The two sites thus alternate in having donor and acceptor functions. The weak point in this model lies in its failure to explain in a satisfactory way the unidirectional movement of template relative to the peptide's growing point.

(c) *Rolling models* (69). Identical sites, A, B, C, ..., are arranged symmetrically around the ribosome. After transfer of the peptide from site A to B, instead of moving back to site A it is transferred on to the next aminoacyl sRNA in site C, from there to site D and so on. The ribosome rolls, as it were, along the template. The problem of template movement is thus solved, but an elaboration of the model would be required to provide a mechanism to keep the motion unidirectional. In its simple form the rolling model predicts the occurrence of multiple sites for interaction with sRNA. However, more sophisticated models can be constructed in which the ribosome undergoes conformational changes such that only one site is open at a time.

(d) *Rotational model.* Combinations of model "a" or "b" with the rolling model are also possible. If the ribosomal subunits can rotate relative to one another (70) then there might, for instance, be a multi-site rolling interaction of template with the 30S subunit and a two-site interaction of sRNA with the 50S subunit.

Each of the above models provides the adjacent sRNA binding sites which are necessary for peptide bond formation. In the absence of further structural information about the ribosome, we favour the translocation model, since it provides the most plausible mechanism for coupling unidirectional template movement to peptide growth. It is also best in conformity with observations which indicate that, in the absence of protein synthesis, there is a unique binding site for sRNA on the 50S subunit (39) and on the 30S subunit (35), and one for chloramphenicol on the 50S subunit (71). As noted above, the presence of unique sites is predicted by the translocation model, whereas the rolling model can be reconciled with unique binding sites only by the addition of special refinements.

It should be emphasized that although the translocation model helps to envisage certain aspects of peptide synthesis, no satisfactory explanation has yet been made for the mechanism of sRNA translocation or mRNA movement. Before these and other questions can be answered, it will be necessary to have further information about the structure and functions of the various components involved.

TERMINOLOGY

The basic model provides a framework in which to develop a terminology of protein synthesis. For the hypothetical enzyme which catalyzes peptidyl transfer from sRNA to aminoacyl sRNA we propose the operational name *peptidyl transferase* (or, perhaps, *polypeptidyl transferase* to distinguish it from other peptidyl transferases). This name is preferable to *peptide polymerase* because peptidyl transferase is more definitive of the actual reaction catalyzed and is in conformity with the conventions of enzyme terminology (72). The name *peptide synthetase* should be avoided, since the balance of evidence at present (42, 66; see also Part II) indicates that the peptidyl transfer reaction does not directly involve GTP. (Synthetase reactions, by definition, involve nucleoside triphosphates).

We define the sRNA molecule, from which the peptidyl group is transferred, the *peptidyl donor–sRNA* or *donor–sRNA* for short. The region on the ribosome which interacts with the donor–sRNA at the time of peptidyl transfer is termed *peptidyl donor–sRNA binding site* or in abbreviated form, *donor–sRNA site*. Aminoacyl sRNA accepts the peptidyl group from donor sRNA upon interaction with the ribosome at a region which we term the *aminoacyl sRNA receptor site*, or more simply, the *receptor site*.

The above definitions are functional and can be applied to most elaborations of the basic model. Further studies are required to relate the functional sites, thus defined, to structural sites deduced from sRNA binding studies.

Other systems of terminology have been proposed for sRNA binding sites. An alternative to the present functional terminology is that of Wettstein, and Noll (43), which supposes three sites, named *entrance* (or *decoding*), *middle* (or *condensing*) and *exit* sites. As another alternative, the simple terms *amino acid site* and *peptide site* are inadequate since it is those sites on the ribosome with which the sRNA moieties interact that we wish to describe. These sites can be termed *aminoacyl sRNA site* and *peptidyl sRNA site* on the structural basis of binding studies. Aminoacyl sRNA site can also be applied in a functional sense, but the term peptidyl sRNA site leads to ambiguity because, subsequent to transfer, peptidyl sRNA resides in the aminoacyl sRNA site.

The terms donor-sRNA site and receptor site apply to any regions of the ribosome which interact with the sRNA molecules at the time of peptidyl transfer. The terminology is versatile and allows for description of binding sites on 30S as well as 50S subunits.

PART II – LOCALIZATION OF PEPTIDYL TRANSFERASE ON THE 50S RIBOSOMAL SUBUNIT BY STUDY OF THE PUROMYCIN REACTION

Puromycin is an analogue of the aminoacyl adenosine moiety of aminoacyl sRNA (Figure 3a and b). It represents that part of the aminoacyl sRNA molecule which participates in peptidyl transfer but lacks those parts con-

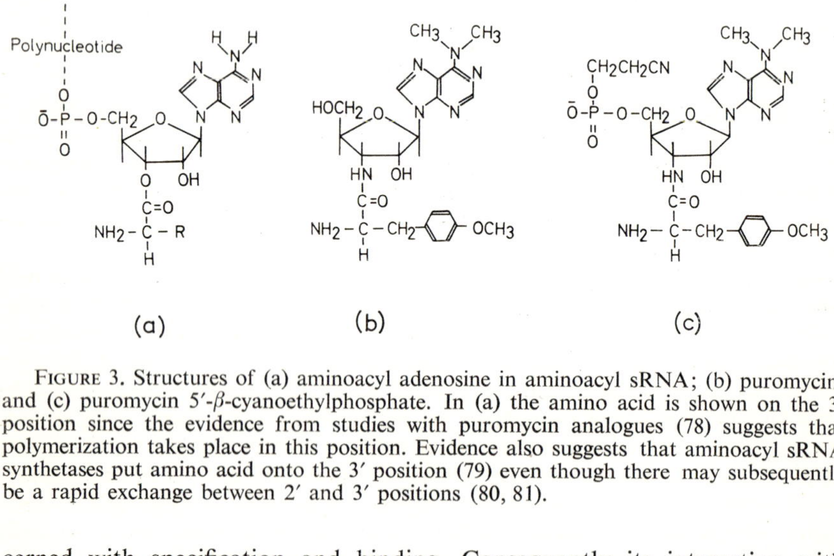

FIGURE 3. Structures of (a) aminoacyl adenosine in aminoacyl sRNA; (b) puromycin, and (c) puromycin 5'-β-cyanoethylphosphate. In (a) the amino acid is shown on the 3' position since the evidence from studies with puromycin analogues (78) suggests that polymerization takes place in this position. Evidence also suggests that aminoacyl sRNA synthetases put amino acid onto the 3' position (79) even though there may subsequently be a rapid exchange between 2' and 3' positions (80, 81).

cerned with specification and binding. Consequently its interaction with the ribosome-sRNA-transferase complex is a great simplification over the natural reaction sequence leading to peptide bond formation. Previous work had indicated that puromycin stimulates the release of growing peptides from ribosomes (73) and from sRNA (38). There was evidence also that in this process the puromycin becomes attached to the released peptides (74, 75). It appeared that the growing peptide is transferred to puromycin instead of aminoacyl sRNA as shown in Equation 2 (cf. Equation 1):

$$\begin{array}{c} R_n \qquad\qquad R' \\ | \qquad\qquad | \\ \text{---NH.CH.CO---sRNA}_n + \text{NH}_2.\text{CH.CO-aminonucleoside} \rightarrow \\ \text{(Puromycin)} \end{array} \qquad (2)$$

$$\begin{array}{c} R_n \qquad\qquad R' \\ | \qquad\qquad | \\ \rightarrow \text{---NH.CH.CO---NH.CH.CO-aminonucleoside} + \text{sRNA}_n \end{array}$$

However, attachment of puromycin had not been demonstrated in the *E. coli* cell-free system. Indeed, there was still the distinct possibility that puromycin could stimulate release of peptides without concomitant attachment (75, 76, 77).

We therefore confirmed the postulated reaction mechanism by characterization of the products of the puromycin reaction in a simple cell-free system. Subsequent work made use of the puromycin reaction to investigate the nature of the peptidyl transfer reaction with particular emphasis on identification of the components involved.

ATTACHMENT OF PUROMYCIN AND OF A LABELLED DERIVATIVE TO RELEASED PEPTIDES

In order to demonstrate the attachment of puromycin to released peptides it was advantageous to label puromycin with a radioactive isotope. Chemical synthesis from [^{3}H] or [^{14}C] components is laborious and expensive and ^{3}H-exchange methods lead to extensive degradation. A phosphate-containing derivative, puromycin 5′-β-cyanoethyl phosphate (Figure 3c) was therefore developed (83). This analogue, which is readily prepared and labelled with ^{32}P, was shown to have activity similar to puromycin both in the inhibition of protein synthesis and in the release of polypeptide chains from sRNA, assayed as described later.

For characterization of the products of the puromycin reaction we studied effects of puromycin on the synthesis of polylysine in a poly A-directed cell-free system from *E. coli*. The oligolysine peptides, which are formed in this system, can be readily fractionated by chromatography (67, 82).

In the uninhibited system the lysine peptides remain bound to sRNA (59, 60). However, incorporation in the presence of puromycin gives rise to products which are unattached to sRNA and behave like free peptides (83). These products were resolved by chromatography into a series of components which did not correspond to normal lysine peptides (Figure 4). Digestion with trypsin followed by re-chromatography showed that these compounds had properties expected of lysine oligopeptides attached through their carboxyl group to puromycin by a peptide linkage.

Further evidence for attachment of lysine peptides to puromycin was obtained by use of the labelled puromycin derivative. Incubation of ^{32}P-labelled puromycin 5′-β-cyanoethyl phosphate with the system using ^{3}H-labelled lysine gave a series of compounds labelled with both ^{32}P and ^{3}H (Figure 5) in ratios expected from lysyl peptides terminated by a single puromycin analogue residue. No significant amounts of lysine peptides unattached to puromycin were released from sRNA in the presence of puromycin. The major products were di- and trilysyl puromycins with smaller amounts of higher homologues.

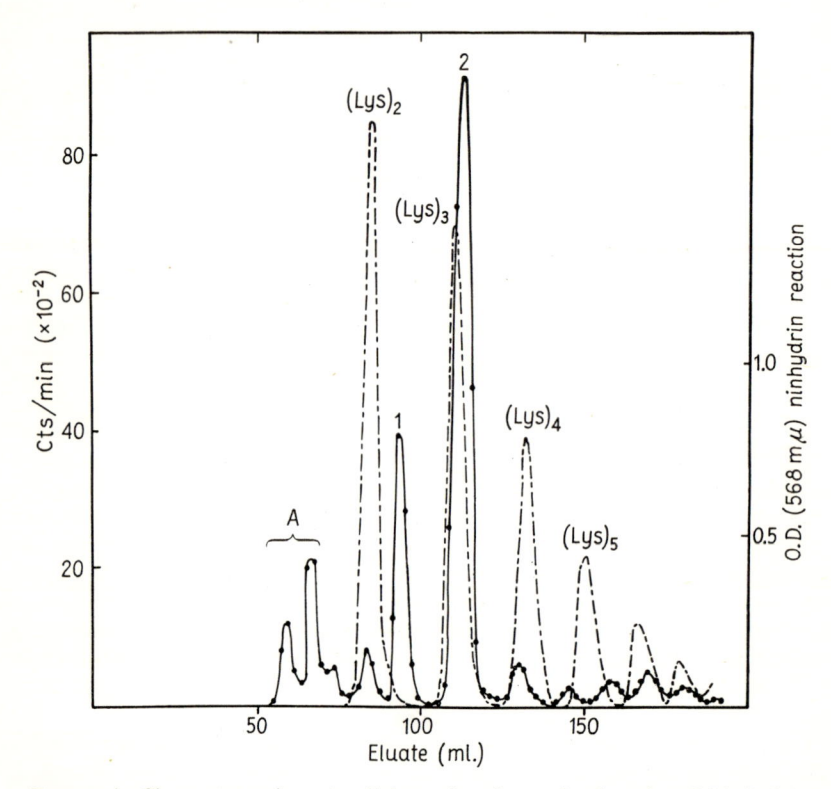

FIGURE 4. Chromatography on cellulose phosphate of released ^{14}C-labelled lysine peptides formed in the presence of poly A and puromycin. Conditions are given in another publication (83) from which the figure is reproduced. ———— Radioactivity; —·—·—, absorbancy (568 mμ) after ninhydrin reaction to locate unlabelled marker peptides added after incubation. Free lysine was eluted in the first 50 ml. (not analyzed). The radioactive peaks designated (A) were present in all incubations irrespective of whether poly A or puromycin were added, and appear to be degradation products of lysine.

The attachment of puromycin by peptide linkage to the C-terminal end of the released peptides clearly indicates that the puromycin reaction is analogous to normal peptide bond formation in protein synthesis. In characterization of the puromycin reaction the poly U-directed polyphenylalanine system has advantages over the lysine system. It is not so easy to characterize the products of the reaction of puromycin with polyphenyl-

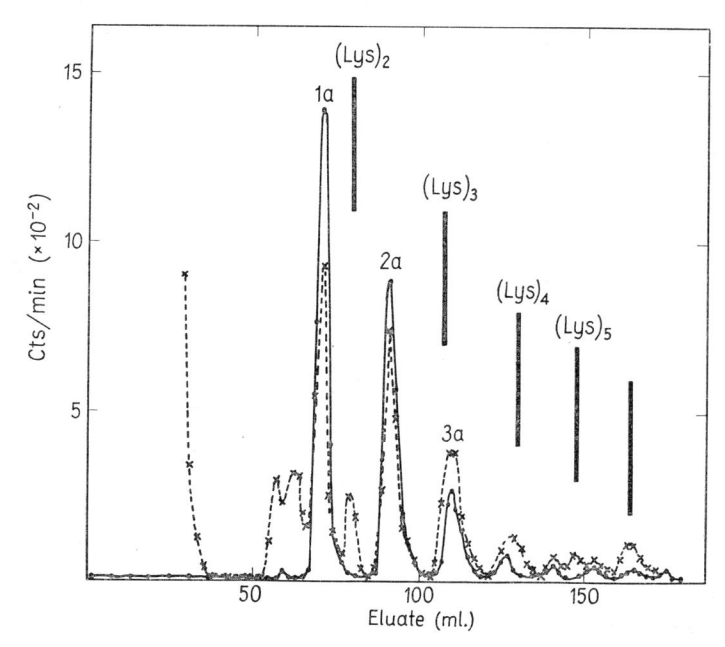

FIGURE 5. Separation on cellulose phosphate of released ³H-labelled lysine peptides formed in the presence of poly A and [³²P] puromycin cyanoethylphosphate. The Figure is reproduced from another publication (83) in which the conditions are given. ——— radioactivity, ³²P; ------ ³H; vertical lines mark the positions of the marker lysine peptides.

alanine owing to the difficulty of fractionating this insoluble material. We have, however, demonstrated that the ³²P-labelled puromycin derivative does become bound to polyphenylalanine during its release from sRNA under a variety of conditions (see below).

ASSAY OF THE PUROMYCIN REACTION

Puromycin action leads to cleavage of the bond between growing peptide chain and the sRNA which binds it to the ribosome (38). Consequently the sRNA (40, 84), and in many cases the nascent protein chain (73), are released from the ribosome. Early assays of the puromycin reaction measured release of nascent protein into the post-ribosomal supernatant fraction. Such assays assume that protein released from sRNA does not stick or adsorb to the ribosome and this is not invariably true. In order to avoid artefacts arising from this possibility we developed other assays.

The *E. coli* poly U-directed polyphenylalanine system (85) has been used throughout the present work. This system is similar to protein synthesis with natural mRNA in its requirements and its behaviour towards puromycin. It is simpler in that the product is a homopolymer; furthermore chain release does not normally occur and nearly all the polyphenylalanine remains attached to sRNA on the ribosome (38, 66).

Ribosomes were charged with polyphenylalanine and then isolated and washed by centrifugation. Such charged ribosomes can be stored for several days at 0°C. Incubation with puromycin under suitable conditions leads to release of polyphenylalanine from sRNA. Using this system we assayed the puromycin reaction in three ways. The first two methods are based on the separation of polyphenylalanyl–sRNA from polyphenylalanine. The third method measures the uptake of radioactive puromycin analogue.

1. *Zonal centrifugation.* Ribosomes are dissociated in the presence of SDS and salt. The resultant polyphenylalanyl–sRNA and free polyphenylalanine remain in solution and can be separated from one another by zonal centrifugation in sucrose gradients (38, 66). Substrate and product of the puromycin reaction can thus be resolved and the polyphenylalanine can be isotopically labelled to determine their relative proportions.

2. *Solubility in cresol.* Although the zonal centrifugation method is well defined and reliable, it is very laborious. We therefore developed a rapid assay (86), which is based on the observation that polyphenylalanine is soluble in *m*-cresol. An excess of *m*-cresol is added to the reaction mixture containing ^{14}C-or ^{3}H-labelled polyphenylalanine-charged ribosomes. The resultant mixture is passed through glass filters. Polyphenylalanine dissolves and passes through the filter; polyphenylalanyl–sRNA forms a precipitate and is estimated by measurement of the radioactivity retained on the filters. Disappearance of substrate can thus be readily followed and percentage release estimated by difference. This method is rapid and gives results in agreement with the zonal centrifugation method.

3. *Uptake of puromycin cyanoethylphosphate.* As described above, this puromycin derivative acts in the same manner as puromycin and can be labelled with ^{32}P. The puromycin reaction can be assayed by incubation of polyphenylalanine-charged ribosomes with [^{32}P] puromycin cyanoethylphosphate followed by precipitation with TCA and estimation of radioactivity in the precipitate by a filtration method (87). Results obtained with assays "1" and "2" have been qualitatively confirmed using this method (87).

CHARACTERIZATION OF THE PUROMYCIN REACTION

Requirements of the puromycin reaction have been studied by incubation of washed polyphenylalanine-charged ribosomes with puromycin under a variety of conditions, followed by assay using one of the above methods. The reaction is not affected by the addition of charged and stripped sRNA or of poly U, is completely dependent on divalent and monovalent cations, and is partially dependent on supernatant fraction and the GTP system (66, 86, 89). It appears that the charged ribosomes can exist in two states, one which does and one which does not require added supernatant factors in order to react with puromycin (66).

Attention has been concentrated on the supernatant independent reaction with a view to characterization and identification of the putative peptidyl transferase enzyme. Investigation of the location of this enzyme was facilitated by the report (38) that polyphenylalanyl–sRNA remains associated with the 50S subunit when the ribosomes are dissociated in low magnesium and separated by zonal centrifugation. We prepared polyphenylalanine-charged 50S ribosomes by this method and found that they would still react with puromycin (66). Figure 6 shows the separation obtained in a typical preparation. The charged 50S subunits sediment slightly faster than uncharged 50S subunits and are well separated from the 30S subunits. When recentrifuged in the same system the charged 50S subunits ran in the same position as before and no 30S subunits could be detected (86). The absence of 30S subunits was further confirmed by electron microscopy, which showed that the charged particles had the same appearance as 50S subunits and that no 30S subunits were observable (88).

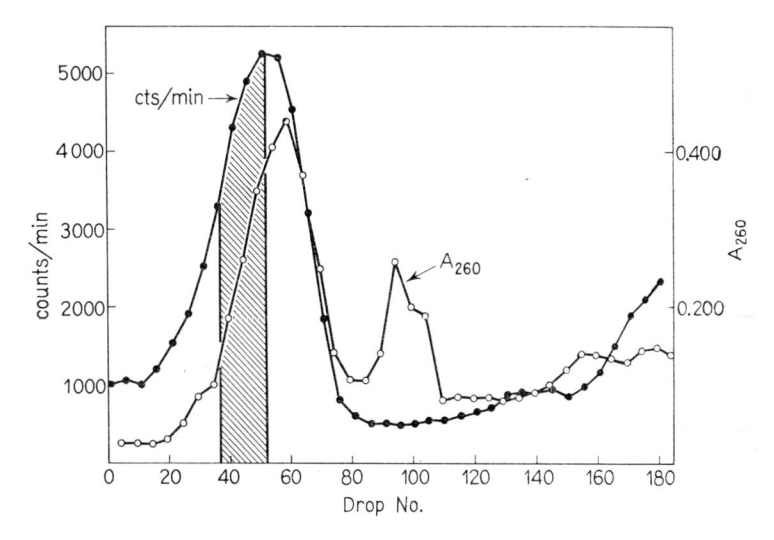

FIGURE 6. Preparation of [^{14}C] polyphenylalanine-charged 50S ribosomes by zonal centrifugation in a sucrose gradient containing 10^{-4} M Mg^{++}. The graph is reproduced from another publication (66) in which conditions are given.

As with charged 70S ribosomes the puromycin reaction would take place to a considerable extent with charged 50S subunits in the absence of supernatant factors (86). Characteristics of the supernatant-independent reaction with 50S subunits are qualitatively similar to those with 70S ribosomes, though in general the rate and the extent of the reaction with 50S subunits is lower. This is thought to be due to instability of the 50S complex in the absence of 30S subunits.

Figure 7 shows progress curves of the puromycin reaction with 70S ribosomes and with 50S ribosomes. With 70S ribosomes approximately 60% of the polyphenylalanine was released from sRNA in 1 min at 30° (Figure 7a). This was followed by the much slower release of a further 20% during the next 30 min.

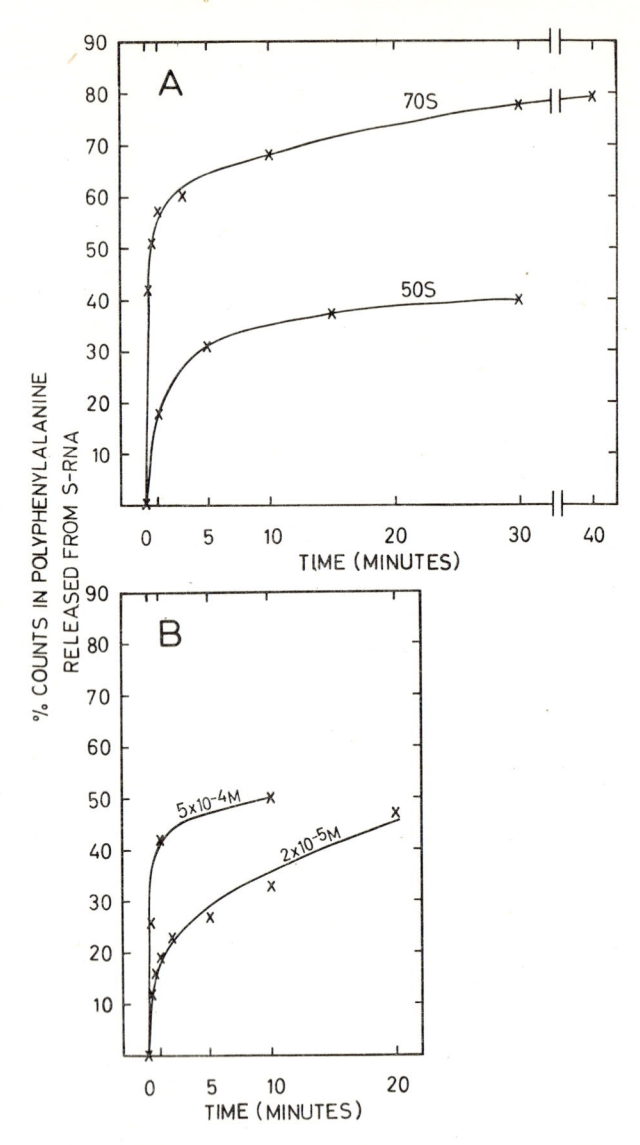

FIGURE 7 a, b. Progress curves of the puromycin reaction. Polyphenylalanine-charged 70S ribosomes were washed three times in 0.01 M Tris pH 7.4, 0.01 M Mg acetate and 0.5 M NH$_4$Cl. 50S ribosomes were prepared by sucrose gradient zone centrifugation and concentrated with Sephadex. Incubation volume was 0.1 ml. Components were: 0.2 mg/ml ribosomes, 0.01 M Tris pH 7.4, 0.01 M Mg acetate and 0.1 M NH$_4$ acetate. Reactions were started by adding puromycin and stopped by addition of 3 ml of *m*-cresol. (a): 70S ribosomes (upper curve) and 50S ribosomes (lower curve). Incubation at 30° with 5×10^{-4} M puromycin; (b): 70S ribosomes at 0°C with 5×10^{-4} M puromycin (upper curve) and 2×10^{-5} M puromycin (lower curve).

Release in the absence of puromycin was about 15% in 30 min. Purified 50S ribosomes reacted in a qualitatively similar way (Figure 7a) but less polyphenylalanine was released and the early stage of the reaction was less rapid. With 70S ribosomes the initial rate is too rapid to follow at 30°. Figure 7b shows that this is also true at 0° with saturating puromycin concentration but that with limiting puromycin (2×10^{-5}M) the kinetics are resolved. None of these progress curves fit exponential kinetics and the relative proportions of rapid to slow reaction vary from one preparation to another.

A number of explanations are possible for the heterogeneous kinetics. It is clear, however, that a considerable percentage of the charged ribosomes react very rapidly with puromycin. In good preparations a detectable reaction takes place in 1 min at 0° with concentrations of puromycin as low as 10^{-7} M. This level of reactivity provides evidence that the reaction takes place by an enzymic mechanism.

It has been generally assumed that peptide bond formation is catalyzed by one of the supernatant enzymes. We were therefore surprised to observe that the supernatant-independent puromycin reaction could not be abolished by repeated washing of the charged ribosomes under a variety of conditions (66, 86). Effects of repeated washing in the presence of salts on the rapid phase of the puromycin reaction are shown in Table 1. These are conditions similar to those which are routinely employed to prepare ribosomes completely dependent upon T and G factors for protein synthesis (47). It can be seen that the first wash leads to an appreciable increase rather than the expected decrease in the extent of the puromycin reaction. Subsequent washes do not lower the extent. As a further check on the possibility that the reaction was taking place as a result of contamination with a supernatant enzyme, the assay was also carried out at low concentrations (Table 1) of the washed ribosomes. This treatment led to an increase rather than a decrease in extent of the reaction. It is thus clear that if the puromycin reaction is catalyzed by a supernatant enzyme this enzyme must be firmly and specifically bound. The alternatives to this possibility are either that the reaction is non-enzymic or that it is catalyzed by an enzyme which is an integral part of the 50S subunit.

TABLE I. Effects of Washing and of Ribosome Concentration on Puromycin Reaction

Polyphenylalanine-charged ribosomes Treatment	Concentration mg/ml	Release of polyphenylalanine %
Unwashed	1	24
3 × washed	1	37
5 × washed	0.8	40
5 × washed	0.4	48
5 × washed	0.2	43
5 × washed	0.1	50

Polyphenylalanine-charged ribosomes were washed by centrifugation in 0.01 Tris, pH 7.4, 0.01 M Mg acetate and 0.5 M NH₄Cl. Assay was carried out under standard conditions (Fig. 7). Incubation was for 10 min at 0°C with 5×10^{-4} puromycin.

Several lines of evidence indicate that the puromycin reaction is catalyzed by an enzyme. (i) It is a defined chemical reaction involving group transfer from one molecule to another. Furthermore, the reaction is specific with respect to substrate. The L-phenylalanine analogue of puromycin is active whereas the D-phenylalanine analogue is inactive (78, 84). Analogues with the amino acid on the 2' or 5' rather than the 3' position of the amino nucleoside are inactive (78). (ii) Aminoacyl or peptidyl sRNA will not react with puromycin in the absence of ribosomes. The reaction is abolished by treatments which disrupt ribosomes or denature proteins (4 M urea, 0.5% SDS, or incubation at 65° (86)). (iii) The charged ribosome complex is stable in the absence of puromycin. Little release of polyphenylalanine from sRNA takes place upon storage of the complex for 7 days at 0°C or upon incubation of the complex for 5 min at 55°. Incubation with puromycin before or after either of these treatments leads to very rapid release. These results show that the reaction with puromycin does not result from non-specific activation of the polyphenylalanyl sRNA.

It is therefore reasonable to suppose that the puromycin reaction is catalyzed by an enzyme. Evidence that this is the same enzyme which catalyzes peptidyl transfer in protein synthesis rests upon the similar nature of the two reactions and the analogy between the structures of puromycin and aminoacyl sRNA. To quote from a previous publication (66), "The existence of a reaction with puromycin, independent of supernatant fraction and GTP, shows that polypeptide sRNA can exist in a state which can react without further modification. This state might have two explanations: (1) a supernatant factor and GTP are required but an intermediate is formed in which they are tightly and specifically bound to the ribosome in a manner permitting the formation of a peptide bond; (2) supernatant fraction and GTP act in a prior step to form a reactive polyphenylalanyl sRNA–50S ribosome complex and the reaction of this intermediate to form a peptide bond is catalyzed by the ribosome itself". In the following sections we shall attempt to distinguish which of these two alternatives is correct.

EFFECTS OF INHIBITORS

As noted in the first part of this paper, GTP is involved in the functioning of the G supernatant factor. Furthermore the GTP analogue, GMP–PCP, specifically inhibits protein synthesis by interference with the GTP reaction. However, when this analogue was tested in the puromycin assay, it was found to affect neither the rate nor the extent of the reaction even at concentrations which give over 90% inhibition of protein synthesis (Table 2). It is therefore probable that peptidyl transfer does not involve GTP or G factor but that these components are involved in some other step.

The T supernatant factor is very labile and loses activity rapidly in the absence of sulphydryl compounds. Both T and G factors are inactivated by sulphydryl inhibitors (47). In contrast polyphenylalanine-charged ribosomes are very stable in the absence of sulphydryl compounds and the puromycin reaction is unaffected by concentrations of parahydroxymercuriben-

TABLE II. Effects of Inhibitors on Protein Synthesis and on the Puromycin Reaction

Inhibitor	Concentration M	% Inhibition Polyphenylalanine synthesis	Puromycin reaction
p-Hydroxymercuri-benzoate	1×10^{-3}	100	0
N-Ethylmaleimide	1×10^{-3}	100	0
Guanylyl methylene-diphosphate	2×10^{-4}	90	0

Polyphenylalanine synthesis was assayed as previously described (66). Conditions for assay of puromycin reaction were as described in Figure 7. Both initial rate and final extent of the reaction were measured.

zoate or N-ethyl maleimide which completely inhibit protein synthesis (Table 2). It follows that, if the puromycin reaction is catalyzed by either the T or the G factor, the enzyme must not only be firmly and specifically bound to the charged 50S ribosome complex but must also be in a protected state.

EVIDENCE FROM THE POLYLYSINE SYSTEM

Rychlik isolated polylysyl sRNA from the poly A-directed *in vitro* system (90). He then studied the reaction of this intermediate with puromycin under various conditions (91). The reaction was assayed by measurement of the conversion of polylysine from an acid-insoluble to an acid-soluble form (equivalent to release from sRNA). From our experiments in the polylysine system (described above), this reaction can be assumed to give rise to peptide bond formation between puromycin and the released lysine peptides. Rychlik found that no reaction took place when polylysyl sRNA was incubated with puromycin alone or in the presence of supernatant. However binding of polylysyl sRNA to ribosomes took place in the presence of poly A and the resultant complex reacted rapidly with puromycin. Both ribosomes and poly A were required for the release of polylysine by puromycin, but the addition of supernatant did not stimulate the reaction.

These experiments provide further information regarding the identity of the peptidyl transferase enzyme. The ribosomes which were employed to catalyze the puromycin reaction were uncharged and were washed by a procedure which makes ribosomes dependent upon both T and G factors for protein synthesis. We therefore infer that binding of peptidyl transferase on ribosomes is not dependent upon the presence of the peptidyl sRNA complex.

The results with polylysyl sRNA, together with those from the polyphenylalanine system, lead us to the conclusion that one or more molecules of peptidyl transferase enzyme are specifically integrated into the structure of the 50S subunit. We can furthermore say that this enzyme is insensitive to sulphydryl inhibitors not only in the presence of polyphenylalanyl sRNA (see above) but also in the absence of peptidyl sRNA substrate. This follows from observations that the 30S but not the 50S subunit is inactivated for incorporation by treatment with sulphydryl inhibitors (92).

The possibility cannot be excluded at present that a molecule of supernatant enzyme is firmly and specifically bound to a site on the ribosomes in a protected form, and that exchange at this site is linked in an obligatory way to peptide synthesis. However, we consider it more probable that peptidyl transferase is a non-exchangeable protein of the 50S ribosomal subunit and that supernatant proteins together with GTP have functions other than catalysis of peptidyl transfer.

SUMMARY OF MAIN POINTS

1. Peptide bond formation takes place by transfer of growing peptide from sRNA to aminoacyl sRNA (basic model).

2. An enzyme for this reaction is postulated and named *peptidyl transferase* (basic model).

3. Catalysis of peptide bond formation by peptidyl transferase has been resolved from other reactions of protein synthesis by use of puromycin.

4. Peptidyl transferase is an integral part of the nascent protein-ribosome complex (washing polyphenylalanine-charged ribosomes does not abolish the supernatant-GTP-independent reaction with puromycin).

5. Peptidyl transferase is also integrated into the structure of uncharged ribosomes (reaction of polylysyl sRNA with puromycin is catalyzed by uncharged ribosomes, washed under conditions which make them dependent upon T and G supernatant factors for protein synthesis).

6. Peptidyl transferase is associated with the 50S subunit (puromycin reaction takes place with charged 50S subunit in absence of 30S subunit).

7. Inhibitor studies provide independent evidence that peptidyl transfer does not involve GTP or G factor (non-inhibition by GMP–PCP; involvement of G factor function with GTP).

8. If peptidyl transferase is identical with T or G factor then it must be in a protected form when bound to the ribosome (puromycin reaction and 50S subunit insensitive to sulphydryl inhibitors; T and G factors inactivated).

9. It follows from "5" and "6" that peptidyl transferase is integrated into the structure of the normal 50S subunit. This conclusion is not incompatible with the possibility (point "10") that the enzyme dissociates from the ribosome in protein synthesis.

10. It follows from "4" and "5" that if peptidyl transferase is identical with one of the supernatant factors, then (a) it must be capable of firm and specific integration into the ribosome structure, and (b) dissociation of the bound enzyme from ribosome must be both dependent upon and obligatory for protein synthesis. Such a mechanism is improbable, and since still further restrictions would have to be applied to such a bound molecule of supernatant protein to explain behaviour towards inhibitors (points "7" and "8"), we consider it more likely that peptidyl transferase is a non-exchangeable ribosomal protein.

CONCLUSIONS

Peptide bond formation is catalyzed by an enzyme, peptidyl transferase, which is integrated into the structure of the 50S ribosomal subunit. The enzyme is situated in a position available to both acceptor and donor sRNA molecules. It is probable that peptidyl transferase is non-identical to the supernatant proteins required for protein synthesis, and that GTP is not directly involved in its action. The translocation model of protein synthesis provides other possible functions for supernatant factors and GTP.

ACKNOWLEDGEMENTS

We should like to thank Dr. F. H. C. Crick and other members of the laboratory for stimulating discussions and criticism of the manuscript. In particular we thank Professor F. M. Dixon for advice on terminology.

REFERENCES

1. Sarabhai, A. S., Stretton, A. O. W., Brenner, S. and Bolle, A., *Nature* **201**, 13 (1964).
2. Yanofsky, C., Carlton, B. C., Guest, J. R., Helinski, D. R. and Henning, U., *Proc. Nat. Acad. Sci. U.S.* **51**, 266 (1964).
3. Jacob, F. and Monod, J., *J. Mol. Biol.* **3**, 318 (1961).
4. Brenner, S., Jacob, F. and Meselson, M., *Nature* **190**, 576 (1961).
5. Gros, F., Hiatt, H. H., Gilbert, W., Kurland, C. G., Risebrough, R. W. and Watson, J. D., *Nature* **190**, 581 (1961).
6. Terzaghi, E., Okada, Y., Streisinger, G., Tsugita, A., Inouye, M. and Emrich, J., *Science* **150**, 387 (1965).
7. Salas, M., Smith, M. A., Stanley, W. M., Wahba, A. J. and Ochoa, S., *J. Biol. Chem.* **240**, 3988 (1965).
8. Smith, M. A., Salas, M., Stanley, W. M., Wahba, A. J. and Ochoa, S., *Proc. Nat. Acad. Sci. U.S.* **55**, 141 (1966).
9. Thach, R. E., Cecere, M. A., Sundararajan, T. A. and Doty, P., *Proc. Nat. Acad. Sci.* **54**, 1167 (1965).
10. Crick, F. H. C., *Progress in Nucleic Acid Research* **1**, 163 (1963).
11. Crick, F. H. C., Barnett, L., Brenner, S. and Watts-Tobin, R. J., *Nature* **192**, 1227 (1961).
12. Khorana, H. G., These meetings.
13. Tsugita., A. *J. Mol. Biol.* **5**, 284 (1962).
14. Weisblum, B., Gonazo, F., von Ehrenstein, G. and Benzer, S., *Proc. Nat. Acad. Sci. U.S.* **53**, 328 (1965).
15. Khorana, H. G., These meetings.
16. Zachau, H. G., Acs, G. and Lipmann, F., *Proc. Nat. Acad. Sci. U.S.* **44**, 885 (1958).
17. Preiss, J., Berg, P., Ofengand, E. J., Bergmann, F. H. and Dieckman, M., *Proc. Nat. Acad. Sci. U.S.* **45**, 319 (1959).
18. Hecht, L. I., Stephenson, M. L. and Zamecnik, P. C., *Proc. Nat. Acad. Sci. U.S.* **45**, 505 (1959).
19. Crick, F. H. C., In Symp. Soc. Exp. Biol., **XII**, 139 (1958).

20. Hoagland, M. B., In Structure and Function of Genetic Elements, Brookhaven Symposia in Biology **12**, 40 (1959).
21. Chapeville, F., Lipmann, F., von Ehrenstein, G., Weisblum, B., Ray, W. J. and Benzer, S., *Proc. Nat. Acad. Sci. U.S.* **48**, 1086 (1962).
22. Petermann, M. L., The Physical and Chemical Properties of Ribosomes, Elsevier, Ch. 7, 113 (1964).
23. Tissieres, A., Watson, J. D., Schlessinger, D. and Hollingworth, B. R., *J. Mol. Biol.* **1**, 221 (1959).
24. Moller, W. and Boedtker, H., *Acides ribonucleiques et polyphospates.* Editions du Centre National de la Recherche Scientifique, Paris, p. 99 (1962).
25. Rosset, R., Monier, R. and Julien, J., *Bull. Soc. Chim. Biol.* **46**, 87 (1964).
26. Sanger, F., Brownlee, G. G. and Barrell, B. G., *J. Mol. Biol.* **13**, 373 (1965).
27. Brownlee, G. G. and Sanger, F., *J. Mol. Biol.* in press (1966).
28. Waller, J. P. and Harris, J. I., *Proc. Nat. Acad. Sci. U. S.* **47**, 18 (1961).
29. Waller, J. P., *J. Mol. Biol.* **7**, 483 (1963).
30. Waller, J. P., *J. Mol. Biol.* **10**, 319 (1964).
31. Lebry, P. S., Cox, E. C. and Flaks, J. G., *Proc. Nat. Acad. Sci. U.S.* **52**, 1367 (1964).
32. Traut, R. R. Unpublished observations.
33. Gilbert, W., *J. Mol. Biol.* **6**, 347 (1963).
34. Takanami, M. and Okamoto, T., *J. Mol. Biol.* **7**, 323 (1963).
35. Kaji, H. and Kaji, A., *Federation Proc.* **24**, 408 (1965).
    Matthaei, J. H., Amelunxen, F., Eckert, K. and Heller, G., *Berichte der Bunsengessellschaft für Physikalische Chemie* **68**, 735 (1964).
36. Suzuka, I. and Kaji, A., *Federation Proc. Abstracts* **25**, 403 (1966).
37. Vazquez, D. and Monro, R. E. To be published.
38. Gilbert, W., *J. Mol. Biol.* **6**, 389 (1963).
39. Cannon, M., Krug, R. and Gilbert, W., *J. Mol. Biol.* **7**, 360 (1963).
40. Cannon, M., *Biochem. J.* **98**, 5 (1966).
41. Warner, J. R. and Rich A., *Proc Nat. Acad. Sci. U.S.* **51**, 1134 (1964).
42. Arlinghaus R., Schaeffer, J. and Schweet, R., *Proc. Nat. Acad. Sci. U.S.* **51**, 1291 (1964).
43. Wettstein, F. O. and Noll, H., *J. Mol. Biol.* **11**, 35 (1965).
44. Watson, J. D., 50th Anniversary Meeting, French Biochemical Society (1964).
45. Nathans, D. and Lipmann, F., *Proc. Nat. Acad. Sci. U.S.* **47**, 497 (1961).
46. Allende, J. E., Monro, R. E. and Lipmann, F., *Proc. Nat. Acad. Sci. U.S.* **51**, 1212 (1964).
47. Nishizuka, Y. and Lipmann, F., *Proc. Nat. Acad. Sci. U.S.* **55**, 212 (1966).
48. Lucas-Lenard, J. and Lipmann, F., *Federation Proc.* **25**, 215 (1966),
49. Keller, E. B. and Zamecnik, P. C., *J. Biol. Chem.* **221**, 45 (1956).
50. Monro, R. E., *Cold Spr. Harb. Symp. Quant. Biol.* **26**, 156 (1961).
51. Nathans, D., von Ehrenstein, G., Monro, R. E. and Lipmann, F., *Federation Proc.* **21**, 127 (1962).
52. Conway, T. W. and Lipmann, F., *Proc. Nat. Acad. Sci. U.S.* **52**, 1462 (1964).
53. Chan, M. and McCorquodale, D. J., *J. Biol. Chem.* **240**, 3116 (1965).
54. Hershey, J. W. B. and Monro, R. E., *J. Mol. Biol.* **18**, 68 (1966).
55. Myers, T. C., Nakamura, K. and Flesher, J. W., J.A.C.S. **85**, 3292 (1963).
56. Simon, L., Myers, T. C. and Mednieks, M., *Biochim. Biophys. Acta* **103**, 189 (1965).

57. Bishop, J., Leahy, J. and Schweet, R., *Proc. Nat. Acad. Sci. U.S.* **46**, 1030 (1960).
58. Dintzis, H. M. *Proc. Nat. Acad. Sci. U.S.* **47**, 247 (1961).
59. Bretscher, M. S., *J. Mol. Biol.* **7**, 446 (1963)
60. Bretscher, M. S., *J. Mol. Biol.* **12**, 913 (1965).
61. Glassmann, E., Allen, E. M. and Schweet, R. S., *J. Am. Chem. Soc.* **80**, 4427 (1958).
62. Lipmann, F., Hülsman, W. C., Hartmann, G., Boman, H. G. and Acs, G., *J. Cell. Comp. Physiol.* **54**, 75 (1959).
63. Berg, P., Bergmann, F. M., Ofengand, E. J. and Dieckmann, M., *J. Biol. Chem.* **236**, 1726 (1961).
64. Borsook, H., *Adv. Protein Chem.* **VIII**, 128 (1953).
65. Kornberg, A., *Adv. Enzymol.* **18**, 191 (1957).
66. Traut, R. R. and Monro, R. E., *J. Mol. Biol.* **10**, 63 (1964).
67. Smith, J. D., *J. Mol. Biol.* **8**, 772 (1964).
68. Schweet, R., Arlinghaus, R., Heintz, R. and Schaeffer, J., M. D. Anderson Symposium (1965).
69. Davis, B. D., *Cold. Spr. Harb. Symp. Quant. Biol.* **XXVIII**, 294 (1963).
70. Zubay, G., In *The Molecular Basis of Neoplasia*, University of Texas Press, Austin (1962).
71. Vazquez, D., *Biochem. Biophys. Res. Comm.* **15**, 464 (1964).
72. Report of the Commission on Enzymes of the International Union of Biochemistry. In Comprehensive Biochem. **13**, 2nd Edition (1965).
73. Morris, A. and Schweet, R., *Biochim. Biophys. Acta* **47**, 415 (1961).
74. Allen, D. W. and Zamecnik, P. C., *Biochim. Biophys. Acta* **47**, 865 (1962).
75. Nathans, D., *Proc. Nat. Acad. Sci. U.S.* **51**, 585 (1964).
76. Nathans, D., Allende, J. E., Conway, T. W., Spyrides, G. J. and Lipmann, F., In *Informational Macromolecules*, ed. by H. J. Vogel, V. Bryson and J. O. Lampen, p. 361, New York, Academic Press (1963).
77. Morris, A., Arlinghaus, R., Favelukes, S. and Schweet, R., *Biochemistry* **2**, 1084 (1963).
78. Nathans, D. and Neidle, A., *Nature* **197**, 1076 (1963).
79. Zachau, H. G. and Feldmann, H., *Progress in Nucleic Acid Research and Molecular Biology* **4**, 217 (1965).
80. Wolfenden, R., Rammler, D. H. and Lipmann, F., *Biochemistry* **3**, 329 (1964).
81. McLaughlin, C. S. and Ingram, V. M., *Federation Proc.* **22**, 350 (1963).
82. Smith, M. A. and Stahman, M. A., *Biochem. Biophys. Res. Comm.* **13**, 251 (1963).
83. Smith, J. D., Traut, R. R., Blackburn, G. M. and Monro, R. E., *J. Mol. Biol.* **13**, 617 (1965).
84. Monro, R. E. Unpublished observations.
85. Nirenberg, M. W., Matthaei, J. H. and Jones, O. W., *Proc. Nat. Acad. Sci. U. S.* **47**, 494 (1962).
86. Monro, R. E., Traut, R. R. and Maden, B. E. H. To be published.
87. Monro, R. E., Blackburn, G. M. and Traut, R. R. Unpublished results.
88. Huxley, H. E. and Monro, R. E. Unpublished results.
89. Maden, B. E. H. and Monro, R. E. To be published.
90. Rychlik, I., Collection Czech. Chem. Commun. **30**, 2259 (1965).
91. Rychlik, I., *Biochim. Biophys. Acta* **114**, 425 (1966).
92. Traut, R. R. To be published.
93. Fox, C. F. and Kennedy, E. P., *Proc. Nat. Acad. Sci. U. S.* **54**, 891 (1965).

94. Lynen, F., *Federation Proc.* **20**, 941 (1961).
95. Hsu, R. Y., Wasson, G. and Porter, J. W., *J. Biol. Chem.* **240**, 3736 (1965).
96. Moore, P. B. Ph. D., Thesis, Harvard University., (November 1965).
97. Allen, D. W. and Zamecnik, P. C., *Biochem. Biophys. Res. Comm.* **11**, 294 (1963).
98. Hosokawa, K., Fujimura, R. K. and Nomura, M., *Proc. Nat. Acad. Sci. U.S.* **55**, 198 (1966).
99. Staehelin, M. and Meselson, M., *J. Mol. Biol.* **15**, 245 (1966).
100. Lermann, M. F., Spirin, A. S., Gavrilova, L. P. and Galor, V. F., *J. Mol. Biol.* **15**, 268 (1966).
101. Kaji, H., *Federation Proc.* **25**, 777 (1966).

DISCUSSION

(Chairman: E. FREDERICQ)

P. LEDER: Since peptidyl sRNA is not removed from ribosomes in the washing procedure is it possible that supernatant enzyme is also not removed? A substrate–enzyme–ribosome complex might have considerable stability.

R. E. MONRO: This is quite possible in the case of the polyphenylalanine experiments and is one of the alternatives considered. However, Rychlik's experiments in the polylysine system show that presence of substrate is not necessary for strong binding of the enzyme to ribosomes.

J. H. MATTHAEI: It would be a desirable experiment to feed back polyphenylalanyl sRNA to the 50S ribosomes after efficient washing to see whether these would still not require a supernatant enzyme for the puromycin reaction.

R. E. MONRO: This experiment cannot be done because polyphenylalanyl sRNA is insoluble. The corresponding experiment could be tried with polylysyl sRNA, but it might not be possible to effect entry to the donor site on the 50S subunit in the absence of 30S subunits and poly A. The advantage of the polyphenylalanine system for implication of the 50S subunit as the site of peptidyl transfer is that polyphenylalanyl sRNA remains stuck to the 50S subunit after synthesis. The need for mRNA and 30S subunit to effect entry to the peptide transfer site is thus obviated. However, with regard to the question of whether or not supernatant enzyme is involved, the evidence from the polylysine experiments is just as valid whether 70S ribosomes or purified 50S subunits are employed.

F. SANGER: Is there any need to assume the presence of an enzyme, peptidyl transferase, or could the puromycin reaction be catalyzed by the ribosome itself?

R. E. MONRO: Evidence is included in the text that the puromycin reaction is specifically catalyzed by a component on the 50S subunit, and that treatments which denature proteins abolish the reaction. Although we have no direct evidence to prove that the catalytic agent is composed of protein, it is reasonable to suppose that this is the case by analogy with other biochemical reactions. A molecule of peptidyl transferase appears to be integrated into the structure of a cell organelle, the ribosome. There may also be other functional proteins integrated into the ribosome structure. Analogous structural integration of functional proteins can be observed elsewhere in the cell as, for example, in muscle, mitochondria, cell membrane (93) and in the multi-enzyme unit involved in fatty acid synthesis (94, 95). In this respect a novel feature about ribosomes is the presence of RNA as a major constituent.

We visualize peptidyl transferase as being similar to a normal, soluble enzyme, and its function as a catalyst could involve cofactors and prosthetic groups. However, it is important to emphasize that the functional unit is the whole ribosome. Access of substrate to peptidyl transferase is controlled by the other macromolecules with which this enzyme is structurally integrated. As distinct from most enzyme catalysis, association of substrate with enzyme involves simultaneous interactions with several macromolecules beside the enzyme.

If peptidyl transferase is a protein molecule, as we think, it should be possible to extract and purify it from ribosomes. It is possible that under suitable conditions the purified protein could catalyze peptidyl transfer or a related exchange reaction. If this could be demonstrated it would provide definitive proof that peptide bond formation in protein synthesis is catalyzed by an enzyme which is part of the ribosome.

E. Fredericq: As regards the binding of sRNA and mRNA by ribosomes, do you know what part is played respectively by the protein moiety and the RNA moiety of the ribosome?

R. E. Monro: Both RNA and protein constituents are essential for activity. RNA amino groups are important for interaction between subunits and for mRNA binding. Treatment of ribosomes with nitrous acid and formaldehyde inhibits both these interactions by a mechanism shown to be due to blocking of nucleic acid amino groups (96). Hydrogen bonding between ribosomal RNA and mRNA and between the RNA moieties of $30S$ and $50S$ subunits is thought to be at least partially responsible for binding reactions.

Mild treatment of ribosomes with T1 ribonuclease decreases incorporation activity and mRNA binding (97). Removal of part of the ribosomal protein causes inactivation. Recombination of protein with the depleted particle restores activity (98, 99, 100). The protein specific reagents, FDNB (96), p-hydroxymercuribenzoate and N-ethyl maleimide (92) cause inactivation of incorporation. Mild treatment with trypsin results in inhibition of sRNA binding to both $30S$ and $50S$ subunits (35, 101).

E. Fredericq: Is it possible at the present time to distinguish whether the binding process occurs on the ribosome surface or in a deeper part, involving partial unfolding of its structure?

R. E. Monro: The ribosome is a rather hydrated structure and it is possible that interactions take place not only on the surface, but also internally. We have no information regarding these alternatives at present.

THE GENETIC CODE

Chairman's Introductory Remarks

F. Sanger (Chairman)

Probably the most exciting and spectacular development in biochemistry during the last few years has been the elucidation of the genetic code. In 1961 Matthaei and Nirenberg were studying the effect of tobacco mosaic virus RNA on the incorporation of amino acids by a cell free system from *E. coli*. As a control experiment they tried the effect of polyuridylic acid. Surprisingly this caused a very large incorporation of phenylalanine and led to the identification of the triplet code word for phenylalanine as UUU. This initial experiment sparked off a veritable explosion of research which has today resulted in the almost complete deduction of the genetic code. The earlier work by Nirenberg and his colleagues and by Ochoa and his colleagues using random mixed polynucleotides made it possible to determine the composition of many of the codons but not the actual sequence of nucleotides in these codons. The next major step forward was the finding by Leder and Nirenberg that a specific binding could be obtained between an amino acid transfer RNA and its corresponding coding nucleotide in the presence of ribosomes. This made it possible to test synthetic trinucleotides of defined sequence for their coding properties. Progress in this work has been very largely dependent on the supply of synthetic nucleotides as is evident from the outstanding results of Khorana and his group. They have been foremost in the development of chemical and enzymic methods for synthesizing nucleotides, and their elegant work has greatly increased confidence in the assignment of the various codons. We are very fortunate to have most of the leading groups of workers in this field well represented in this symposium, which will cover most aspects of the genetic code.

Polynucleotide Synthesis and the Genetic Code—II*

H. Gobind Khorana

Institute for Enzyme Research, The University of Wisconsin, Madison, Wisconsin, U.S.A.

Introduction

I should begin by referring to five major landmarks in the development of the chemistry and biochemistry of the nucleic acids, all made within the space of the last 14 years and the landmarks which to my way of thinking have provided firm foundations for the present day structure of the field of molecular biology. There was the classical paper in 1952 by Brown and Todd (1) which elucidated the nature of the covalent bonds in nucleic acids and this really represented the climax of the efforts of about three generations of chemists and biochemists, the most prominent among whom was P. A. Levene. In 1953, came the Watson–Crick (2) structure which described the macromolecular organization of DNA and focussed attention, in particular, on the biological significance of this physical structure. In 1956, Kornberg and coworkers (3) discovered the enzyme, DNA polymerase, which showed that DNA could be replicated to yield more DNA in the test tube. In 1960, several groups of workers (4–8) almost simultaneously discovered the enzyme, DNA-dependent RNA polymerase, which clarified the manner in which information from DNA may be transcribed to a ribopolynucleotide (messenger RNA). The fifth landmark in our story is the development of the cell-free amino acid incorporating systems, which again has a long history but one thinks, in particular, of the work of Zamecnik and Hoagland (9), of Berg, of Lipmann and, in regard to the bacterial system, that of Tissieres and Watson (10) and, in 1961, of the important refinement made by Nirenberg and Matthaei (11).

As early as 1953, we began our efforts to develop chemical methods for the construction of ribo- and deoxyribopolynucleotide chains containing the internucleotide bonds which occur in nature (For recent reviews of the synthetic work see 12–14). By 1962 the total chemical methodology had developed to the point that we posed the question: Can chemical synthesis make a contribution to the study of the fundamental and biologically vital process of information flow, DNA → RNA → Protein? Specifically, the reaction sequences we proposed to investigate were as illustrated in Chart 1. Early work in this laboratory had encouraged the hope that short deoxypolynucleotides containing repeating nucleotide sequences would serve

---

* A recent review (Ref. 12) with the same title is regarded as Part I.

CHART 1. Proposed Sequence of Enzymatic Reactions for the Synthesis of Specific Polypeptides using Chemically Synthesized Deoxyribopolynucleotides

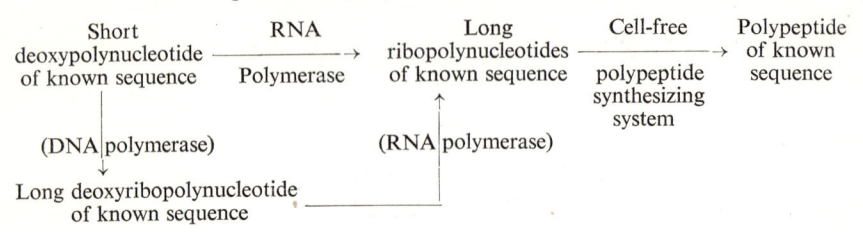

```
   Short                RNA            Long             Cell-free        Polypeptide
deoxypolynucleotide  ─────────→   ribopolynucleotides  ─────────→       of known
of known sequence    Polymerase   of known sequence    polypeptide       sequence
       │                                 ↑             synthesizing
                                                          system
   (DNA polymerase)               (RNA polymerase)
       ↓                                 │
Long deoxyribopolynucleotide             │
   of known sequence       ──────────────┘
```

as templates for the polymerases and that therefore a combination of chemistry and enzymology would produce polynucleotide messengers of known sequences. The latter would be of potential interest in the study of the genetic code because these would be expected to direct the synthesis of polypeptides of known sequences. It is recalled that a direct correlation of the polynucleotide and polypeptide sequences was the central goal of molecular biology during these years. In a nutshell, all of these expectations have in fact been realized and accounts of different phases of this work have recently appeared (see e.g. Ref. 12 and 13). In the following, attention is to be focused primarily on recent work and references will be made to earlier work insofar as these illustrate the principles of the approaches involved.

CHEMICAL SYNTHESIS OF POLYNUCLEOTIDES

The synthesis of all of the sixty-four ribotrinucleotides (14) and the earlier work on the synthesis of deoxyribopolynucleotides containing repeating dinucleotide sequences (15) and repeating trinucleotide sequences (16–18) has all been published. More recent efforts have been devoted (a) to the development of rapid procedures for the polymerization of suitably protected deoxyribotrinucleotides and of the isolation of the corresponding oligomers and (b) to the synthesis of specific deoxyribopolynucleotides by stepwise condensation of preformed di-, tri- and tetranucleotide blocks. Satisfactory progress has resulted in both these areas and the synthesis of the deoxyribopolynucleotides listed in Table I has been accomplished through the enthusiastic and heroic efforts of Drs. S. A. Narang, Hans Kössel, Eiko Ohtsuka, T. M. Jacob and Henry Büchi during the past year (19). The important general point in the list given in Table I to be noted is that the syntheses all comprise complementary (in the antiparallel base-pairing sense) sets of deoxyribopolynucleotides. This general requirement of synthesis of segments of both strands was forced upon us by the failure of DNA polymerase to respond to deoxyribopolynucleotides corresponding to one strand only (19). The second general point regarding the synthesis of deoxypolynucleotides containing repeating trinucleotide sequences is that because earlier work had shown the dependability of the approach as illustrated in Chart 1 for rigid assignment of codon sequences, it was clearly desirable to *prove* the total structure of the genetic code by this method. For a comprehensive attack on the codon assignments and for comprehen-

TABLE I. Synthetic Deoxyribopolynucleotides with Repeating Sequences

Repeating Trinucleotide Sequences

$\begin{Bmatrix} d(TTC)_4 \\ d(AAG)_4 \end{Bmatrix}$ $\begin{Bmatrix} d(TAC)_{4-6} \\ d(TAG)_{4-6} \end{Bmatrix}$

$\begin{Bmatrix} d(TTG)_{4-6} \\ d(CAA)_{4-6} \end{Bmatrix}$ $\begin{Bmatrix} d(ATC)_{3-5} \\ d(ATG)_{3-5} \end{Bmatrix}$

$\begin{Bmatrix} d(CCT)_{3-5} \\ d(GGA)_{3-5} \end{Bmatrix}$ $\begin{Bmatrix} d(GGA)_{3-5} \\ d(CCT)_{3-5} \end{Bmatrix}$

$\begin{Bmatrix} d(CGA)_{3-5} \\ d(CGT)_{3-5} \end{Bmatrix}$

Repeating Tetranucleotide Sequences

$\begin{Bmatrix} d(TTTC)_3 \\ d(AAAG)_{3-4} \end{Bmatrix}$

$\begin{Bmatrix} d(TATC)_3 \\ d(TAGA)_2 \end{Bmatrix}$

$\begin{Bmatrix} d(TTAC)_4 \\ d(TAAG)_2 \end{Bmatrix}$

sive physical studies of DNA-like polymers, the synthesis of a variety of repeating trinucleotide sequences was undertaken. It should be pointed out that the theoretical number of double-stranded DNA-like polymers containing all possible repeating trinucleotide sequences derived *from more than one type of base* is only ten. Of these, seven DNA-like polymers can be generated from the list given in Table I.

Justification for the selection of the repeating tetranucleotide sequences is given in a later section.

DNA POLYMERASE-CATALYZED SYNTHESIS OF DNA-LIKE POLYMERS CONTAINING REPEATING TRI- AND TETRANUCLEOTIDE SEQUENCES

An extremely fortunate and pratically useful feature of the action of DNA-dependent RNA polymerase and of DNA polymerase on synthetic deoxyribopolynucleotides containing repeating sequences is that the products, while obeying the Watson–Crick base-pairing rules, are much longer in chain lengths than those of the synthetic templates (20–22). Of the two polymerases, DNA-polymerase has proved to be much the preferred tool for the amplification and gross multiplication of the chemically generated repeating nucleotide sequences and our recent preoccupation has therefore been mostly with the DNA polymerase, the progress in this phase of the work being due to the determined efforts of Dr. R. D. Wells. Table II lists the types of reactions so far elicited from DNA-polymerase. Reaction (1) leading to the synthesis of DNA-like polymers with repeating dinucleotide sequences has been fully documented previously (20, 21) and attention is here devoted to reactions (2) and (3). As an example of the reaction of type (2), we may cite the kinetics of the synthesis of poly d-TTC:GAA (Figure 1). The most direct means for showing the chemical structure of the polymer

TABLE II. Types of Reactions Catalyzed by DNA-Polymerase

(1) $d(TG)_6 + d(AC)_6 + \begin{Bmatrix} dTTP \\ dATP \\ dCTP \\ dGTP \end{Bmatrix} \longrightarrow$ Poly d-TG:CA†

(2) $d(TTC)_4 + d(AAG)_3 + \begin{Bmatrix} dTTP \\ dATP \\ dCTP \\ dGTP \end{Bmatrix} \longrightarrow$ Poly d-TTC:GAA†

(3) $d(TATC)_3 + d(TAGA)_2 + \begin{Bmatrix} dTTP \\ dATP \\ dCTP \\ dGTP \end{Bmatrix} \longrightarrow$ Poly d-TATC:GATA†

† All of the DNA-like polymers are written so that the colon separates the two complementary strands. The complementary sequences in the individual strands are written so that antiparallel base-pairing is evident.

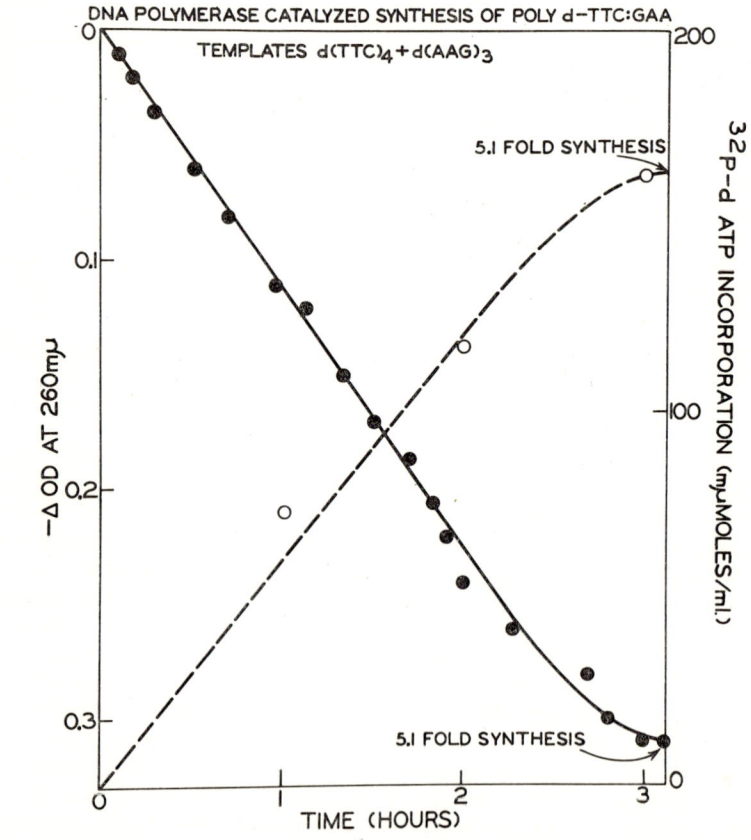

FIGURE 1. Kinetics of DNA-polymerase catalyzed synthesis of Poly d-TTC:GAA

is by nearest neighbor analysis using one α-^{22}P-labeled deoxynucleoside triphosphate at a time and degradation of the isolated polymer to deoxyribonucleoside 3'-phosphates. The results which are shown in Table III show *complete* accord with theoretical expectation for a polymer with strictly

TABLE III. Nearest Neighbor Frequency Analysis of Poly d-TTC:GAA
Templates: d(TTC)₄+d(AAG)₃

α-^{32}P-labeled Triphosphate	Radioactivity in deoxynucleoside 3'-phosphates							
	dAp		dGp		dCp		dTp	
	cpm	%	cpm	%	cpm	%	cpm	%
dATP	12,836	50.0	12,851	50.0	0	0	0	0
dGTP	13,684	100	0	0	0	0	0	0
dCTP	0	0	0	0	0	0	9,623	100
dTTP	0	0	0	0	12,860	50.6	12,565	49.4

repeating trinucleotide sequences in the complementary strands. As an example of a DNA-like polymer with repeating tetranucleotide sequence we show the kinetics of the incorporation of isotope from α-^{32}P-labeled TTP into acid-insoluble product using d(TATC)₃+d(TAGA)₂ as templates

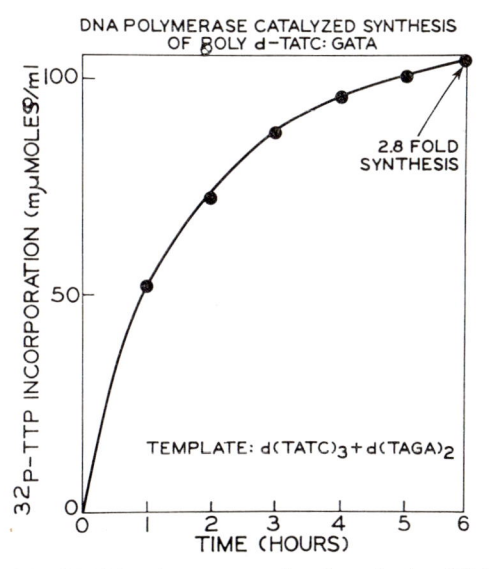

FIGURE 2. Kinetics of DNA-polymerase catalyzed synthesis of Poly d-TATC:GATA.

(Figure 2). Here again, the reaction starts off without any lag period and leads to net synthesis. Nearest neighbor frequency analysis as seen in Table IV gives within experimental error excellent agreement between experimentally observed and theoretically expected results for repeating tetranucleotide

TABLE IV. Nearest Neighbor Frequency Analysis of Poly d-TATC:GATA Templates: d(TATC)₃+d(TAGA)₂

α-^{32}P-labeled Triphosphate	Radioactivity in deoxynucleoside 3′-phosphates							
	dAp		dGp		dCp		dTp	
	cpm	%	cpm	%	cpm	%	cpm	%
dATP	0	0	6425	35.0	0	0	11,938	65.0
dGTP	22,467	100	0	0	0	0	0	0
dCTP	34	0.2	0	0	0	0	16,682	99.8
dTTP	15,920	66.6	0	0	7.985	33.4	0	0

sequence. It should be further added that with every repeating DNA-like polymer that has been prepared, the individual strands have been copied by RNA polymerase (see below) and the ribopolynucleotide products when subjected to the nearest neighbor analysis technique give results in *complete* accord with theory. A double check is thus provided for the conclusion that all of the DNA-like polymers prepared have the strictly repeating patterns of nucleotide sequences originally provided in the chemically synthesized templates.

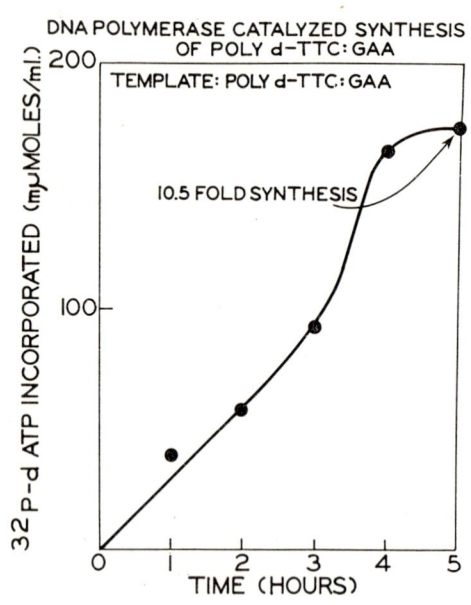

FIGURE 3. Reutilization of poly d-TTC:GAA as template for DNA-polymerase.

A property of the DNA-like polymers, which is of paramount significance, is their ability to re-seed the synthesis by the DNA polymerase of more of the same product. An example of this type of reaction leading to more than 10-fold net synthesis is shown in Figure 3. The same feature has previously been established for repeating dinucleotide polymers (20). The importance of this finding can be hardly overstressed: once the specific sequences have been put together by well-defined and unambiguous chemical synthesis, DNA-polymorase ensures their permanent availability, an expected but, nevertheless, dramatic feature of DNA-structure, namely, its ability to guide its own replication.

TABLE V. New DNA-like Polymers with Repeating Sequences

Repeating Trinucleotide Sequences	Repeating Tetranucleotide Sequences
Poly d-TTC:GAA	Poly d-TTTC:GAAA
Poly d-TTG:CAA	
Poly d-TAC:GTA	
Poly d-ATC:GAT	Poly d-TATC:GATA

The total DNA-like polymers prepared so far are listed in Table V and without going into individual details we can summarize the total characteristics of the DNA polymerase catalyzed reactions studied to date: (1) Chemically synthesized segments corresponding to both strands are required for reactions to proceed; (2) Minimal size of the two complementary segments used as primers varies between 8 and 12 nucleotide units; (3) Synthesis is extensive; (4) Products are high mol.wt. and are double-stranded with sharp melting transitions; (5) Nearest neighbor analyses invariably show the individual strands to contain appropriate repeating tri- or tetranucleotide sequences; (6) High mol.wt. products can be reutilized as primers for more synthesis.

SINGLE-STRANDED RIBOPOLYNUCLEOTIDES WITH REPEATING TRI- AND TETRANUCLEOTIDE SEQUENCES

All of the DNA-like polymers prepared contain a maximum of three different bases in every strand. It is therefore possible by providing in the reaction mixture only three appropriate ribonucleoside triphosphates, to restrict the action of RNA polymerase so as to transcribe only one strand at a time. The same principle was used previously in the preparation of single-stranded ribopolynucleotides containing two nucleotides in alternating sequence (23). This expectation has been fully realized in the case of all of the DNA-like polymers listed in Table V. Thus, for example, poly d-TAC:GTA (Chart 2) when exposed to RNA polymerase in the presence of UTP+ATP+GTP gives poly UAG and when this enzymatic reaction is carried out in the presence of UTP+ATP+CTP, the result is the formation of poly UAC. The two ribopolynucleotides have been shown to be polymers of the trinucleotides UpApGp and UpApCp respectively by the technique of nearest neighbor analysis, the results of which are given in Figures 4 and 5. As seen from these data, the transfer of the label in each experiment is exactly as would

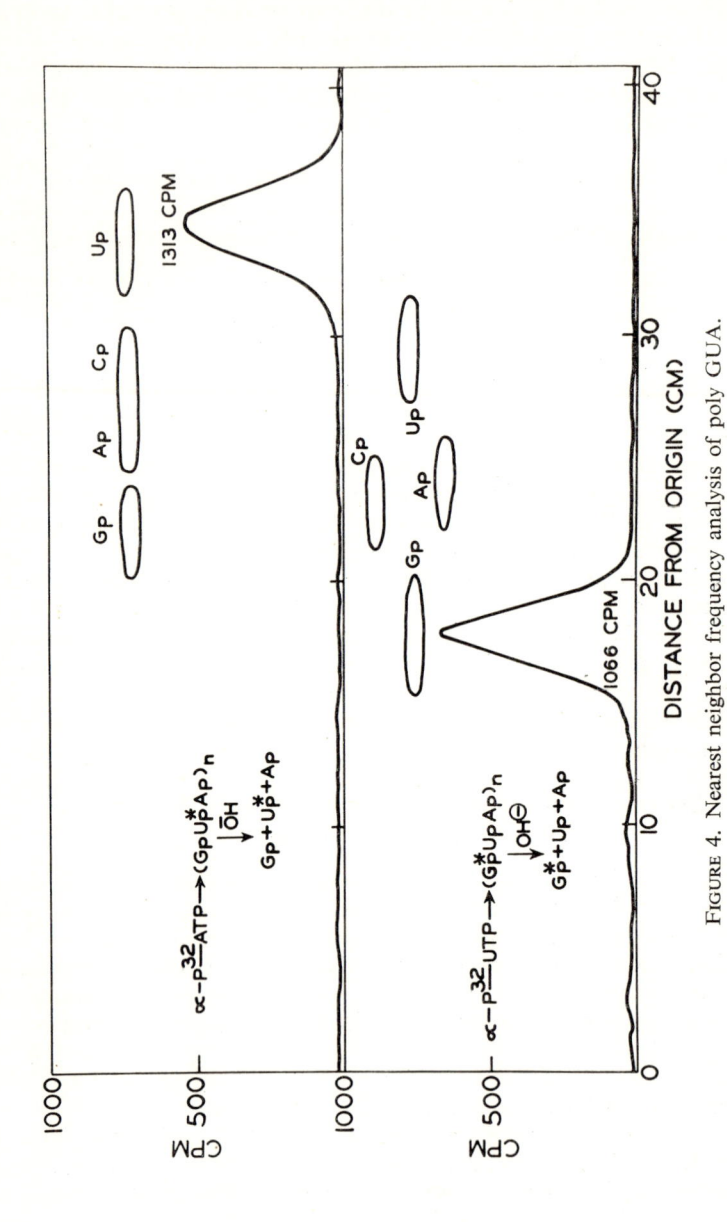

FIGURE 4. Nearest neighbor frequency analysis of poly GUA.

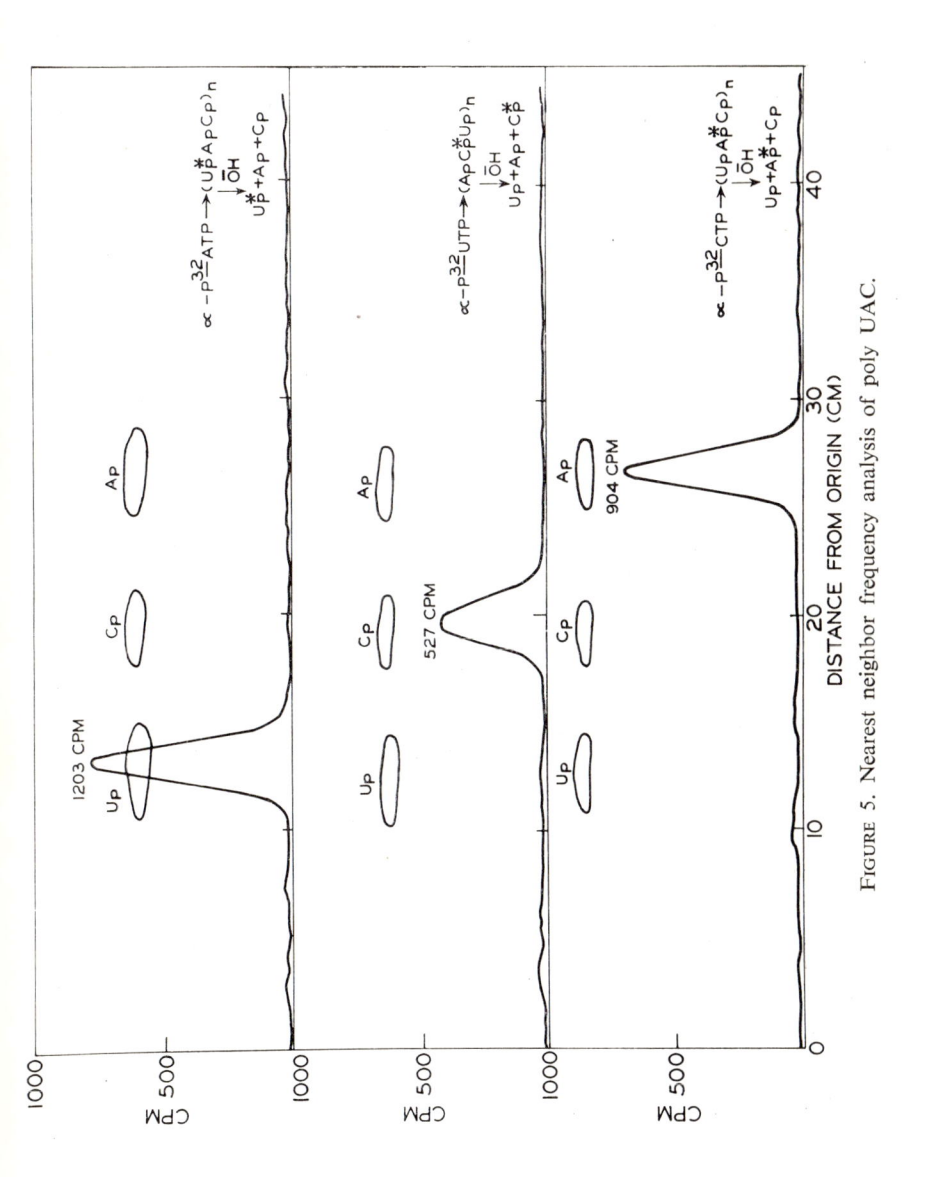

POLYNUCLEOTIDE SYNTHESIS AND THE GENETIC CODE 217

FIGURE 5. Nearest neighbor frequency analysis of poly UAC.

be theoretically expected. There is also evident in these results a further confirmation of the antiparallel base-pairing mechanism in the transcription by RNA polymerase.

The above observations apply exactly to the DNA-like polymer poly d-TATC:GATA and from it either of the two ribopolynucleotides containing the repeating tetranucleotide sequence, poly GAUA and poly UAUC, may be prepared (Chart 2). The results of nearest neighbor frequency analysis shown in Figures 6 and 7 confirm the presence of the repeating tetranucleotide sequence. A further point of practical interest is that the synthesis

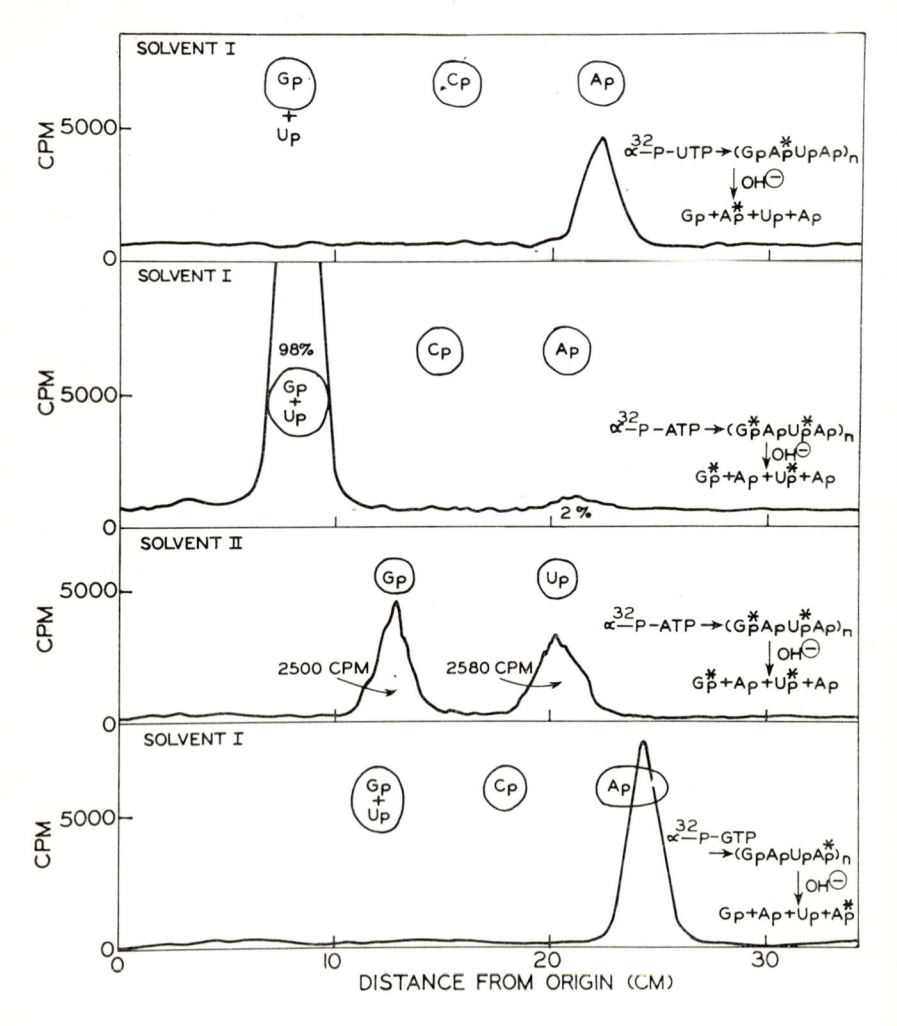

FIGURE 6. Nearest neighbor frequency analysis of poly GAUA.

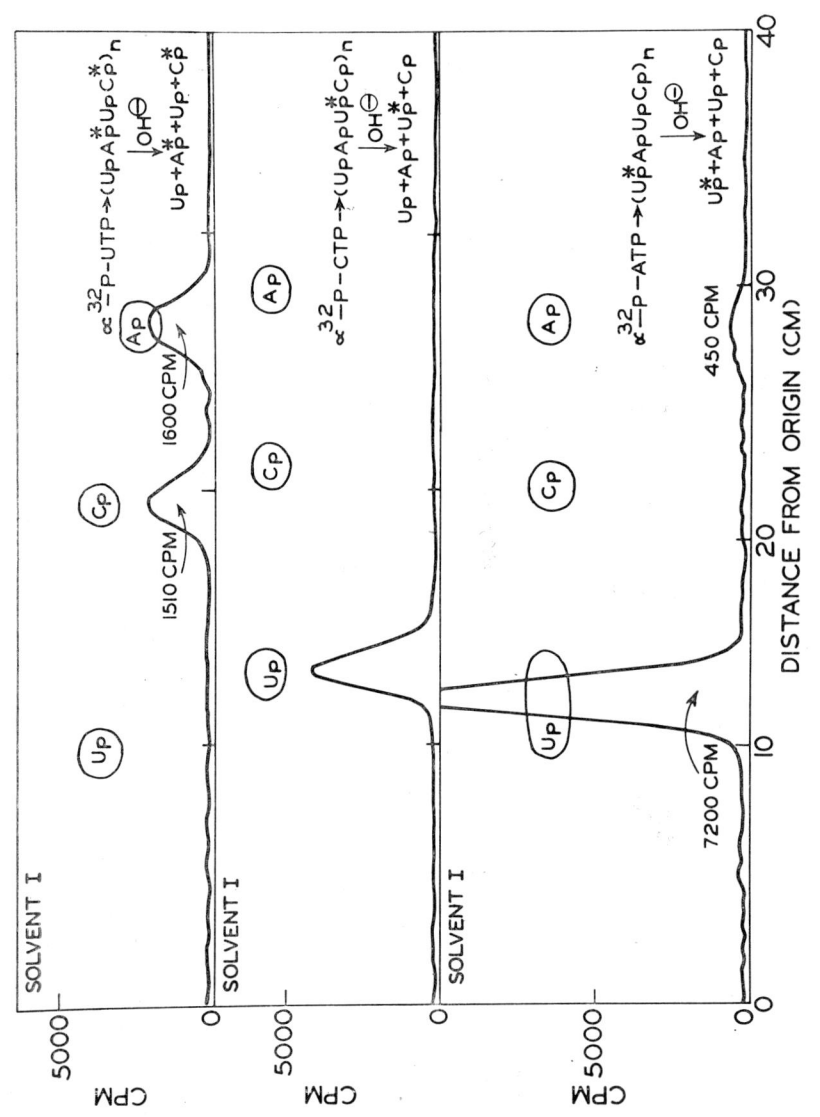

FIGURE 7. Nearest neighbor frequency analysis of poly UAUC.

of the ribopolynucleotides is usually extensive (3–10 fold that of the DNA-template), amount of synthesis showing an increase with an increase in the amount of the enzyme.

CHART 2. Preparation of Single-stranded Ribopolynucleotides from DNA-like Polymers Containing Repeating Tri- and Tetra-nucleotide Sequences

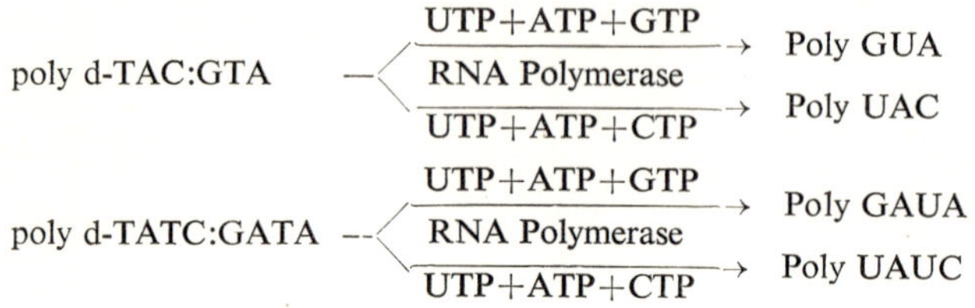

```
                         UTP+ATP+GTP
                        ─────────────→  Poly GUA
    poly d-TAC:GTA   ─⟨  RNA Polymerase
                         UTP+ATP+CTP
                        ─────────────→  Poly UAC

                         UTP+ATP+GTP
                        ─────────────→  Poly GAUA
    poly d-TATC:GATA ─⟨  RNA Polymerase
                         UTP+ATP+CTP
                        ─────────────→  Poly UAUC
```

CELL-FREE AMINO ACID INCORPORATIONS AND THE CODON ASSIGNMENTS

From a combination of the results obtained with poly AAG (formation of polylysine, polyarginine and polyglutamate) and with polynucleotides containing repeating dinucleotide sequences (formation of four copolypeptides, Table VI) direct proof for the three letter and non-overlapping nature

TABLE VI. Polypeptide Synthesis Directed by Polynucleotides of Alternating Sequence

Template Ribopolynucleotide	Copeptide formed	Probable Codon sequence
Poly UC	(Ser-Leu)$_n$	Ser. UCU Leu. CUC
Poly UG	(Cys-Val)$_n$	Val. GUG Cys. UGU
Poly AG	(Arg-Glu)$_n$	Arg. AGA Glu. GAG
Poly AC	(Thr-His)$_n$	Thr. ACA His. CAC

of the genetic code has previously been provided (see e.g. Ref. 20, 23). For assignment of codon sequences to different amino acids, the binding technique developed by Nirenberg and Leder (24) was extensively used by Nirenberg and coworkers (see e.g. 25) and by Söll et al. (26). The total structure of the code which emerged from (a) cell-free polypeptide synthesis of defined sequence, (b), trinucleotide-stimulated binding of aminoacyl-sRNA to ribosomes and (c) other genetical and mutagenic evidence, has been presented and discussed recently. The simple and elegant binding technique which could, in principle, have given information on all of the code, proved, in fact, not to be completely reliable. Often the effects observed were too small and, in certain cases, the results were ambiguous. In other situations, authentic codons were shown not to give any detectable effect in the binding assay (26, 27). It was therefore concluded (13, 26) that further information on those sections of the codon assignments, where uncertainty prevailed, was desirable by the use of repeating polymers. The studies on cell-free polypeptide syntheses as directed by the messengers that are afforded by the DNA-like polymers of Table V are not as yet comprehensive. The results available to date (work of Drs. Richard Morgan, R. D. Wells and Hans Kössel) are presented in Table VII. These when incorporated into

TABLE VII. Amino Acid Incorporations Using Polymers with Repeating Tri- and Tetranucleotide Sequences

RNA	Amino acid	Incorporation (mµmoles) −template	+template	Codon assignment
Poly r-UUC	Phe	0.010	0.15	UUC
,,	Ser	0.044	0.16	UCU
,,	Leu	0.042	0.40	CUU
Poly r-CAA	Gln	0.028	0.11	CAA
,,	Thr	0.11	2.4	ACA
,,	Asn	(deamidation to aspartic acid)		AAC
Poly r-UUG	Cys	0.25	0.79	UGU
,,	Leu	0.068	2.0	UUG
,,	Val	0.050	3.3	GUU
Poly r-GUA	Val	0.024	9.3	GUA
,,	Ser	0.034	0.15	AGU
Poly r-UAC	Tyr	0.052	0.44	UAC
,,	Thr	0.035	4.7	ACU
,,	Leu	0.043	0.20	CUA
Poly r-AUC	Ileu	0.027	3.6	AUC
,,	Ser	0.037	0.17	UCA
,,	His	0.045	0.35	CAU
Poly r-GAU	Asp	0.07	1.770	GAU
Poly r-UAUC	Tyr	0.12	0.88	UAU
,,	Ileu	0.14	0.65	AUC
,,	Ser	0.18	0.70	UCU
,,	Leu	0.22	0.82	CUA

the previous data on codon assignments, lead to the codon catolog shown in Table VIII. The following especially interesting points may be noted. (1) Poly UAG stimulates, in the incorporating system prepared from *E. coli* B, a non-permissive strain, the incorporation of valine (GUA) and serine (AGU). AGU is clearly a codon for serine, the only reliable previous evidence on this point being that of Okada and Streisinger (25, 28). (2) CUA and CUU are both now shown to be codons for leucine, the binding experiments in these cases again failed to give any detectable effects. (3) Poly GAU stimulated the incorporation of methionine and of aspartic acid. These findings confirm the codon assignments AUG and GAU for methionine and aspartic acid respectively. A homopolypeptide corresponding to the third repeating codon (UGA) present in poly GAU should also have been observed. Incorporation of no other amino acid has, however, been detected so far. Since AUG is also a codon for peptide chain initiation (29) and may be expected to fix the reading frame, poly GAU might have stimulated the incorporation of methionine only. However, as mentioned above, under the conditions used in the present experiment, aspartic acid (0.01 M Mg^{++}) was also in-

TABLE VIII. Codon Assignments from Polypeptide Synthesis and/or Stimulation of Aminoacyl-sRNA Binding to Ribosomes (March 1966)

1st	U (2nd)	C	A	G	3rd
U	Phe	Ser	Tyr	Cys	U
	Phe	Ser	Tyr	Cys	C
	?	Ser	Ochre	?	A
	Leu	Ser	Amber	Try	G
C	Leu	Pro	His	Arg	U
	Leu	Pro	His	Arg	C
	Leu	Pro	Gln	Arg	A
	Leu	Pro	Gln	Arg	G
A	Ileu	Thr	Asn	Ser	U
	Ileu	Thr	Asn	Ser	C
	?	Thr	Lys	Arg	A
	Met	Thr	Lys	Arg	G
G	Val	Ala	Asp	Gly	U
	Val	Ala	Asp	Gly	C
	Val	Ala	Glu	Gly	A
	Val	Ala	Glu	Gly	G

' The assignments not underlined are on the basis of binding experiments only. The assignments singly underlined are on the basis of copolypeptide and/or homopolypeptide syntheses and gave essentially no binding. The assignments doubly underlined are derived from both polypeptide synthesis and binding experiments.

corporated and the product of this incorporation has been characterized as poly-aspartic acid. The failure to detect so far the incorporation of a third amino acid in the presence of poly GAU is, therefore, probably not due to the phasing effect of AUG (4). The repeating tetranucleotide polymer, poly UAUC, stimulated the incorporation of four amino acids in agreement with the theoretical expectation that a repeating tetrapeptide should be formed. While detailed sequential analysis of the polypeptidic product has yet to be performed, the indications are that the four amino acids are, *in fact*, present so as to give the repeating tetrapeptide sequence.

The repeating polymers derivable from the DNA-like polymers of Table V are potentially capable of establishing with complete certainty the remainder of the genetic code. Prominent trinucleotides for which assignments at this moment are completely uncertain are UUA, AUA and UGA. In addition, confirmation of UAA and UAG as the chain-terminating triplets in non-permissive strains would be highly desirable. All of these answers may be expected in the near future from the repeating polymers in Table V.

Specificity of sRNA for Recognition of Codons

Many of the features of the genetic code as they have become apparent have been commented on before (see e.g. 13, 26). One major comment relevant to the present discussion that can be made is that most of the possible sixty-four triplets are meaningful codons. This conclusion has focussed attention on the question: Is there invariably a discrete sRNA species for the recognition of every codon or can one sRNA species, under certain conditions, recognize more than one codon? That the latter may be the case is suggested, for example, by the work on alanine–sRNA where purification and structural analysis has shown only one sRNA species in yeast for this amino acid. In our laboratory, this question has been studied principally by Dr. Dieter Söll using the trinucleotide-stimulated binding to ribosomes of aminoacyl–sRNA's prepared from purified sRNA fractions of yeast and *E. coli* (30). The results which are briefly described below, show that one sRNA can often recognize multiple codons provided they differ in the third letter only. The observed patterns for multiple codon recognition are consistent with the postulates of the Wobble hypothesis recently proposed by Francis Crick (31).

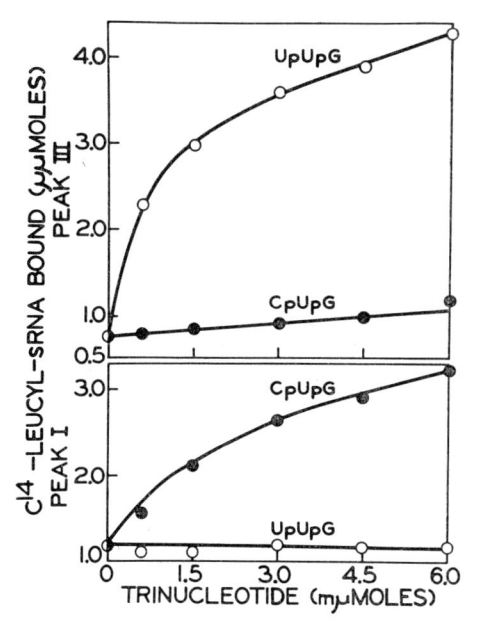

FIGURE 8. Stimulation of the binding of *E. coli* ^{14}C-leucyl-sRNA species I and III to ribosomes by the trinucleotides UpUpG and CpUpG.

In the table of codon assignments (Table VIII), leucine and arginine codons fall in two "boxes" and thus contain two sets of codons differing in first letter. (Serine takes an exceptionally big jump, with two sets of codons differing in first and second letter.) We first studied the question of sRNA-specificity for codons differing in first letter. The results of binding experiments (Figure 8) using leucine–sRNA–I and leucine–sRNA–III (*E. coli*) show striking specificity for CUG and UUG respectively. Results obtained using purified arginine sRNA (yeast) fractions (Figure 9) showed similarly striking specificity, one sRNA fraction recognizing codons beginning with the letter A and another sRNA fraction recognizing the set of codons beginning with the letter C. So, we conclude that whenever a change in the first letter of the codon occurs, a new sRNA species would be required to recognize it.

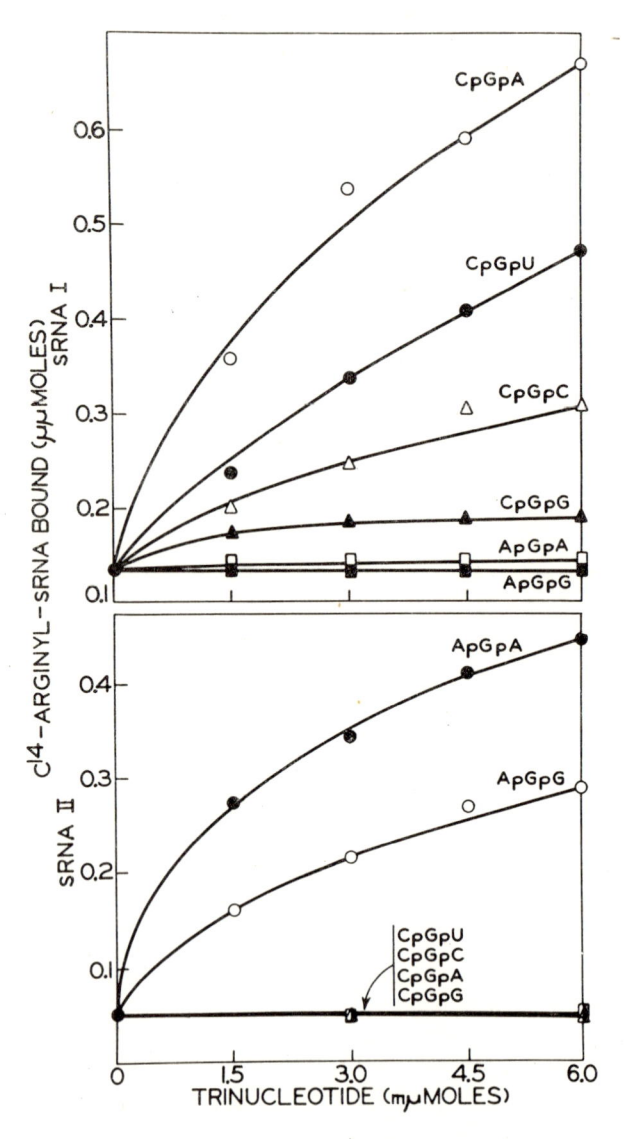

FIGURE 9. Stimulation of the binding of yeast ^{14}C-arginyl-sRNA's I and II to ribosomes by the trinucleotides CpGpU, CpGpC, CpGpA, CpGpG, ApGpA and ApGpG.

On the other hand, several examples of multiple codon recognition when a change in the third letter is involved have been uncovered. In fact, the data of Figure 9 also show clearly that one arginine–sRNA species can recognize AGA and AGG and another sRNA species recognizes the set of codons, CGA, CGU and CGC. Figure 10 shows again the ability of one purified alanine–sRNA species (yeast) to recognize three codons, GCU, GCA and GCC, no response being observed for GCG. The patterns of multiple recognition evident from these examples are (1) that the letters U, C and A in the third position may be recognized as a group and (2) that the letters A and G may similarly be recognized as a group. Further examples of multiple recognition are provided by the two purified peaks of serine–sRNA (*E. coli*).

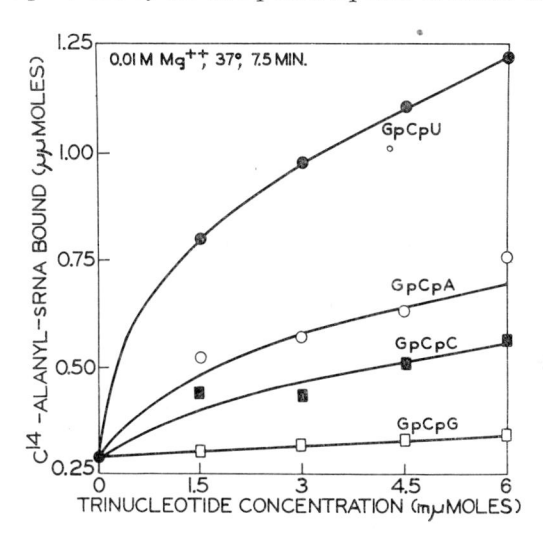

FIGURE 10. Stimulation of the binding of yeast ^{14}C-alanyl-sRNA to ribosomes by the trinucleotides GpCpU, GpCpC, GpCpA, and GpCpG.

The results shown in Figure 11 show that one peak can recognize UCU and UCC while the second peak recognizes well UCG and UCA mainly. The conclusion that in the third place pyrimidines can be exchanged is also supported by results with purified phenylalanine sRNA (yeast) (30) which can recognize both UUC and UUU. Finally, Figure 12 shows the recognition patterns of the multiple peaks for glycine sRNA (yeast). Peak I recognizes GGA and probably GGG but not the codons with the third letter U or C. Peak II recognizes GGC, GGU and perhaps GGA but not GGG. Peak III best recognizes GGG and appears to be specific for this codon, the additional response to other codons probably being the results of contamination by peak II. Peak IV gave results very similar to those of peak II. (Peak IV appears to be an aggregate of peak II, Ref. 30).

Table IX summarizes the total results on the specificity of sRNA for recognition of codons. One important consequence of these results is that

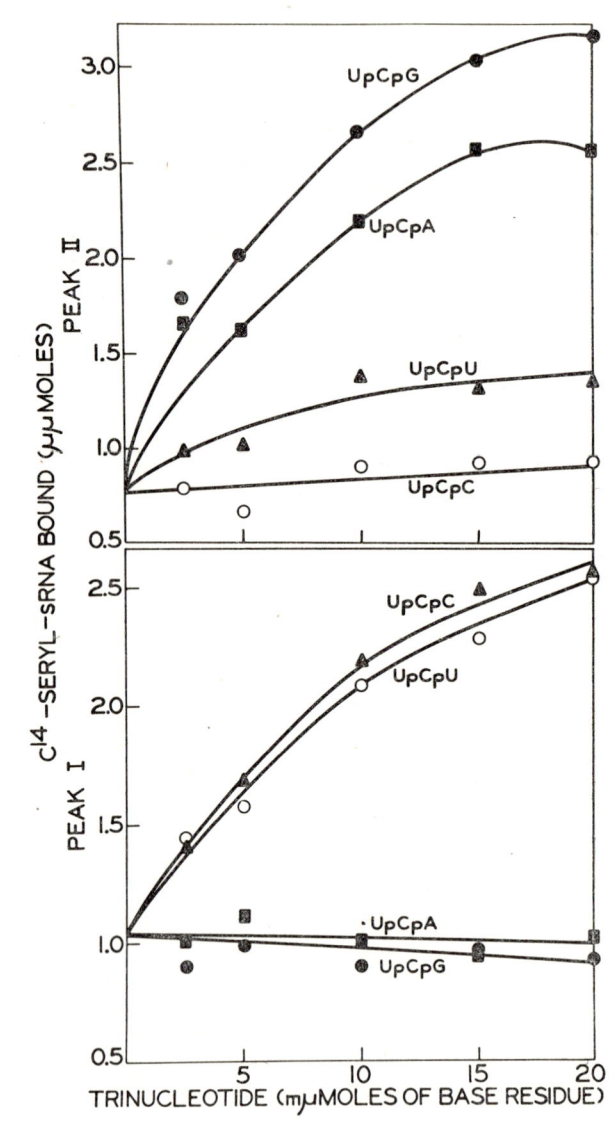

FIGURE 11. Stimulation of the binding of *E. coli* ^{14}C-seryl-sRNA peaks I and II to ribosomes by the trinucleotides UpCpU, UpCpC, UpCpA and UpCpG.

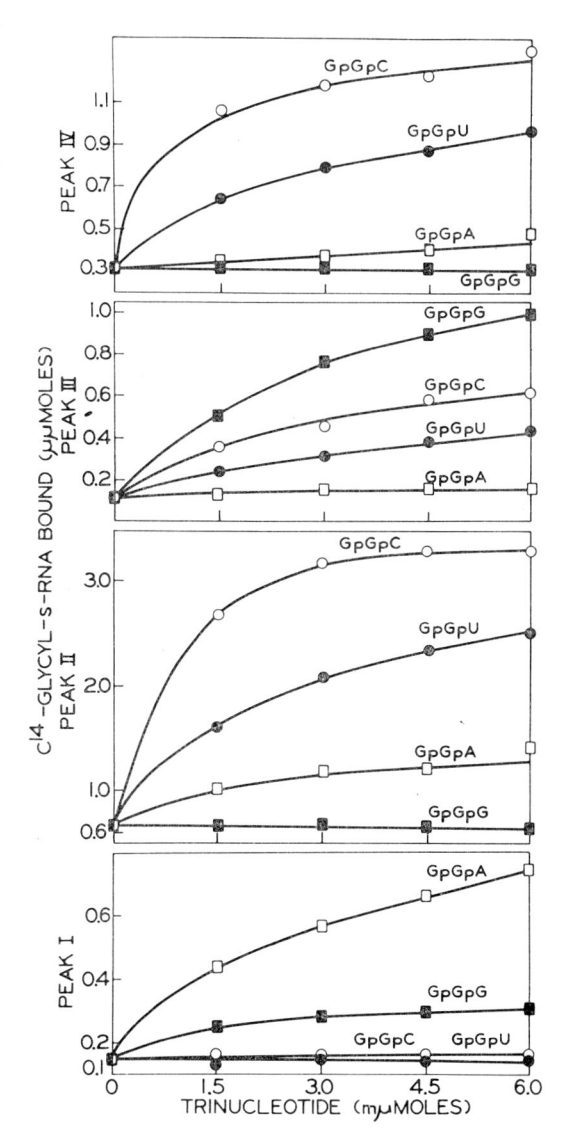

FIGURE 12. Stimulation of the binding of yeast ^{14}C-glycyl-sRNA species to ribosomes by the trinucleotides GpGpU, GpGpC, GpGpA and GpGpG.

TABLE IX. Specificity of sRNA for Codon Recognition

Type of specificity	sRNA	Source	Codons
First letter change	Arg-I	Yeast	CG(U, C, A)
	Arg-II	Yeast	AG(A, G)
	Leu-I	E. coli	CUG
	Leu-III	E. coli	UUG
Third letter; U or C	Gly-II	Yeast	GG(U, C)
	Ileu	E. coli	AU(U, C)
	Phe	Yeast	UU(U, C)
	Ser-I	E. coli	UC(U, C)
Third letter; A or G	Arg-II	Yeast	AG(A, G)
	Gly-I	Yeast	GG(A, G)
	Ser-II	E. coli	UC(A, G)
Third letter; G	Gly	Yeast	GGG
	Leu-I	E. coli	CUG
	Leu-III	E. coli	UUG
	Met	E. coli	AUG
	Try	E. coli	UGG
Third letter; U, C or A	Ala	Yeast	GC(U, C, A)
	Arg-I	Yeast	CG(U, C, A)

the minimum number of sRNA species necessary for the recognition of the total codons is quite small and the total number of sRNA species that is actually found in any organism easily exceeds the above number. The biological significance, especially in regard to control mechanisms, of this important finding of sRNA redundancy has been discussed elsewhere (30).

CHAIN TERMINATION

A large body of information recently appearing from different laboratories has shown that certain triplets can cause termination of polypeptide chains. The sequences UAA and UAG have been concluded for two chain terminating triplets. As mentioned above, confirmation of these sequences would be desirable and, furthermore, polynucleotides with repeating tetranucleotides containing these triplets, commonly designated, respectively, Ochre and Amber triplets, might provide simple and defined systems for further studies of the mechanism of chain termination. It is with these considerations in mind that the chemical synthesis of the deoxypolynucleotides (Table 1) d-(TATC)$_3$ and d-(TTAC)$_4$ were undertaken. When transcribed by RNA polymerase, these would give the polynucleotides (GAUA)$_n$ and (GUAA)$_n$. It is seen that the former contains the Amber triplet and the latter contains the Ochre triplet, these triplets repeating every fourth time. The maximum size of the polypeptides that one would hope to get using these polymers and the amino acid incorporating system from non-permissive strains such as *E. coli* B would be tripeptides.

CHAIN INITIATION

N-Formylmethionine-sRNA is now known to be a chain initiator sRNA in *E. coli* (29) and the chain initiating triplets appear to be of the general sequence XUG. That AUG may fix the frame for translation on the ribosomal surface has been observed in our present work and commented on above. In our repeating polymers, whenever chain initiator triplet is absent, such as in poly AC or poly UAC, the reading can apparently start anywhere along the polynucleotide chain. Further interesting results have recently been obtained with poly UG containing the two nucleotides in alternating sequence. The results obtained by Dr. Hara Ghosh and given in Table X

TABLE X. Stimulation of ^{3}H-formyl and Methionine Incorporation by Poly UG

	Expt. I—Poly UG (Cys-Val) ^{3}H-Formyl mµmoles	^{14}C-Met mµmoles	Expt. II—Poly AC (Thr-His) ^{3}H-Formyl mµmoles
Complete	1.9	0.16	0.26
–Template	0.3	0.04	0.26
–Met	0.38	—	
–Val	0.3	0.06	

demonstrate that poly UG, but not poly AC, stimulates the incorporation of the label from formate into acid-insoluble polypeptide. Poly UG also stimulates the incorporation of methionine and, presumably, this amino acid goes into the N-terminal position of the copolypeptide of valine and cysteine. These results provide proof for the first time from incorporation data in cell-free system that GUG is a chain initiating triplet in addition to its being a codon for valine. It is the objective of our current work to prove by peptide sequential analysis that (1) methionine is present at the amino terminal position and (2) whether the peptide sequence then is H$_2$N-met-cys-val-(cys-val)$_n$———.

MECHANISM OF GENETIC SUPPRESSION

A further example of the use of ribopolynucleotides containing repeating dinucleotide sequences as messengers has been on the mechanism of genetic suppression in some of the suppressed strains of *E. coli* mutants which normally make only a defective (CRM) protein A of tryptophane synthetase. From the work of Yanofsky and coworkers it is known (Table XI) that in one case one glycine residue in A protein is replaced by cysteine (mutant A-78) and in another case (A-23) another glycine is replaced by arginine. Suppressed mutants (A-78-Su-9 and A-23-Su-6) restore to a small extent the original glycine in place of cysteine and arginine respectively. In collaboration with Dr. Yanofsky, Dr. Naba Gupta has been able to show that the amino acid incorporating system prepared from *E. coli* B when supple-

TABLE XI. Tryptophane Synthetase (A Protein) (C. Yanofsky et al.)

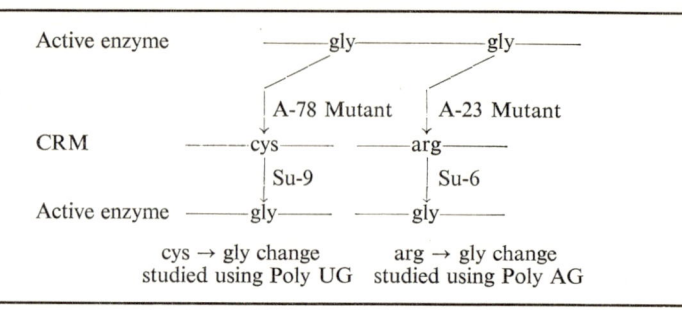

Active enzyme	———gly———	———gly———
	A-78 Mutant ↓	A-23 Mutant ↓
CRM	———cys———	———arg———
	Su-9 ↓	Su-6 ↓
Active enzyme	———gly———	———gly———
	cys → gly change studied using Poly UG	arg → gly change studied using Poly AG

mented with sRNA from A-78-Su-9 brings about the synthesis of valine-glycine copolypeptide under the direction of poly UG. (The latter polymer normally directs the synthesis of valine-cysteine copolypeptide.) Valine–glycine copolypeptide formed *specifically* in the presence of sRNA from A-78-Su-9 strain has been thoroughly characterized. Similarly Drs. J. Carbon and P. Berg have shown that a copolypeptide of glutamic acid–glycine is formed under the direction of poly AG (this normally directs the formation of arginine–glutamic acid copolypeptide) when *E. coli* B system is supplemented by sRNA from A-23-Su-6. The results clearly demonstrate that the genetic suppression in these strains is acting at the level of sRNA.

CONCLUDING REMARKS

The combined use of the methods of organic chemistry and of enzymology has provided a variety of messenger RNA's of completely defined sequence. The use of these in cell-free amino acid incorporating system has provided direct proof for the three letter and non-overlapping properties of the genetic code. Furthermore, the complete characterization of defined polypeptide products has led to unequivocal assignments of a large number of codons to amino acids. It seems very likely that in the very near future extension of the present methods will establish the total structure of the genetic code on a completely certain basis.

The availability of a variety of DNA-like polymers is expected to facilitate further studies of the chemistry and enzymology of DNA and, in particular, of the mechanism of the transcription process.

REFERENCES

1. Brown, D. M. and Todd, A. R., in *Nucleic Acids*, New York, Academic **1**, p. 409 (1955).
2. Watson, J. D. and Crick, F. H. C., *Nature* **171**, 737 (1953).
3. Lehman, I. R., Bessman, M. J., Simms, E. S. and Kornberg, A., *J. Biol. Chem.* **233**, 163 (1958).
4. Weiss, S. B., *Proc. Natl. Acad. Sci. U.S.* **46**, 1020 (1960).
5. Stevens, A., *J. Biol. Chem.* **236**, PC43 (1961).

6. Hurwitz, J., Furth, J. J., Anders, M. and Evans, A., *J. Biol. Chem.* **237**, 3752 (1962).
7. Chamberlin, M. and Berg, P., *Proc. Natl. Acad. Sci. U.S.* **48**, 81 (1962).
8. Burma, D. P., Kroger, H., Ochoa, S., Warner, R. C. and Weill, J. D., *Proc. Natl. Acad. Sci. U.S.* **47**, 749 (1961).
9. Zamecnik, P. C., *Harvey Lectures*, **LIV**, 256 (1958–1959).
10. See e.g. Tissieres, A., Schlessinger, D. and Gros, F., *Proc. Natl. Acad. Sci. U.S.* **46**, 1450 (1960).
11. Matthaei, J. H. and Nirenberg, M. W., *ibid* **47**, 1580 (1961).
12. Khorana, H. Gobind, *Federation Proc.* **24**, 1473 (1965).
13. Khorana, H. G., Jacob, T. M., Moon, M. W., Narang, S. A. and Ohtsuka, E., *J. Am. Chem. Soc.* **87**, 2954 (1965).
14. Lohrmann, R., Söll, D., Hayatsu, H., Ohtsuka, E. and Khorana, H. G., *J. Am. Chem. Soc.* **88**, 819 (1966).
15. Ohtsuka, E., Moon, M. W. and Khorana, H. G., *J. Am. Chem. Soc.* **87**, 2956 (1965).
16. Jacob, T. M. and Khorana, H. G., *J. Am. Chem. Soc.* **87**, 2971 (1965).
17. Narang, S. A. and Khorana, H. G., *J. Am. Chem. Soc.* **87**, 2981 (1965).
18. Narang, S. A., Jacob, T. M. and Khorana, H. G., *J. Am. Chem. Soc.* **87**, 2988 (1965).
19. Wells, R. D., Jacob, T. M., Kossel, H. R., Morgan, A. R., Narang, S. A., Ohtsuka, E. and Khorana, H. G., *Federation Proc.* **25**, 404 (1966).
20. Wells, Robert D., Ohtsuka, E. and Khorana, H. G., *J. Mol. Biol.* **14**, 221 (1965).
21. Byrd, C., Ohtsuka, E., Moon, M. W. and Khorana, H. G., *Proc. Natl. Acad. Sci. U.S.* **53**, 79 (1965).
22. Nishimura, S., Jacob, T. M. and Khorana, H. G., *Proc. Natl. Acad. Sci. U.S.* **52**, 1494 (1964).
23. Nishimura, S., Jones, D. S. and Khorana, H. G., *J. Mol. Biol.* **13**, 302 (1965).
24. Nirenberg, M. W. and Leder, P., *Science* **145**, 1399 (1964).
25. Nirenberg, M., Leder, P., Bernfield, M., Brimacombe, R., Trupin, J., Rottman, F. and O'Neal, C., *Proc. Natl. Acad. Sci. U.S.* **53**, 1161 (1965).
26. Söll, D., Ohtsuka, E., Jones, D. S., Lohrmann, R., Hayatsu, H., Nishimura, S. and Khorana, H. G., *Proc. Natl. Acad. Sci. U.S.* **54**, 1378 (1965).
27. Brimacombe, R., Trupin, J., Nirenberg, M., Leder, P., Bernfield, M. and Jaouni, T., *Proc. Natl. Acad. Sci. U.S.* **54**, 954 (1965).
28. Okada, Y. and Streisinger, G., Unpublished work.
29. Clark, B. F. C. and Marcker, K. A., *J. Mol. Biol.* **17**, 394 (1966).
30. Söll, D., Jones, D. S., Ohtsuka, E., Faulkner, R. D., Lohrmann, R., Hayatsu, H. and Khorana, H.G.; Cherayil, J.D., Hampel A. and Bock, R.M., *J. Mol. Biol.* **19**, 556 (1966).
31. Crick, F. H. C., *J. Mol. Biol.* (In press).

DISCUSSION

(Chairman: F. SANGER)

PHILIP LEDER: In view of the GUG codeword for methionine and your proposal about the existence of a new species of sRNA for a GUG–AUG type degeneracy, do you feel that there will be more than one methionine initiator sRNA?

H. G. KHORANA: All we can say at this time from our polypeptide experiments using poly-UG is that GUG is an initiator triplet and that there is a specific formyl–methionine–sRNA which recognizes this triplet. If AUG can be shown by similar amino acid incorporation experiments also to initiate polypeptide chains in the same way, then certainly our present data on codon-anticodon recognition patterns predict the presence of a second initiator sRNA.

H. MATTHAEI: Poly dinucleotides with complementary bases are inactive templates in polypeptide synthesis. How much inhibition of template activity is observed in polynucleotides, if 2 of 3 bases are complementary?

H. G. KHORANA: Many of the ribopolynucleotide polymers with repeating trinucleotide sequences that we have prepared have the possibility of making 2/3rd of total possible hydrogen bonds by forming hair-pin structures. This possibility of course was of great concern to us at the outset of our present work. But as you have seen from our data, for example with poly-UAC, no difficulty was observed in getting all of the expected amino acid incorporations. This is fortunate because it is now possible to prove all of the code by specific polypeptide synthesis using our approach.

B. F. C. CLARK: How did the chain length of the polypeptide product directed by poly-$(UG)_n$ in the presence of formyl–methionyl–sRNA compare with that of the product (as already published) made in the absence of formyl–methionyl–sRNA.

H. G. KHORANA: We have no quantitative data on the chain length of polypeptides directed by poly-UG in the presence and in the absence of formyl–methionyl–sRNA. It is our impression, however, that in the presence of the initiator sRNA, shorter polypeptide chains are being made. Using ^{3}H-formate and ^{14}C-valine in the same tube and measuring the ratio of the incorporation of the two labels, the average chain length of polypeptide chains formed is about 12 units.

ADDENDUM (added in proof): (1) In regard to the total codon assignments, (Table VIII), the assignment UUA for leucine is confirmed from more recent work with repeating polymers and the assignment AUA for isoleucine is also now established [Khorana, H. G., Büchi, H., Ghosh, H., Gupta, N., Jacob, T. M., Kössel, H., Morgan, R., Narang, S. A., Ohtsuka, E. and Wells, R. D., Cold Spring Harbor Symp. Quant. Biol., June 1966 (In press)]. The only triplet for which an assignment is lacking at this time is UGA.

(2) In more recent work with repeating polymers poly UG, poly GAU, and poly UUG, the assignments of AUG and GUG as chain initiating triplets has been deduced.

Analysis of the Genetic Code by Amino Acid Adapting

H. MATTHAEI, G. HELLER, H. P. VOIGT, R. NETH[*], G. SCHÖCH AND
H. KÜBLER

From the Max-Planck-Institute for Experimental Medicine, Göttingen, Germany

The sequential arrangement of building blocks in bio-macromolecules is guided directly or indirectly by genes. The decoding system of the cell builds the amino acid sequences of the proteins. All primary gene products known, however, are ribonucleic acids (RNA), which in turn have essential functions in the synthesis of the secondary gene products, the proteins. The biochemical specificities of nucleic acids controlling these synthetic processes in

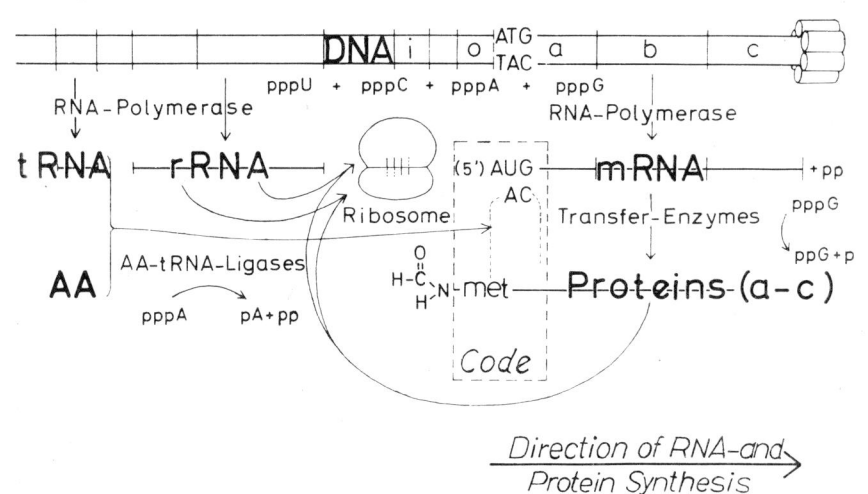

FIGURE 1. Diagram illustrating the role of nucleotides in the cellular decoding system for the synthesis of primary (RNA) and secondary gene products (proteins) (15).

a template translating mechanism appear to reside mainly in properties of their nucleic acid bases yielding the complementary pairs A:U (T) and G:C, and perhaps some others[†]. Figure 1 indicates the role of nucleotides in the decoding system and also the enzymatic devices belonging to it. The set of rather indirect correspondences between nucleotide triplets in mRNA

[*] On leave from Universitäts-Kinderklinik, Hamburg-Eppendorf.
[†] Abbreviations used are listed in Table III.

and amino acids coded thereby, is referred to in a narrow sense as the genetic code (16). After the recent finding of a base complementary antiparallel sequence standing in analogous position in each tRNA nucleotide sequence established (6, 28, 1), the successful adaptor-hypothesis of Crick is very likely to be true once again.

Deciphering the code started five years ago. A first coding unit was found in an assay system for mRNA (8) polymerizing phenylalanine under the direction of polyuridylic acid (10, 19). Two years later, extensive use of random polynucleotides in this polypeptide synthesizing system led to the knowledge of 49 nucleotide compositions of coding units for 20 amino acids. All nucleotide sequences of these units remained unknown, however. The search for approaches using again functional biochemical analysis seemed more promising by that time than a direct analysis of mRNA nucleotide sequences (9). Several laboratories independently and simultaneously developed a simpler system in which mRNA-directed binding of aaRNA to ribosomes occurred (12, 21, further references in 16). To characterize the two binding processes going on in this system, it was called the "adapting" system (12). Nirenberg and Leder first reported the enormous technical simplification and improvement of the assay through adsorption of the ribosomes, together with mRNA and labeled aaRNA bound to them, on a filter of cellulose nitrate (21). Figure 2 shows the steps occurring on the ribosome in the adapting system, and indicates the subsequent steps as they probably follow in peptidizing systems (13). The adapting position seems to be on the 30S (12); the transfer position, however, is on the 50S subunit of the ribosome (in Ref. 13).

Extensive syntheses of all 64 sequences possible among the 4 major RNA nucleotides were essential, besides availability of the adapting system. Figure 3 characterizes in Eq. 3 the enzymic synthesis of polynucleotides started with a dinucleoside monophosphate by polynucleotide phosphorylase. This method was used for trinucleoside diphosphates (2, 22) as well as for longer chains (14, 16). With the poly-oligonucleotides synthesized by Khorana with RNA polymerase on chemo-synthetic oligodeoxynucleotide templates (Figure 3, Equation 4), it has been possible to confirm by peptidizing experiments many triplet assignments in internal rather than terminal position within the mRNA chain (7). Poly-trinucleotides directed the synthesis of three poly-amino acids. For establishing triplet assignments, therefore, it was again useful to consult the adapting system. In this case, 64 chemo-synthetic trinucleoside diphosphates were tested (24) against the amino acid likely to be coded, because of the codon base compositions known. It is satisfactory to have the cell-free decoding results thus based on a variety of different approaches. Complete results of testing all 64 triplets against all 20 amino acids are still to be published, however, in order to demonstrate the sufficiently high degree of unambiguous specificity reached in the cell-free systems used.

The purpose of this paper is to extend previous reports (14, 16) by adding the full table of molar amounts of amino acid adapted to 61 of the 64 triplets, on which amino acid assignments could be based. Many improvements

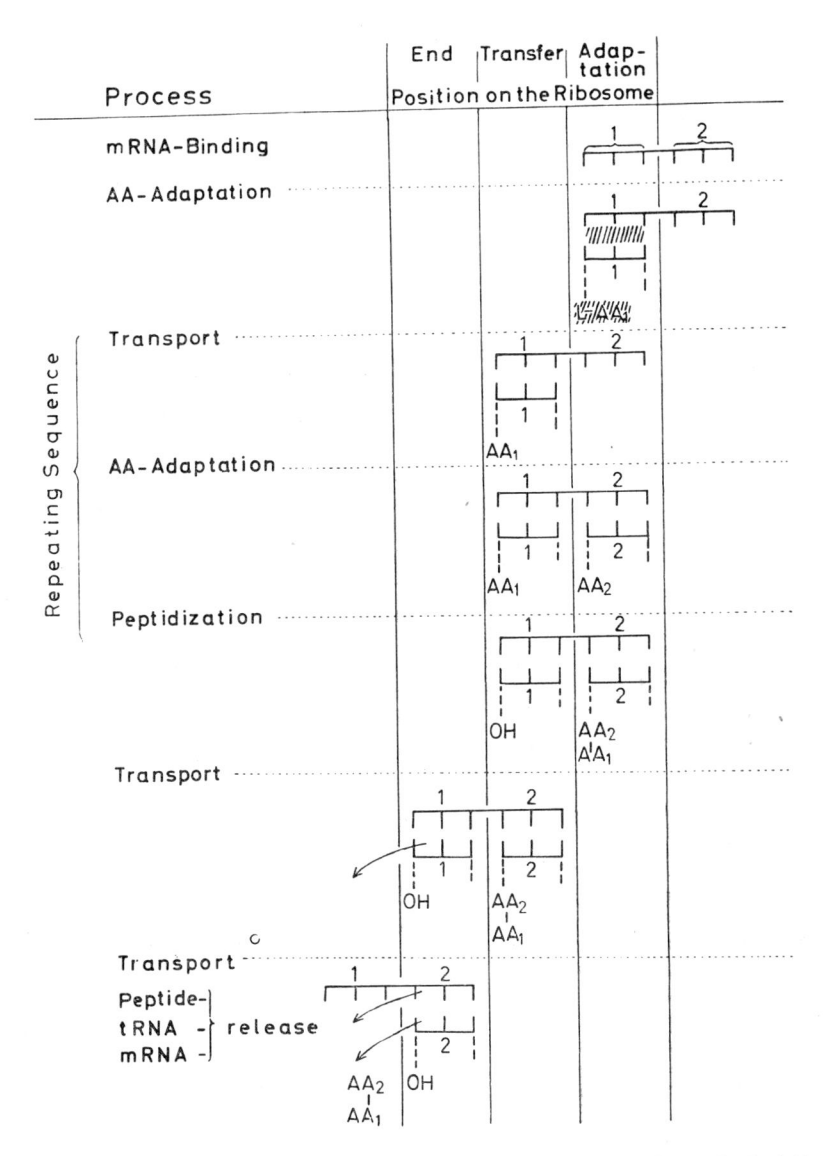

FIGURE 2. Our current working hypothesis on subsequent steps in protein synthesis (13). In the adapting system, only the first two processes occur: binding of mRNA and of aminoacyl RNA dependent thereon to the ribosome which, for assaying, is bound to a cellulose nitrate filter (21).

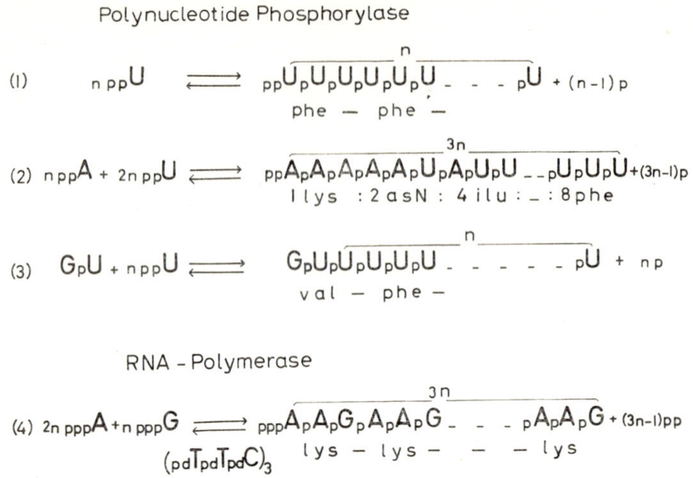

Polynucleotide Phosphorylase

$$(1) \quad n\,ppU \rightleftharpoons \overbrace{ppU_pU_pU_pU_pU_pU\_\_\_\_pU}^{n} + (n-1)p$$
$$\quad\quad\quad\quad\quad\quad\quad phe - phe -$$

$$(2) \quad n\,ppA + 2n\,ppU \rightleftharpoons \overbrace{ppA_pA_pA_pA_pA_pU_pA_pU_pU\_\_pU_pU_pU}^{3n} + (3n-1)p$$
$$\quad\quad\quad\quad\quad\quad\quad 1\,lys\;:2\,asN:\;4\,ilu:\_:8phe$$

$$(3) \quad GpU + n\,ppU \rightleftharpoons \overbrace{G_pU_pU_pU_pU_pU\_\_\_\_\_pU}^{n} + np$$
$$\quad\quad\quad\quad\quad\quad\quad val - phe -$$

RNA - Polymerase

$$(4) \quad 2n\,pppA + n\,pppG \rightleftharpoons \overbrace{pppA_pA_pG_pA_pA_pG\_\_\_pA_pA_pG}^{3n} + (3n-1)pp$$
$$\quad\quad (pdTpdTpdC)_3 \quad lys - lys - \quad - lys$$

FIGURE 3. Enzymatic syntheses of polynucleotides used for decoding experiments.

in the methodology are described, as they are essential for further interpretation, and may also explain why we succeeded in deciphering a number of codons unassigned so far by other laboratories. Finally, we wish to add a contribution to the problem of universality on the level of interactions between ribosomes and tRNA.

MATERIALS

E. coli cells A19 (RNase I$^-$; strain isolated and kindly supplied by Drs. Gesteland and Watson) were grown as before (9) to 0.7 OD$_{500}$ under limiting, logarithmically increased aeration resulting at 32° in a generation time of 35–40 min. They were chilled with ice-water from outside and harvested, washed and extracted for ribosomes (P100) and 100,000g supernatant (S100) as previously described (8, 9), or dried after washing for extraction of sRNA.

Micrococcus lysodeikticus cells, spray-dried, were purchased from Miles Chemical Comp., Elkhart, Ind., USA.

Nucleoside diphosphates were obtained from Boehringer (UDP, ADP, GDP) and Schwarz (CDP, IDP). Li-salts had to be converted into Na-salts, after elution from DOWEX 50×2.

Dinucleoside monophosphates were mainly products of the Waldhof Company (Mannheim, Germany). IpI and UpU were obtained by deamination of ApA and CpC with NaNO$_2$ at pH 3.5.

Labeled amino acids from Dr. Hempel (H) at the University of Köln, Amersham (A), Calbiochem (C), and Schwarz (S), had the specific activities indicated in Table 1.

Standard salts: Trizma Base reagent grade (Sigma) was neutralized to pH at 20°C with HCl p.a. (E. Merck), redistilled from glass. KCl, NaCl,

(NH₄)₂SO₄, and MnCl₂ were p.a. grades from Merck, recrystallized twice in the presence of EDTA (Fluka), MnCl₂ likewise in the presence of Desferal (Ciba), and once from water. Mg-acetate and NH₄Cl p.a. were from Riedel, MgCl₂ El from Merck, all diamines purest grades from Fluka, except 1,7-diaminoheptane from Schuchardt.

Methods

PURIFICATION OF RIBOSOMES

Assay for exonuclease activity: The system reported previously (16) was sensitized by increasing the specific activity of the ^{14}C-polyuridylic acid substrate to 4 µc/µM nucleotide. The amount of ribosomes tested for exonuclease activity was chosen for zero order kinetics of product released. Descending chromatography (16) was routinely performed for about 30 min

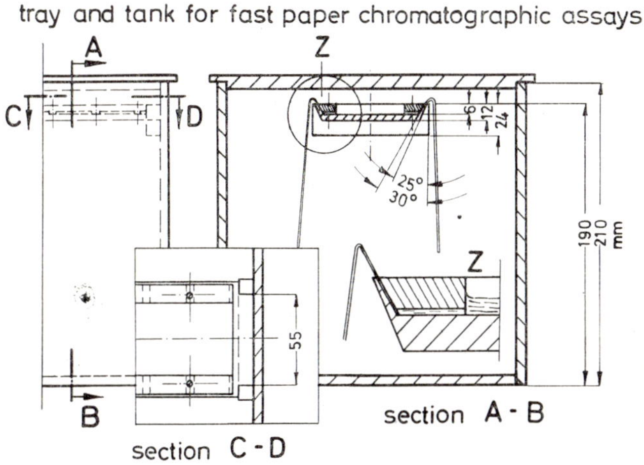

FIGURE 4. Convenient setup for fast separation of labelled products by descending chromatography.

until the solvent front had exceeded the starting spot by 6 cm. The tank and tray described in Figure 4 were used. This is a very useful lab bench design for quick routine chromatographic analysis of many products like polynucleotides, mono- and oligonucleotides, aaRNA, aminoacyl adenosine and amino acids (5). The migrating mononucleotide released was detected under UV and counted in a Tricarb-Spectrometer, using the toluene–PPO–POPOP-scintillation system (12). Counting efficiency was 53% for ¹⁴C on Whatman 3MM paper.

Assay for adapting activity: To determine the adapting activity of ribosomes after each step in the purification procedure, 2.4 A₂₆₀-units (approx. 50 µµ-moles) of 70S ribosomes were incubated for 15 min at 33°C with the saturating amount of 10 µg polyuridylic acid (Takamine) and 800 µµmoles of sRNA charged with 23 µµmoles of ³H-phenylalanine and 19 nonlabeled amino acids in 30 mM Tris-HCl 7.2—6 mM KCl—16 mM MgCl$_2$. Ribosomes from the reaction mixtures were adsorbed on cellulose nitrate filters (Millipore HAWP 0.25μ, 25 mm Ø), washed with 10 mM Tris-HCl 7.2—15 mM MgCl$_2$, dried under infrared light and counted as described.

Washing, MnCl$_2$-dialysis and chromatography on G75 of ribosomes: Details of the washing procedure were described before (14, 16). The media required

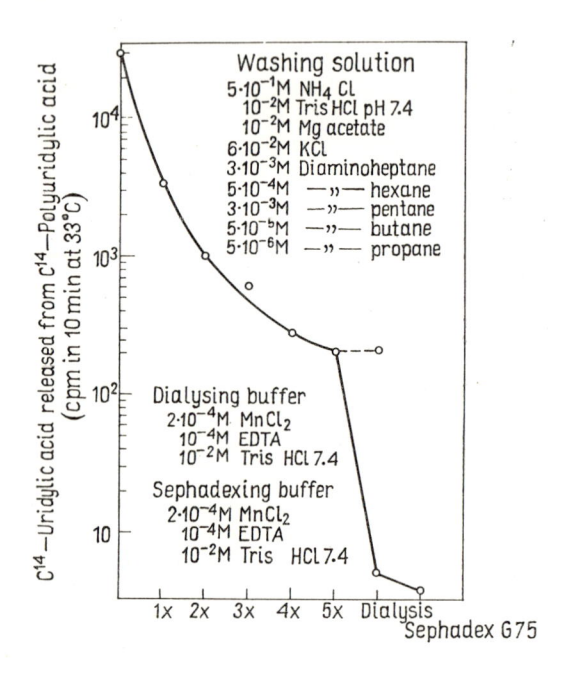

FIGURE 5. Purification of *E. coli* A19 ribosomes from nucleolytic activity in five washing steps, dialysis against MnCl$_2$ and chromatography on SEPHADEX G75. Nuclease was measured by ^{14}C-uridylic acid (sp. act. = 4 μc/μM) released from ^{14}C-polyuridylic acid during 10 min at 33°C. Cpm found in the uridylic acid spot after chromatography of a 25 μl aliquot were corrected for a unit amount of 2.4 OD$_{260}$-units per 100 μl reaction mixture (approx. 50 $\mu\mu$moles) of 70S ribosomes or equivalent subunits.

for washing and a subsequent 20 hour dialysis of the ribosomes redissolved—as after each washing step—in standard buffer (9) to 240 OD$_{260}$-units are indicated in Figure 5. Exonuclease activity of the 5×washed ribosomes is reduced further by a factor of 10 to 20 when it was previously reduced to the level of 1/1000 to 1/500 by the previous washings. In all cases, this washing procedure and dialysis together resulted in only 10^{-4} times the original nucleolytic activity measured (see Figure 5). 1200 OD$_{260}$-units in 5 ml of dialysed ribosome solution were then passed over SEPHADEX G75 (2.5× 30 cm²) with little further purification at this point. Figure 6 shows an initial rise in phenylalanine adapting activity to a level kept nearly constant during 5 washes, MnCl₂ dialysis, and passage over SEPHADEX. The blank adapting in the absence of polynucleotide added is minimized during washing. The specific adapting activity for lysine was equally maintained after this ribosome purification. In the polyphenylalanine synthesizing system (19), these pure ribosomes are also most highly active. Incorporation was linear with time for over 30 min at 37°C and went on for three hours.

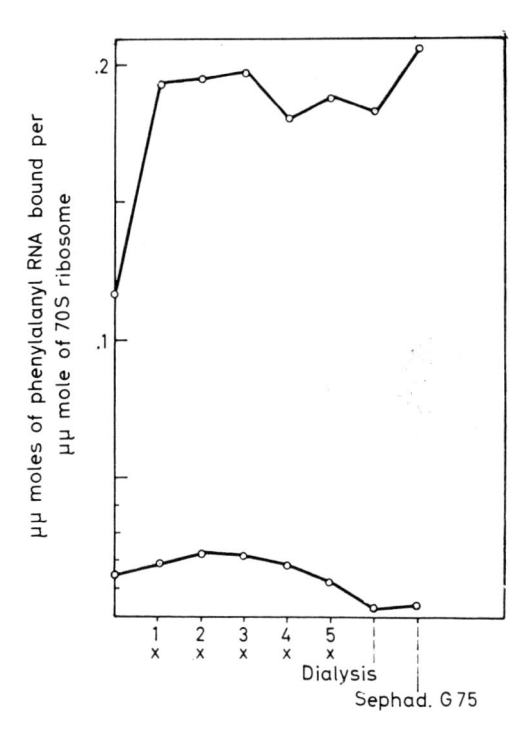

FIGURE 6. Phenylalanine adapting activity after the steps in the ribosome purification described. Curves show binding of phenylalanyl RNA in presence and absence of a saturating amount of polyuridylic acid (for assay conditions see text).

Preparation of human ribosomes: Fetal material (3–4 months) was stored and pulverized under liquid nitrogen in a stainless steel mortar. After thawing, it was gently homogenized further in an ice-chilled motor-driven Potter-Elvehjem type homogenizer by mixing with 3 tissue weights of ice-cold 0.25 M sucrose—1mM MgCl₂—1 mM K·EDTA—30 µg/ml streptomycine sulfate. All subsequent operations were carried out below 4°C. The homogenate was centrifuged for 10 min at 10,000 g, the decanted supernatant again for 20 min at 30,000 g to obtain the supernatant fraction (S30). This S30 was finally spun for 90 min at 133,000 g. The aspirated supernatant (S133) was stored at −35°C. The ribosomal pellets (P133) gave a yield of about 600 OD₂₆₀ units from 50 g of fresh material. P133 was suspended to 100 OD$_{260}$-units per ml in 20 mM Tris/HCl pH 7.4—20 mM NaCl—70 mM KCl—6 mM MgCl$_2$—0.5 mM K·EDTA pH 7—30 μg/ml streptomycine sulfate, after washing pellet surface and tube with this buffer. Ribosomal adapting activity was completely stable in this buffer during storage at −35°C and −45°C for at least a week. Attempts are being made to purify these ribosomes.

Preparation of polynucleotides XpYpZ...pZ. The synthesis of dinucleoside monophosphate primed polynucleotides was previously described (3,16). Using a 100-fold purified polynucleotide phosphorylase preparation, however, the fraction of primer incorporated was substantially lower than the fraction of substrate, particularly in the case of polyadenylic and -inosinic acids. Therefore, we have developed a method for preparing 800–1000-fold purified polynucleotide phosphorylase. The new procedure is also time-saving compared to the method of Singer (23) which has been very helpful to molecular biologists for a number of years. For polymer syntheses with the purer enzyme, the ionic environment had to be changed towards low concentrations in NaCl. Li-salts are very inhibitory to this enzyme preparation.

Purification of polynucleotide phosphorylase. The procedure will only be described in principle here. It involves the following steps:

1. Standard extraction after lysis of 2×20 g spray dried *Micrococcus lysodeikticus* cells with 33% saturated ammonium sulfate (23).
2. Conventional precipitation with 66% saturated ammonium sulfate, sedimentation and redissolving in 40 ml of 10 mM Tris-HCl pH 8.0—1 mM MgCl$_2$—1 mM β-Mercaptoethanol (MEt; "buffer 1"). Purification with this step is 2-fold.
3. Chromatography on three columns of SEPHADEX G200 (each 3.5× ×15 cm²), pumping buffer 1 upwards from one into the next. Purification is 30–40-fold.
4. Centrifugation (3°) for 90 min at 105,000 g and aspiration of the lower three quarters of the supernatant, to purify from ribosomal and topfloating material.
5. Chromatography of 20 ml aliquots on a column of DEAE-SEPHADEX A25 (beads; 1.5×10 cm²) with a linear gradient of 0.05 to 0.40 M NaCl in 2×100 ml of 50 mM Tris-HCl 8.0—1.5 mM MgCl$_2$—0.5 mM K·EDTA pH 7.0—10 mM MEt (buffer 2) and reconcentration after dialysis against buffer 2 on a small pad of SEPHADEX A25 and elution with 0.3 M NaCl in this buffer. Purification is 10–20-fold. The overall purification was 800–1000-fold measured by the UDP-polymerization assay described previously (16), using ApG as a primer in saturating concentrations. Protein was estimated after precipitation from 10% TCA by a microprocedure derived from the method of Lowry using crystalline bovine serum albumin as standard. After the fifth step, the enzyme does not contain nucleolytic activities detectable by the exonuclease procedure described above or by the endonuclease assay of Neu and Heppel (in Ref. 16). Overall yields are around 10 per cent. Further purification was obtained by carrier-free high voltage electrophoresis according to Hannig (see Addendum).

Preparation of polynucleotides. The 800-fold purified polynucleotide phosphorylase preparation Fraction V used in this system had 1200 UDP-polymerization units per mg protein in 1 ml of buffer 1. It was stable in ice for several weeks. Figure 7 shows the complete primer dependence of UDP, CDP, ADP, and IDP polymerization. The substrate incorporation can be adjusted to the pace of primer uptake by rather low concentrations in NaCl

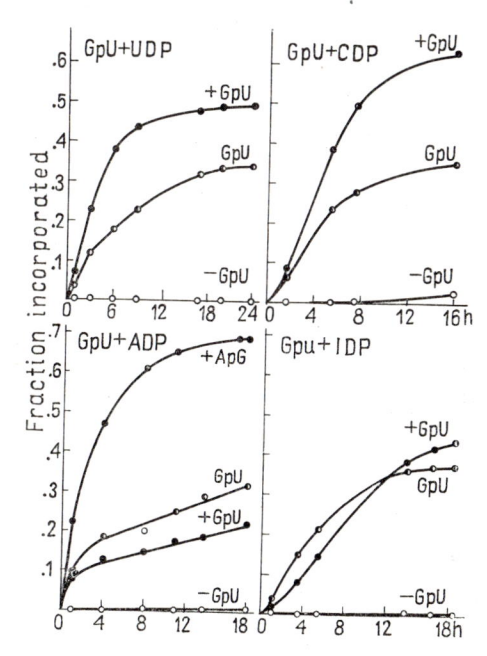

FIGURE 7. Dinucleoside monophosphate primed syntheses of polynucleotides. Primer incorporation was measured by label in G-^{3}H-pU (GpU), and primer dependent substrate incorporation by addition of traces of ^{14}C-UDP (+GpU or +ApG). Circles close to abscissae indicate complete primer dependence of polymerization (for conditions and assay see text).

indicated below. In contrast to the less pure enzyme used before, this preparation polymerizes at different rates in the presence of equal concentrations of different dinucleoside monophosphate primers. This calls for control by accurate nucleotide and chain length analyses, to be published. With the larger fraction of primer incorporated, the new set of Poly-A and -I preparations became more active in coding by the 5′-terminal triplet.

Reaction mixtures for polymer syntheses were incubated at 37° during the time periods, followed by control assays, shown in the curves of Figure 7. They contained in μmoles per actual total volume of 1 ml: 150 Tris 8.8–10 Mg Cl$_2$—0.4 K·EDTA—300 (100 for UDP) NaCl—2.1 μM dinucleoside monophosphate and 42 μM K-UDP or Na-CDP, -ADP, or -IDP, adjusted to pH 7 with KOH—7.5 μg of Fraction V protein and 50 μg of cryst. bovine serum albumine (Serva). All primers and substrates were analysed as described before. At the end of the incubation periods, samples were deproteinized by 4 phenol extractions as described elsewhere (16). The chromatography on columns of SEPHADEX G50 (coarse, beadform; 2×60 cm²) was performed, however, with 0.1 M Tris-HCl pH 7.2 rather than ammonium bicarbonate used earlier, because of hydrolytic effects on aminoacyl RNA, due to impurities in the latter salt (5). The peak fractions containing slightly retarded polynucleotide chains were used for decoding experiments after adjustment with the same buffer to 0.6 μmoles mononucleotide per ml.

Synthesis of amino acyl RNA. We have been improving our preparation and charging methods for sRNA for years, in order to reach complete charging. This may have enabled us last year to easily detect even those coding units which simultaneously were not observed by other laboratories (12, 16). For the first time, we prepared 99% charged sRNA. The RNA was isolated by the method of Berg and purified further as described earlier (16). By the use of Mg·EDTA (5), it could be kept stable. We added a SEPHADEX-step to the procedure described (16), in order to remove contaminating ATP and CTP (5): Charging of 5 mg sRNA was done in 3 ml of the reaction mixture reported (16) with the enzyme preparation described (5). Thereafter, Mg Cl₂ and K·EDTA were added to 10 mM. The chilled samples were deproteinized by 4 shakings with water-saturated phenol at 4°. After one precipitation from 70% ethanol at 0°, the RNA preparations were dissolved in quarz-distilled water and passed over columns of SEPHADEX G25 (fine beads, 1.6×25 cm²), chilled to 4°. SEPHADEX was purified with fresh distilled aqueous phenol and 10 mM K·EDTA pH 7, and washed with 10 mM KCl (suprapure, Merck), which was left at its pH of about 5. Before applying 5 mg aaRNA in 1.6 ml of water to the column, 2.5 ml of K·EDTA were soaked in. The aaRNA was eluted with 10 mM KCl. Table I shows the values calculated for each labeled amino acid esterified in the presence of the 19 nonlabeled other amino acids in percent of sRNA measured by absorption at 260 mμ. Radioactivity was counted after drying 10 μl of the aaRNA eluted from the column onto Whatman GFA glass fibre filters, and converted into molar terms on the basis of steadily controlled counting efficiencies of a Tricarb-liquid-scintillation spectrometer. Eluted aaRNA was adjusted to 0.4 mg (10 A_{260})/ml, and stored frozen in small aliquots at −45°.

The ionic environment of the adapting system. The system for amino acid adapting became necessarily refined after its first development in 1963/1964 (12, 21). In addition to further purification of the macromolecular constituents reported above, the ionic environment characterized in Table II was improved. The original concentrations in Tris⁺ and Mg⁺⁺ (16) were maintained except for the extra amount of Mg⁺⁺ required to complex the EDTA introduced here. These concentrations were chosen in order to observe even the adapting of isoleucine to AUU, a triplet rather difficult to detect in this system (see Table III). They were likewise appropriate for deciphering more than 60 triplets. The temperature of 33° was also optimal as judged by both the parameters of specificity and efficiency of a number of triplets tested. The time curves, however, were not quite satisfactory for a number of amino acids, whose adapting products apparently started to decay—after 5 to 10 minutes. This was shown to result from non-enzymatic hydrolysis of the amino acid-tRNA esters, particularly in the presence of certain divalent cations. In the system reported before (16), ammonium carbonate p.a. (Merck) was the major source of such impurities. But other chemicals including Tris-HCl, and MgCl$_2$ to be used in this system were also somewhat

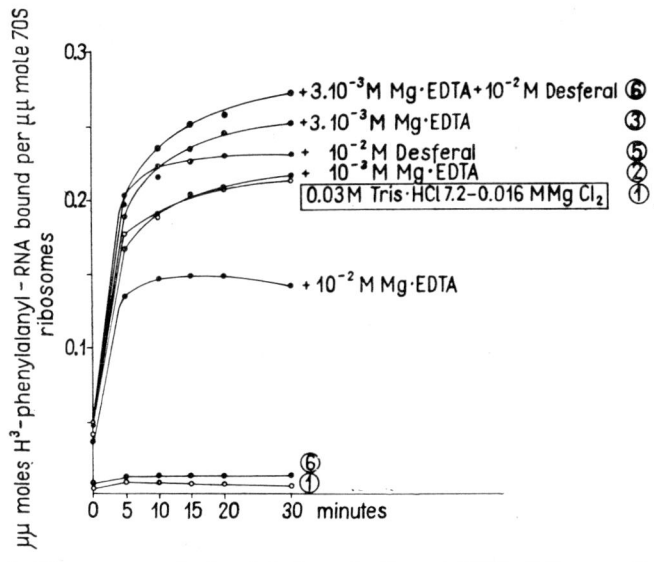

FIGURE 8. Time curvse of phenylalanine adapting at 33°C: Influence of chelating agents (5). The lower cures 1 and 6 are controls, without poly U.

destructive, even when highly purified (see MATERIALS). We therefore add 3 mM MgCl$_2$ and 3 mM K·EDTA pH 7, which, even in case of the relatively stable ester of phenylalanine, improves the time course of adapting (see Figure 8). An extensive study of the hydrolytic effect by many divalent cations on the esters of all 20 protein amino acids is to be published elsewhere (5). We are by no means sure whether this little effort spent so far can reasonably be considered as an optimum reached for both most specific and efficient adapting. Considering the many codons and adaptors to fit, the system is of a complexity which allows for only a gradual approach towards perfection.

RESULTS AND DISCUSSION

Coding nucleotide sequences. The system described in Table II was shown to contain limiting concentrations of ribosomes and mRNA, whereas aaRNA was saturating in all cases tested (12, 16). Although unfavorable, and li-

TABLE I. Charging of soluble RNA from *E. coli* A19 expressed in % RNA with 20 protein amino acids (of specific activities given in c/m). Radioactivity was measured after elution of the aaRNA from SEPHADEX G25 (see text). Calculations are based on 25 A_{260}-units per mg RNA and a molecular weight of 25,000 for sRNA. * = 0.81% of the RNA was formyl-methionine charged.

Alanine	(H)	(^{3}H : 5000)	6.10%	Leucine	(A)	(^{14}C : 155)	13.50%
Arginine	(A)	(^{14}C: 293)	6.82	Lysine	(H)	(^{3}H :3000)	5.10
Asparagine	(A)	(^{14}C: 30)	.98	Methionine*	(A)	(^{35}S : 122)	2.92
Aspartic acid	(A)	(^{14}C: 106)	4.35	Phenylalanine	(H)	(^{3}H :3000)	2.86
Cysteine	(A)	(^{35}S: 24)	.72	Proline	(S)	(^{3}H :1150)	4.25
Glutamic acid	(A)	(^{14}C: 148)	5.00	Serine	(A)	(^{14}C : 96)	4.54
Glutamine	(C)	(^{14}C: 19)	9.94	Threonine	(A)	(^{14}C : 100)	5.80
Glycine	(A)	(^{14}C: 67)	7.46	Tryptophan	(A)	(^{3}H : 875)	3.21
Histidine	(S)	(^{14}C: 132)	1.17	Tyrosine	(H)	(^{3}H :3000)	3.22
Isoleucine	(A)	(^{14}C: 174)	3.28	Valine	(A)	(^{14}C : 160)	8.05
						RNA charged	99.25%

miting the adapting system, the equilibrium in the binding of mRNA to ribosomes is reached within the first 2 minutes at room temperature, showing a maximum initial velocity only during the first 50 milliseconds (17). On the other hand, the binding of aminoacyl RNA approaches its much more favorable equilibrium only after 15–30 minutes, even at 33°C (Figure 8). The utilization of the 5'-terminal triplet in our mRNA models XpYpZ...pZ$_{\sim 30}$ for adapting of the corresponding amino acid is in the order of 0.5 to 1.5% in many cases, shown in Table III. CpGpC...pC$_{\sim 30}$ coded arginine with an efficiency of up to nearly 3 per cent (16). In view of the widely differing efficiency of different triplets, this rather low yield in adapting product, obtained with a certain amount of triplets added, demands accurate analyses of polynucleotides. We may point out, however, that on the average this yield is 30 times as high as in the coding experiments using trinucleoside diphosphates (2, 22, 24).

TABLE II. System for amino acid adaptation stimulated by polynucleotides of the type XpYpZ...pZ.

Ribosomes†	50 μμmoles
Polynucleotide chains‡, approx.	200 μμmoles
Soluble RNA§	800 μμmoles
99% charged with 20 amino acids carrying per labeled amino acid§§	40 μμmoles
per total volume††	100 μl
MgCl$_2$ (EL, Merck)	19 mM
Tris-HCl pH 7.2 (Sigma)	30 mM
KCl (Suprapure, Merck)	11 mM
K.EDTA pH 7.0 (Fluka)	3 mM
incubated for 15 min at 33°C	

† calculated at 16 A$_{260}$-units : 1mg : 0.356 mμmoles of 70S.
‡ average chainlength = 30 nucleotides; 6 mμmoles of nucleotide.
§ calculated at 25 A$_{260}$-units : 1mg : 40 mμmoles of sRNA.
§§ extreme values to this average were 6 μμmoles of cysteine and 108 μμmoles of leucine in 800 μμmoles of sRNA.
†† The XpYpI...pI-polymers were tested in 1.0 ml of the same ionic environment.

In Table III the stimulations observed in our system due to addition of polynucleotides XpYpZ...pZ$_{\sim 30}$, over the blank, which was obtained in the presence of the homopolynucleotide, are given in μμmoles. On these stimulations we have based the amino acid assignments indicated (12, 16). The values in parentheses express the same stimulations in percent. By use of various polynucleotides synthesized with 800-fold purified polynucleotide phosphorylase under the conditions described in this communication, substantially higher stimulations were obtained in all cases tested. These values (marked ") are given in additional lines of Table III. In a subsequent communication, we shall report the full results of testing all these new polynucleotide preparations in order to draw definite conclusions, based on their base and chain length analyses, concerning the efficiency of the different codons in this system (17). However, a few corrections must be made, which have resulted from this work. Last year we described some stimulation of leucine by the triplets UAA and UAG, and of alanine by AUC, reproducibly observed in the system from *E. coli* A19 (12, 16). These stimulations amounted to 0.27 (= +27%) and 0.54 (= +72%) μμmoles leucine with UAA and UAG, and to 0.92 (= +98%) μμmoles alanine with AUC. With the newly synthesized polynucleotides, these stimulations could not be observed in adapting systems derived from either strain A19 or strain B. Instead, AUC gave the rather small effect on isoleucine, shown in Table III, which was to be expected from the base composition of coding units determined several years ago (11, 20, 26). All triplet assignments made during the last one and a half years in *E. coli* A19 (12, 16) are in agreement now with results from adapting on trinucleoside diphosphates, and with conclusions from peptide synthesizing systems translating the same triplets located in internal rather than terminal position (7, 11, 20, 26), and from amino acid exchanges observed *in vivo* (in Ref. 16).

Located at the 5'-end of polynucleotide chain, all coding units are active in this system. Considering various possibilities for the determination of a starting point for reading mRNA, the question arose as to whether ribosomes have a specificity or preference for binding to the 5'-end, possibly regardless of its nucleotide sequence. Under the ionic conditions reported here and earlier (16), adapting was directed also by the homopolymer part of our polymers, at least in the case of polyuridylic and polycytidylic acids. Moreover, the 5'-terminal triplets not coding phenylalanine or proline, respectively, gave small or undetectable inhibitions of this adapting (16). However, we have shown a number of polymers XpYpZ ... pZ, in which the triplet XpYpZ, but hardly the triplet YpZpZ, was read. In other cases, triplet YpZpZ was read as well, or even better, than triplet XpYpZ. To rule out a possibly non-statistical influence of a 5'-exonuclease, we tested some of the latter polymers with ribosomes at their different stages of purification reported here. No decrease in the possible "pseudo-shifted" reading from the second letter was observed even at 10^{-4} times the original level of exonuclease (27). The residual exonuclease at this stage would have removed no more than 5% of the 5'-terminal nucleotide, even if it were a mere 5'-exonuclease. We must conclude that in principle, ribosomes would adapt amino acids all over the polynucleotide chain in this system. It seems to depend on the efficiency of the individual triplet, or the adaptor fitting it or both, whether more or less, but so far never absolute, preference is observed for reading of the 5'-terminal triplet. The efficiencies of several triplets given in Table III seem to be in agreement with observations by Smith et al. (25) on the poor incorporation of asparagine (AAC) and threonine (ACA) into the amino terminal position of polylysine synthesized under the direction of polynucleotides of the series $(Ap)_n Cp(Ap)_m$. We feel, however, that in spite of quite large differences in efficiency seen for different triplets in this system, the many rather active triplets could not all be used for determining the origin of reading mRNA. Otherwise, for a number of amino acids, no codon might be left for directing this amino acid into internal

ANALYSIS OF THE GENETIC CODE 247

TABLE III. The mRNA code found in Escherichia coli A19 by amino acid adapting on the 5'-terminal triplet in polynucleotides of type XpYpZ...pZ~30 are given in $\Delta\mu\mu$moles, and in percent (values in parentheses). For system components see Reference 16, and for the values marked", Table II. + = Stimulations by the homopolymers are given in percent above the minus polymer (ribosome-) blanks. Only the small filter blanks were subtracted throughout.
† = this G was substituted by I; ‡ = synthesis of this polymer was started with IpI; " = this value was obtained with the new set of polynucleotides and the ionic environment described in this paper. B = in agreement with previous conclusions from polypeptide syntheses on random polynucleotides (11, 20, 26). M = this assignment was made by only one group (14). K or N = this triplet was assigned (14) in agreement with either Dr. Khorana's (K) or Dr. Nirenberg's (N) laboratory, respectively (2, 24). For all other triplets, results of the three laboratories support the same conclusion (2, 7, 14, 16, 24).

Abbreviations used: (pp) A = adenosine (diphosphate), C = cytidine, G = guanosine, I = Inosine, T = thymidine, U = uridine. Amino acids (AA): ala = alanine, arg = arginine, asN = asparagine, asp = aspartic acid, cys = cysteine, glN = glutamine, glu = glutamic acid, gly = glycine, his = histidine, ilu = isoleucine, leu = leucine, lys = lysine, met = methionine, phe = phenylalanine, pro = proline, ser = serine, thr = threonine, try = tryptophan, tyr = tyrosine, val = valine. DNA = deoxyribonucleic acid; (m, r, t) RNA = (messenger, ribosomal, transfer) ribonucleic acid.

1st			2nd Nucleotide					3rd
	U		C	A		G		
U	[phe 5.08+(3150)B	ser	.29 (17)" 1.28 (38)"	tyr	.96 (117)B	cys	.06 (49)B .17 (1800)"	U
	phe 1.06 (180)B	ser	.39 (23) 1.24 (23)"	tyr	1.20 (200)B	cys	.26 (93) .17 (1900)"	C
	leu .67 (20)B, M .98 (71)"	ser	.09 (9)K, M			cys	.09 (88)" M	A
	leu .48 (64)B	ser	.51 (83)			try	.09 (37)B, M, K	G†
C	leu .56 (30)B,M,N	pro	.40 (47)	his	.36 (220)B	arg	2.85 (93)B	U
	leu .16 (17)B	pro	2.09+(820)	his	.05 (81)B	arg	2.54 (158)B	C
	leu .85 (33)M	pro	.27 (60)	glN	.40 (45)B	arg	.54 (41)	A
	leu .60 (29)	pro	.02 (27)K, M	glN	.17 (25)	arg	.34 (14)	G†
A	ilu .20 (74)B	tre	.44 (323)	asN	.16 (212)B	ser	.55 (25)	U
	ilu .10 (16)"B	tre	1.02 (123)	asN	.08 (42)B	ser	.28 (26)B	C
	met .12 (29)M	tre	.08 (57)	lys	10.50+(256)B	arg	.60 (42)B	A
	met .06 (55)B	tre	.04 (316)	lys	.19 (17)	arg	.28 (23)M	G†
G	val 1.00 (66)B	ala	1.07 (121)	asp	1.70 (237)B	gly‡	.02 (147)B	U
	val 1.18 (266)	ala	.86 (91)	asp	1.26 (300)B	gly‡	.24 (26)B, K, M	C
	val .17 (41)	ala	.14 (45)	glu	.05 (11)B .11 (73)"	gly‡	.17 (15)B, K, M	A
	val .25 (35)	ala	.42 (84)	glu	1.13 (120)M, K	gly‡	1.88+(170)K, M	G†

position. The question of some positive influence on the activity of a given triplet by its terminal position will be the subject of another publication. There may be certain initiator triplets if not only one that have a much more exaggerated relative efficiency, probably in a different and more natural ionic environment. The *in vitro* adapting system showing us the coding by all or nearly all triplets may cause the ribosomes to bind and adapt to additional unphysiological sites of the mRNA molecule.

On the one hand, further investigation must be directed towards finding the environment adequate for binding of the ribosomes to only the natural initiator nucleotide sequence in mRNA. On the other hand, we must also try to improve the ionic environment of the adapting system for decoding experiments, in order to get still better specificity and possibly higher efficiency.

TABLE IV. Specific binding of amino acyl RNA from *E. coli* by human ribosomes. Values in $\mu\mu$moles of *E. coli* amino acyl RNA bound after 15 min at 33° in a 200 μl reaction mixture containing 2.0 A_{260}-units of ribosomes, approximately 400 $\mu\mu$moles of polynucleotide chains (12 mμmoles of nucleotide), and 1600 $\mu\mu$moles of sRNA 99 percent charged with 1 labeled and 19 nonlabeled amino acids (see Table I). For ionic environment see METHODS

Amino acid	Minus polymer	Poly-U	Poly-C	Poly-A
Alanine	.45	.37	.42	.44
Arginine	1.11	1.16	1.34	1.45
Aspartic acid	.53	.47	.66	.59
Asparagine	.07	.08	.24	.14
Cysteine	.40	.38	.47	.38
Glutamic acid	.81	.71	.52	.91
Glutamine	.44	.29	.39	.42
Glycine	.75	.63	.65	.70
Histidine	.11	.11	.05	.12
Iso-leucine	.27	.28	.37	.29
Leucine	1.41	1.34	1.65	1.33
Lysine	.55	.28	.49	4.09
Methionine	.22	.17	.22	.17
Phenylalanine	.34	2.18	.35	.24
Proline	.30	.22	.68	.21
Serine	.73	.72	.74	.64
Threonine	.97	.90	.77	.78
Tryptophan	.29	.33	.30	.40
Tyrosine	.63	.58	.58	.76
Valine	.35	.50	.30	.54

Universality on the level of ribosome-aaRNA interactions. The code would be completely universal if all organisms translated the same coding units into the same amino acids. This is universality on the level of codon-adaptor-interactions necessarily occurring on ribosomes, taking aaRNA and ribosomes from the same organism. We have taken human ribosomes for adapting experiments with *E. coli* aaRNA to see whether universality exists also on the level of ribosome–aaRNA interactions. In the simplest terms, of course, this could mean no more than the universality among ribosomal specificities for interaction with codon-anticodon pairs. Table IV contains the results of a first adapting experiment with human ribosomes and sRNA from *E. coli* esterified with 20 labeled amino acids by the *E. coli* activating enzymes. As one would expect from *E. coli* ribosomes, the human ribosomes adapt the amino acids lysine to poly-A, phenylalanine to poly-U, and proline to poly-C. Whereas the minus polymer blank in the *E. coli* system is always highest for lysine (16), human ribosomes show a noteworthy low blank for this amino acid. We are trying to prepare an endogenous human adapting system, in order to systematically compare the code in such a different organism.

ACKNOWLEDGMENT

We thank Karin Eckert, Renate Obermeier, Inge Röber, Wiltrud Ludewig, Erika Gaertner, and Jörg Schmidt for excellent assistance in the preparative and assay work. Advice by Dr. Andrea Parmeggiani in the preparation of the manuscript is gratefully appreciated. The Deutsche Forschungsgemeinschaft has supported this investigation.

SUMMARY

1. A cell-free system from *E. coli* A19 for mRNA-directed binding of aminoacyl RNA to ribosomes (= amino acid adapting) was improved:
   (a) Ribosomes were purified to 10^{-4} times their original content in nucleolytic activity without loss of specific activity in both adapting and peptide synthesizing systems.
   (b) 64 polynucleotides of type XpYpZ ... pZ$_{\sim 30}$ were synthesized by 800-fold purified polynucleotide phosphorylase, and subsequently purified and fractionated.
   (c) 20 batches of aminoacyl RNA were synthesized by charging each time with a different labeled and the remaining 19 non-labeled amino acids. Under the conditions described, the sum of the 20 amino acids esterified amounted to 99% of the sRNA.
   (d) The ionic environment of the adapting system was supplied with 3 mM MgCl$_2$ and K·EDTA to avoid non-enzymatic hydrolysis of the amino acid esters during the incubation time.
2. 61 nucleotide triplets were deciphered in this system.
3. The efficiency of the different triplets varied widely in this system, thereby offering a possible means for determination of the start point for reading of the genetic message.
4. Human ribosomes did bind specifically lysine, phenylalanine, and proline adaptors from *E. coli* upon supply of poly-adenylic, -uridylic, and -cytidylic acids, respectively, demonstrating universality on the level of interactions between ribosome and adaptor.

REFERENCES

(See Ref. 16 for complete coverage of pertinent literature)

1. Bayev, A. A., Venkstern, T. V., Mirzabekov, A. D., Krutilina, A. I., Axelrod, V. D., Li, L. and Engelhardt, V. A. This Volume, page 286.
2. Brimacombe, R., Trupin, J., Nirenberg, M., Leder, P., Bernfield, M. and Jaouni, T., *Proc. Nat. Acad. Sci.* **54**, 954 (1965).
3. Cramer, F., Küntzel, H. and Matthaei, J. H., *Angewandt. Chem.* (Int. Ed.) **3**, 589 (1964).
4. Eck, R. V., *Science* **140**, 477 (1963).

5. Heller, G. and Matthaei, H. (In preparation).
6. Holley, R. W., Agpar, J., Everett, G. A., Madison, J. T., Marquisee, M., Merrill, S., Penswick J. R. and Zamir, A., *Science* **147**, 1462 (1965).
7. Khorana, H. G. This Volume, page 209.
8. Matthaei J. H. and Nirenberg, M. W., *Biochem. Biophys. Res. Comm.* **4**, 404 (1961).
9. Matthaei, J. H. and Nirenberg, M. W., *Proc. Nat. Acad. Sci.* **47**, 1580 (1961).
10. Matthaei, J. H., Jones, O. W., Martin, R. G. and Nirenberg, M. W., *Proc. Nat. Acad. Sci.* **48**, 666 (1962).
11. Matthaei, J. H., *Nova Acta Leopoldina* **26**, 45 (1963).
12. Matthaei, J. H., Amelunxen, F., Eckert K. and Heller, G., *Ber. Bunsenges. Phys. Chem.* **68**, 735 (1964).
13. Matthaei, J. H., *Zentralbl. Bakteriol.* **198**, 65 (1965).
14. Matthaei, J. H., Heller, G., Voigt, H.-P., Kleinhauf, H., Küntzel, H., Vogt, M. and Matthaei, H., *Naturwiss.* **52**, 653 (1965).
15. Matthaei, J. H., *Jahrb. d. Akad. d. Wiss. in Göttingen.* 1965, p. 14.
16. Matthaei, J. H., Kleinhauf, H., Heller, G., Voigt, H.-P. and Matthaei, H., *Proc. Mendel Centennial, Genetics Soc. of America*, Fort Collins, Colorado, Sept. 1965, p. 104.
17. Matthaei, H., Voigt, H.-P., Heller, G., Neth, R., Schöch, G., Kübler, H., Amelunxen, F. and Parmeggiani, A., *Cold Spring Harbour Symp.*, June 1966 (In press).
18. Matthaei, H. and Eckert, K., *Biochem. Biophys. Res. Comm.* (In preparation).
19. Nirenberg, M. W. and Matthaei, J. H., *Proc. Nad. Acad. Sci.* **47**, 1588 (1961).
20. Nirenberg, M. W., Jones, O. W., Leder, P., Clark, B. F. C., Sly, W. S. and Pestka, S., *Cold Spring Harb. Symp. Quant. Biol.* **28**. 549 (1963).
21. Nirenberg, M. and Leder, P., *Science* **145**, 1399 (1964).
22. Nirenberg, M., Leder, P., Bernfield, M., Brimacombe, R., Trupin, J., Rottman, F. and O'Neal, C., *Proc. Nat. Acad. Sci.* **53**, 1161 (1965).
23. Singer, M. F. and O'Brien, B., *J. Biol. Chem.* **238**, 328 (1963).
24. Söll, D., Ohtsuka, E., Jones, D. S., Lohrman, R., Hayatsu, H., Nishimura, S. and Khorana, H. G., *Proc. Nat. Acad. Sci.* **54**, 1378 (1965).
25. Smith, M. A., Salas, M., Stanley jr., W. M., Wahba, A. J. and Ochoa, S., *Proc. Nat. Acad. Sci.* **55**, 141 (1966).
26. Speyer, J. F., Lengyel, P., Basilico, C., Wahba, A. J., Gardner, R. S. and Ochoa, S., *Cold Spring Harb. Symp. Quant. Biol.* **28**, 559 (1963).
27. Voigt, H.-P. and Matthaei, H. Unpublished results.
28. Zachau, H. G., Dütting, D. and Feldmann, H. This Volume, page 271.

ADDENDUM (added in proof): Hannig-electrophoresis of polynucleotide phosphorylase carried through step 5 as described above gave 5-fold purification with 50% yields of 2000 to 5000-fold purified, absolutely primer dependent UDP polymerizing activity. These preparations also polymerized ADP, CDP, IDP, and some GDP, but only at and above pH 9.5 (17).

Studies on the Translation of the Genetic Message with Synthetic Polynucleotides*

M. A. Smith[†], M. Salas[‡], M. B. Hille[†], W. M. Stanley, Jr., A. J. Wahba and S. Ochoa

Department of Biochemistry, New York University School of Medicine, New York, New York, U.S.A.

It was shown (1, 2) that, in a cell-free system of protein synthesis consisting of purified *E. coli* ribosomes and *L. arabinosus* supernatant, a system of low nuclease activity, oligonucleotides of the type ApApApApApAp ... pApApC[§] (AAAAAA ... AAC) with an AAC codon at the 3′-end of the chain (Table I, type II) directed the synthesis of oligopeptides of the structure Lys-Lys-Lys ... Lys-Asn with NH_2-terminal lysine and COOH-terminal asparagine. Since the biological assembly of the polypeptide chains of proteins proceeds from the NH_2-terminal to the COOH-terminal amino acids, these results showed that the genetic message is translated by reading the messenger in the direction from the 5′- to the 3′-end of the polynucleotide chain. More recent work (3) with oligonucleotide messengers having the AAC (asparagine) or the ACA (threonine) (4) codon in the vicinity of the 5′-end fully substantiated this conclusion.

Oligonucleotides of the type ApApCpApApAp ... pApApA (Table I, type III) were synthesized and tested for lysine (AAA codon), asparagine (AAC codon) and threonine (ACA codon) incorporation in the *E. coli–L. arabinosus* system. However, these polynucleotides promoted negligible or no incorporation of asparagine or threonine into tungstic acid-insoluble material relative to the amount of lysine incorporated. Thus, the triplet ApApC at the 5′-end of the chain was not read. The possibility that this was due to the lack of a 5′-phosphate residue was excluded, for oligonucleotides of the structure pApApCpApApAp ... pApApA (Table I, type IV) behaved exactly as the type III polymers. In an effort to overcome this difficulty, oligonucleotides (Table I, type V) of the type ApApApApCpAp ... pApApA (A_4CA_n) and ApApApApApCpAp ... pApApA (A_5CA_n) were

---

* Aided by grants from the National Institutes of Health, U.S. Public Health Service, and E. I. Du Pont de Nemours and Company, Inc.

[†] Postdoctoral Fellow of the National Institutes of Health, U.S. Public Health Service.

[‡] International Postdoctoral Fellow of the National Institutes of Health, U.S. Public Health Service. Permanent address, Instituto Marañón, Centro de Investigaciones Biológicas, C.S.I.C., Madrid, Spain.

[§] Shorthand writing of polynucleotides and abbreviations for nucleotides, amino acid residues in polypeptide chains, etc., are as recommended by the Journal of Biological Chemistry and previously used (1, 2).

TABLE I. Polarity of Translation with Synthetic Oligonucleotides

	Oligonucleotide messenger type		Peptides synthesized
(I)	A_n	(5') AAAAAAAAA...AAAAAA (3') (NH_2)	Lys-Lys...Lys-Lys(COOH)
(II)	A_nC	AAAAAAAAA...AAAAAC	Lys-Lys...Lys-Asn*
(III)	A_2CA_n	AACAAAAAA...AAAAAA	Lys-Lys...Lys-Lys
(IV)	pA_2CA_n	pAACAAAAAA...AAAAAA	Lys-Lys...Lys-Lys
(V)	A_4CA_n	AAAACAAAA...AAAAAA	Thr-Lys...Lys-Lys†
(V)	A_5CA_n	AAAAACAAA...AAAAAA	Asn-Lys...Lys-Lys†
	(NH_2) ———— Direction of peptide synthesis ———→ (COOH)		

* NH_2-terminal lysine, no NH_2-terminal asparagine; asparagine released by carboxypeptidase A, not by carboxypeptidase B.
† Threonine (or asparagine) released by carboxypeptidase B, not by carboxypeptidase A.

synthesized. The former (A_4CA_n) promoted the incorporation of lysine, threonine and traces of asparagine, whereas the latter (A_5CA_n) promoted the incorporation of lysine, asparagine and small amounts of threonine. Carboxypeptidase assays showed that the bulk of the threonine in the lysine-containing oligopeptides synthesized with A_4CA_n and the bulk of the asparagine in those synthesized with A_5CA_n messengers was in NH_2-terminal position. These results provide strong additional support for the conclusion that the genetic message is read in a 5' → 3' direction. The same conclusion has recently been reached by other workers (5, 6). The results of Streisinger and collaborators (6) based on studies of the effect of proflavin-induced mutations on phage T_4 lysozyme are particularly noteworthy.

This paper will be mainly concerned with our work with type V polymers (3) and some recent results with similar polynucleotides. The studies with type II polymers were reported last year (1).

PREPARATION AND CHARACTERIZATION OF OLIGONUCLEOTIDES

Type III oligonucleotides ($A_2C^*A_n$) with the cytosine residue labeled with ^{14}C, were prepared with polynucleotide phosphorylase using ADP as substrate and ApApC* as primer (Figure 1). The polymers were isolated by exclusion chromatography on Sephadex G-100 at 25° in 8.0 M urea, 0.5 M ammonium bicarbonate, and recovered by evaporation and lyophilization after exhaustive dialysis against distilled water. The primer was prepared by digestion of random poly AC* (2:1) with pancreatic ribonuclease, phosphomonesterase treatment, and size fractionation by chromatography on DEAE cellulose in ammonium bicarbonate (7). Type IV polymers (p*$A_2C^*A_n$) were prepared and isolated in the same manner as the type III ones but with p*ApApC* as primer. This was prepared by enzymatic phosphorylation of ApApC* with ATP (with the γ phosphate labeled with ^{32}P) using the enzyme described by Richardson (8).

Type V polymers ($A_4C^*A_n$ and $A_5C^*A_n$) with the cytosine residue labeled with ^3H, were prepared by the addition of one single cytidylic acid residue

Type III (A₂C*Aₙ)

ApApC* + n ADP $\xrightarrow{\text{Polynucleotide phosphorylase}}$ ApApC*pApApAp...pApApA + nP_i

Type V(A₄C*Vₙ)

(a) ApApApA + n C*DP $\xrightarrow{\text{Pol. phosphoryl. RNase}}$ ApApApApC*p + nP_i + n−1 C*MP

(b) ApApApApC*p $\xrightarrow{\text{Phosphatase}}$ ApApApApC* + P_i

(c) ApApApApC* + nADP $\xrightarrow{\text{Polynucleotide phosphorylase}}$ ApApApApC*pApAp...pApApA + nP_i

FIGURE 1. Outline of preparation of type III and V oligonucleotides. The asterisk denotes ¹⁴C or ³H label.

(from ³H-labeled CDP) to ApApApA(A₄) or ApApApApA (A₅) with polynucleotide phosphorylase in the presence of pancreatic ribonuclease (Figure 1, step a), to form ApApApApC*p (A₄C*p) or ApApApApApC*p (A₅C*p). This was followed by removal of the 3′-terminal phosphate with phosphomonoesterase (Figure 1, step b). The resulting A₄C* and A₅C*, isolated by DEAE cellulose chromatography in ammonium bicarbonate, were used as primers for the addition of adenylic acid residues from ADP with polynucleotide phosphorylase (Figure 1, step c), and the A₄C*Aₙ and A₅CPAₙ polymers thus formed were isolated as above by Sephadex G-100 chromatography. The A₄ and A₅ used as starting products were prepared from poly A by hydrolysis in 0.1 M ammonium carbonate at 100° for an appropriate length of time, followed by acid hydrolysis of residual cyclic phosphate ends, removal of the resulting terminal 2′- and 3′-phosphate residues with phosphomonoesterase, and chromatography on DEAE cellulose in ammonium bicarbonate.

A purified preparation of *M. lysodeikticus* polynucleotide phosphorylase, virtually free of nuclease and with a requirement for primer, was used for the addition of adenylic acid residues to A₂C*, A₄C* and A₅C* primers and for the addition of one (³H-labeled) cytidylic acid residue to A₄, and A₅, essentially by the procedure of Thach and Doty (9).

The various polymers were characterized by determining the location of the radioactivity following digestion with pancreatic ribonuclease. In each case, all of the radioactivity was recovered in the species expected, A₂C*Aₙ, A₄C*Aₙ and A₅C*Aₙ yielded A₂C*p, A₄C*p and A₅C*p, respectively. Phosphomonoesterase digestion of these products yielded the original primers. Ribonuclease digestion of p*A₂C*Aₙ (type IV) yielded p*A₂C*p, with the 5′-phosphate labeled with ³²P and the cytidylic acid residue labeled with ³H. Phosphomonoesterase hydrolysis of p*A₂C*p released ³²P-orthophosphate and yielded ³H-labeled A₂C* (ApApC*) which, on chromatography, was identical with the primer used for preparation of A₂CAₙ (type III) oligonucleotides.

The average molecular weight and molecular weight distribution were determined both from the ratio of total nucleotide material (measured by absorbancy) to primer (determined by radioactivity) and from their chromatographic behavior on Sephadex G-100 previously calibrated with poly A, poly C and poly U of known and uniform degree of polymerization. There was good agreement between the two methods indicating that all polymer chains were of the desired character and were free of poly (oligo) A chains.

AMINO ACID INCORPORATION WITH $A_nCA_{n'}$ OLIGONUCLEOTIDES

The effect of these oligonucleotides and, as a control, that of random poly AC (15:1) on the incorporation of lysine (AAA codon), asparagine (AAC codon), and threonine (ACA codon) is shown in Table II. The incorporation of glutamine was not investigated because, as previously noted (2), the *L. arabinosus* supernatant has low glutaminyl ~t RNA synthetase activity. All of the polymers promoted the incorporation of lysine.

TABLE II. Amino Acid Incorporation with Poly AC and $A_nCA_{n'}$ Oligonucleotides†

Polymer	Amino acid incorporation				
	$\mu\mu$moles/sample			Per cent of total	
	Lysine	Asparagine	Threonine	Asparagine	Threonine
None	(115)	(23)	(48)		
Poly AC(15:1)	9490	552	576	5.2	5.4
$A_2CA_{\overline{17}}$	570	0	0	0	0
$pA_2CA_{\overline{17}}$	568	0	0	0	0
$A_5CA_{\overline{10}}$	302	35	6	10.2	1.7
None	(61)	(30)	(19)		
$A_5CA_{\overline{14}}$	954	95	20	8.9	1.9
$A_4CA_{\overline{19}}$	725	3	36	0.4	4.7

† Methods as previously described (2,3). Actual incorporation values (blanks without added polynucleotide subtracted from values with polynucleotide) expressed in $\mu\mu$moles/sample.

Random poly AC promoted the incorporation of equal amounts of asparagine and threonine. This was as expected for each amino acid has a 2A1C codon. There was no asparagine or threonine incorporation above the blank values without added oligonucleotide, in the presence of $A_2CA_{\overline{17}}$ or $pA_2CA_{\overline{17}}$, despite the presence of a 5'-phosphate residue in the latter. The incorporation of lysine was promoted by these polymers to about the same extent. In other experiments with A_2CA_n polymers there was some incorporation of both asparagine and threonine but it was negligible relative to the amount of lysine incorporated.

A_4CA_n oligonucleotides directed the incorporation of asparagine but threonine was incorporated to a much lesser extent. This result, suggesting that the artificial messengers were not read randomly, prompted the preparation and testing of A_4CA_n oligonucleotides. The results were quite conclusive, as these polymers directed the incorporation of threonine but only traces of asparagine. While only a few typical experiments are given in Table II, several experiments were carried out with polymers of the A_5CA_n series, varying in length from an average of 16 to one of 25 nucleotide residues, and with polymers of a A_4CA_9 series, varying in length from an average of 16 to one of 34 nucleotide residues, with similar results. The *E. coli–L. arabinosus* system used in these experiments was found to be free of 5'-exonuclease activity. This was assayed by using as substrates oligonucleotides of the type A*pCpAp ... pApApA, with the adenine residue at the 5'-end labeled with ^{14}C. These were prepared in the same way as the type II (Table I) polymers, using A*pC as primer for the addition of adenylic acid residues with polynucleotide phosphorylase.

Position of Asparagine and Threonine in Peptide Chains

Peptides containing ^{12}C-lysine and either ^{14}C-asparagine or ^{14}C-threonine were prepared with $A_5CA_{\overline{19}}$ and $A_4CA_{\overline{29}}$ messengers, respectively, and isolated by carboxymethylcellulose chromatography as previously described (2). As shown in Figure 2, after treatment with carboxypeptidase A the distribution of the peptides containing ^{14}C-threonine or ^{14}C-asparagine remained essentially unchanged. This result indicates that neither of these amino acids was in COOH-terminal position in the lysine peptides for, as shown previously (2), COOH-terminal asparagine is rapidly released from lysine peptides by carboxypeptidase A. Carboxypeptidase A is also known to hydrolyze off COOH-terminal threonine (10). On the other hand, treatment with carboxypeptidase B resulted in the release of about 75 per cent of the radioactivity in each of the two peptides as free ^{14}C-threonine or ^{14}C-asparagine. Previously (2) it had been shown that carboxypeptidase B has no effect on the size distribution of lysine peptides with COOH-terminal asparagine and the same is to be expected of lysine peptides with COOH-terminal threonine, for carboxypeptidase B requires COOH-terminal basic amino acids for activity (11). These results indicate that the bulk of the asparagine and threonine in the lysine peptides investigated was in NH$_2$-terminal position.

Frame Setting

It is apparent from the above results that the triplet at the 5'-end of the chain, i.e. ApApC or pApApC is not read. However, the next triplet down the chain, either pApCpA or pApApC is correctly read with a high frequency as either threonine or asparagine. This suggests that, with the oligonucleotides used, translation begins at the 5'-end, with the first triplet being apparently used as a threading triplet, and that this start sets the reading frame (cf. Table III). Faulty starts due to "jumping" one or two bases and resulting in shifts of reading frame (as indicated in the (b) series of frame settings of Table III) occurred rarely. It should be noted that these shifts must be the result of base "jumping" rather than that of enzymatic removal of 5'-terminal residues from the messengers for, as already pointed out, the cell fractions used for these experiments were devoid of 5'-exonuclease activity.

TABLE III. Translation of A_2CA_n, A_5CA_n, and A_4CA_n Oligonucleotides

Oligonucleotide	Frame setting	Peptides synthesized
Poly A_2CA_n	a. ↑ ApApC pApApA pApApA......	Lys-Lys-Lys...Lys; probably traces of Asn-Lys-Lys...Lys
	b. A ↑ pApCpA pApApA pApApA...	Probably traces of Thr-Lys-Lys...Lys
Poly A_5CA_n	a. ↑ ApApA pApApC pApApA......	Asn-Lys-Lys...Lys; probably traces of Lys-Asn-Lys...Lys
	b. A ↑ pApApA pApCpA pApApA...	Probably small amounts of Lys-Thr-Lys...Lys
Poly A_4CA_n	a. ↑ ApApA pApCpA pApApA......	Thr-Lys-Lys...Lys; probably traces of Lys-Thr-Lys...Lys
	b. ApA ↑ pApApC pApApA......	Probably traces of Asn-Lys-Lys...Lys

The vertical arrows mark the initial point of attachment of ribosomes to the messenger. The underlined triplets at the 5'-ends are not read. "Jumping" of one or two bases with ensuing frame shift (as in (b) series of frame settings) occurs infrequently.

^{12}C-LYSINE–^{14}C-THREONINE PEPTIDES

^{12}C-LYSINE–^{14}C-ASPARAGINE PEPTIDES

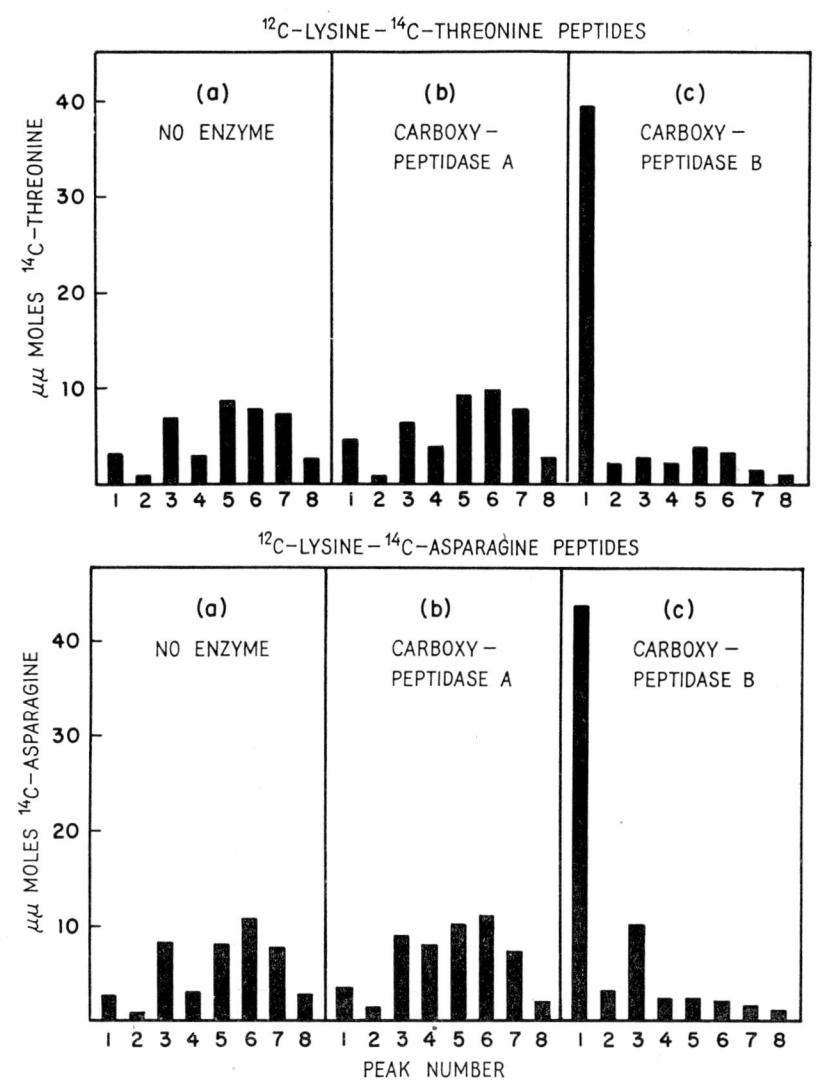

FIGURE 2. Effect of carboxypeptidases A and B on ^{12}C-lysine- ^{14}C-threonine and ^{12}C-lysine- ^{14}C-asparagine peptides. Peptides containing 56 $\mu\mu$moles of ^{14}C-threonine or 60 $\mu\mu$moles of ^{14}C-asparagine, prepared with $A_4CA_{\overline{29}}$ or $A_5CA_{\overline{19}}$ oligonucleotide messengers, respectively, and isolated as previously described (2), were fractionated by chromatography on carboxymethylcellulose before (a) or after incubation for 30 min at 37° with either 13 μg of carboxypeptidase A (b) or 1.8 μg of carboxypeptidase B (c). The procedures for carboxypeptidase treatment and chromatography have been described (2). The effluent was monitored continuously at 220 mμ. Fractions, 0.85 ml, eluted from the column were collected and their radioactivity was measured. Free threonine or asparagine, not retained by the column, are recovered in peak 1. The succeeding peaks, 2, 3, etc., correspond to the dipeptide (1 Thr or 1 Asn, 1 Lys), tripeptide (1 Thr or 1 Asn, 2 Lys), etc., respectively. Free lysine would be recovered in peak 2.

Use of Type V Polymers for Determination of Codon Base Sequence

The fact that type V oligonucleotides are read in frame (at least in the neighborhood of the 5'-end) permits the use of these polymers for unequivocal determination of codon base sequence. The results of Table II show that AAC is an asparagine codon (in agreement with previous results (1, 2) with ApApApApApAp ... ApApC oligonucleotides) and ACA a threonine codon. With use of oligonucleotides of the structure ApApApApApA ... ApApU (2) it had been shown that AAU is another asparagine codon.

Oligonucleotides of the type $A_nUA_{n'}$ were prepared in the same way as the $A_nCA_{n'}$ polymers, using UDP with ^{3}H-label in the uracil moiety instead of ^{3}H-CDP at step a (see Figure 1). Similar oligonucleotides containing one ^{3}H-labeled guanylic acid residue, namely A_3GA_n, have been prepared using ^{3}H-labeled GDP and substituting ribonuclease T_1 for pancreatic ribonuclease at step a of a preparation of type V polymers. Amino acid incorporation experiments with these oligonucleotides as messengers are shown in Table IV. They are compared with random AU and AG copolymers. It may be seen that oligo $A_4UA_{\overline{14}}$ promoted the incorporation of lysine and isoleucine but not that of asparagine. Earlier work with random copolynucleotides indicated that both asparagine and isoleucine had 2AlU codons. The present results show that AUA is an isoleucine codon. The results with oligo $A_3GA_{\overline{18}}$ show that this polymer promoted the incorporation of lysine and glutamic acid but not that of arginine. Both arginine and glutamic acid were known to have 2AlG codons. The results indicate that GAA is a glutamic acid codon. The exclusive stimulation of the incorporation, besides lysine, of either isoleucine (with oligo $A_4UA_{\overline{14}}$) or glutamic acid (with $A_3GA_{\overline{18}}$) shows that, as with the $A_nCA_{n'}$ oligonucleotides, beginning of translation at the 5'-end sets the reading frame.

TABLE IV. Amino Acid Incorporation with poly AU, poly AG and $A_nUA_{n'}$ or $A_nGA_{n'}$ Oligonucleotides*

Polymer	Amino acid incorporation				
	$\mu\mu$moles/sample			Per cent of total	
	Lysine	Asparagine	Isoleucine	Asparagine	Isoleucine
None	(158)	(20)	(38)		
Poly AU (15:1)	12806	456	610	3.3	4.4
$A_4UA_{\overline{14}}$†	538	0	22	0	4.0
	Lysine	Arginine	Glutamic ac.	Arginine	Glutamic ac.
None	(226)	(56)	(106)		
Poly AG (5:1)	8770	844	354	8.5	3.6
$A_3GA_{\overline{18}}$‡	1182	0	26	0	2.1

* Conditions as in Table II.
† AAAAUAAAA...AAA.
‡ AAAGAAAAA...AAA.

Codon base sequence assignments to date from experiments with type V oligonucleotides are summarized in Table V. They are compared with earlier assignments (see Ref. 4) based on the use of Nirenberg's aminoacyl∼tRNA ribosomal binding method and the studies of Khorana with synthetic polyribonucleotides of alternating base sequence (block polymers). Most of the codon base sequence assignments to date are based on binding data and, as pointed out by Khorana (4), it would be important to corroborate them by experiments based on amino acid incorporation and polypeptide synthesis. Of the five codons listed in Table V, our results confirm the binding data for the two asparagine codons, AAC and AAU. The assignment of AUA to isoleucine is a new one because (a) binding results with the trinucleotide ApUpA have been negative, and (b) polynucleotides with an alternating AU sequence are inactive as messengers because of their high degree of secondary structure (4). On the other hand, the ACA assignment for threonine and the GAA assignment for glutamic acid had been previously made both on the basis of binding experiments and of experiments with block polynucleotides (4). Our results with type V oligonucleotides are in agreement with these assignments.

TABLE V. Base Sequence Assignment of Some Codons by Several Methods†

Codon	Method		
	Binding	Block polymers	End triplet
AAC	Asparagine		Asparagine
AAU	Asparagine		Asparagine
ACA	Threonine	Threonine	Threonine
AUA			Isoleucine
GAA	Glutamic ac.	Glutamic ac.	Glutamic ac.

† Data by use of binding and block polymers are from Khorana (4).

Studies with other oligonucleotides, now being prepared in our laboratory, such as $A_n U_{n'}$, $U_n GA_{n'}$, $AUGA_{n'}$, and other $A_n UA_{n'}$ polymers, should permit further codon base sequence studies and might throw some light on the mechanisms of initiation and termination of translation.

SUMMARY

In a system of purified *E. coli* ribosomes and *L. arabinosus* supernatant, of low nuclease content and devoid of 5'-exonuclease activity, A_5CA_n oligonucleotides directed mainly the synthesis of lysine oligopeptides with NH$_2$-terminal asparagine, whereas A_4CA_n polymers directed predominantly the synthesis of lysine oligopeptides with NH$_2$-terminal threonine. The initial ApApA triplet in these polymers was not read. This was not due to the lack of a 5'-phosphate residue, for the introduction of such a residue did not change the translation features of the oligonucleotides. Together with earlier work these results provide conclusive evidence for the 5' → 3' polarity of messenger translation. They also show that the above oligonucleotides are not read randomly by the cell-free system.

Similar results were obtained with $A_nUA_{n'}$ and $A_nGA_{n'}$ oligonucleotides. The former (A_4UA_n) directed the incorporation of lysine and isoleucine. The latter (A_3GA_n) promoted the incorporation of lysine and glutamic acid. To date, these studies have allowed the assignment of the following codon base sequences: asparagine, AAC and AUU; threonine, ACA; isoleucine, AUA; glutamic acid, GAA.

REFERENCES

1. Ochoa, S., 2nd Annual Meeting, *Federation of European Biochemical Societies*, Vienna (April 1965).
2. Salas, M., Smith, M. A., Stanley, W. M., Jr., Wahba, A. J. and Ochoa, S., *J. Biol. Chem.* **240**, 3988 (1965).
3. Smith, M. A., Salas, M., Stanley, W. M., Jr., Wahba, A. J. and Ochoa, S., *Proc. Natl. Acad. Sci. U.S.* **55**, 141 (1966).
4. Khorana, H. G., *Federation Proceedings* **24**, 1473 (1965).
5. Thach, R. E., Cecere, M. A., Sundararajan, T. A. and Doty, P., *Proc. Natl. Acad. Sci. U.S.* **54**, 1167 (1965).
6. Terzaghi, E., Okada, Y., Streisinger, G., Tsugita, A., Inouye, M. and Emrich, J., *Science* **150**, 387 (1965).
7. Staehelin, M., *Biochim. Biophys. Acta* **49**, 11 (1961).
8. Richardson, C. C., *Proc. Natl. Acad. Sci. U.S.* **54**, 158 (1965).
9. Thach, R. E. and Doty, P., *Science* **147**, 1310 (1965).
10. Folk, J. E., Brannberg, R. C. and Gladner, J. A., *Biochim. Biophys. Acta* **47**, 595(1961).
11. Folk, J. E. and Gladner, J. A., *J. Biol. Chem.* **231**, 379 (1958).

DISCUSSION

(Chairman: F. SANGER)

P. N. CAMPBELL: In connection with the problem of starting a chain and utilizing the first triplet in the m-RNA, has not Doty shown that a hexamer leads to the production of a dipeptide?

S. OCHOA: I should like to point out that the promotion of peptide synthesis by the hexanucleotide used by Thach and Doty (*Proc. Natl. Acad. Sci. U.S.*, **54**, 1167 (1965)) was exceedingly small; a fraction of 1 $\mu\mu$mole. Our own experiments did not exclude marginal reading of the triplet at the 5'-end of our nucleotide messengers, but this was certainly negligible by comparison with the reading of the second triplet. I do not believe, therefore, that the results you mentioned invalidate, or conflict with our conclusions.

J. P. SLATER: The control using a poly(AC), 15:1, gave a very much larger incorporation than the polymers A_4CA_{10}. Would Dr. Ochoa consider that the very long length of AAAA ... induces some strain into the ribosomal system employed?

S. OCHOA: I doubt it, because long poly A's give very high incorporation of lysine, as high or higher than that given by AC copolymers. We have found with poly A's of increasing chain length that the total incorporation of lysine increases with increasing chain length. The relatively small incorporation brought about by $A_nCA_{n'}$ copolymers, with 15 to 30 nucleotide residues, appears therefore to be due to a smaller degree of effectiveness of the oligonucleotides as compared to polynucleotides.

P. SZAFRAŃSKI: Do the ribosomes free of ribonuclease bind more oligonucleotides, as compared to mRNA, than ribosomes containing ribonuclease? Is it possible that ribonuclease blocks ribosomal active sites which may be occupied by mRNA?

S. OCHOA: We have not tested either the purified, nuclease-low or the crude ribosomes for binding of polynucleotide messengers. I cannot, therefore, answer this question. I would not expect such interference.

ADDENDUM (added in proof): More recent studies of the translation of natural messenger RNAs and of oligonucleotides having the methionine codon AUG at or near the 5'-end of the chain have shown that AUG is an initiator codon and that two hitherto unknown factors normally associated with the ribosomes are specifically concerned with the initiation of translation. The presence of an initiator codon at or near the 5'-end of synthetic oligonucleotides considerably speeds up translation (Stanley, W. M.,Jr., Salas, M., Wahba, A.J. and Ochoa, S., *Proc. Natl. Acad. Sci. U.S.*, **56**, 290 (1966).

Invited Comments

P. Leder* and B. F. C. Clark[†]

P. Leder: "Synonymous Codewords and Messenger RNA Phasing Mechanisms."

Resume: I should like to describe briefly an experiment, performed together with Dr. M. W. Nirenberg in his laboratory at the National Institutes of Health, which I believe is pertinent to certain points raised by Professor Khorana.

As Prof. Khorana has pointed out, codeword assignments fall largely into groups or sets in which two to four triplets specify a single amino acid. Differences between codewords in each synonymous set are generally, but not always, restricted to the 3'-terminal base which in certain cases may be either purine, in others, either pyrimidine, and in still others, any of the four common bases. It was, therefore, particularly important to determine whether a single molecular species of tRNA would recognize all the codewords in its synonym set or whether a separate species of tRNA would correspond to each of the individual codewords in the set. Of course, settling this question is especially important if one is to understand the basis for recognition between codeword and tRNA.

Prof. Robert Holley and his co-workers at Cornell University have, as you all know, been brilliantly successful in purifying and completely determining the primary structure of *Yeast* alanine tRNA. From work in the laboratories of Dr. Nirenberg and of Prof. Khorana using *E. coli* tRNA, it was known that alanine corresponded to a codeword set consisting of GpCpU, GpCpC, GpCpA and GpCpG. Thus, assuming that, at least in part, Watson–Crick type anti-parallel complementary base pairing is involved in tRNA-codeword recognition, the sequence ...IpGpC..., occurring approximately in the middle of the molecule, appears as the most promising anticodon sequence. An analogous sequence, also containing insosine, occurs in serine tRNA, the complete structure of which will be reported by Prof. H. Zachau at these meetings.

Prof. Holley kindly provided us with *Yeast* alanine tRNA purified by counter current distribution. This material was acylated with ^{14}C-alanine using partially purified yeast alanyl-tRNA synthetase. Each of the four alanine codewords, GpCpU, GpCpC, GpCpA and GpCpG was tested in the conventional ribosomal binding system (1) at 0.02 M (Mg^{++}) for induction of the binding of purified ^{14}C-ala-tRNA to ribosomes. It was found that GpCpU, GpCpC and GpCpA readily induce ^{14}C-ala-tRNA binding,

---

* Biochemistry Division Weiczmann Institute, Rehovoth, Izrael. † Medical Research Council Laboratory of Molecular Biology, Cambridge, England.

whereas GpCpG is least effective. Unfractionated *Yeast* ^{14}C-ala-tRNA responded similarly. On the other hand, when unfractionated *E. coli* ^{14}C-ala-tRNA was tested against these same triplets, the order of binding efficiency was greatly altered, though the same synonymous codeword set induced binding. GpCpG and GpCpA were most efficient; GpCpU, less; and GpCpC, least. Studies were carried out in a range of (Mg^{++}) between 0.01 and 0.03 M with similar results.

Thus it would appear that a single molecular species of tRNA, in this case Holley's purified ala-tRNA, can recognize more than one codon in its synonym set—in concurrence with other data presented earlier this morning by Prof. Khorana—and among species recognizing universal coding assignments, in this case *Yeast* and *E. coli*, codeword preferences do exist. Crick has put forward one mechanism, in addition to the conventional Watson–Crick type base pairing, by which acceptable base pairs could be formed permitting the recognition by a single anticodon sequence of more than one codeword. It is possible, in the light of this "Wobble hypothesis", and in view of the codeword preferences observed, that the *Yeast* alanine anticodon will differ from the *E. coli* alanine anticodon at the base corresponding to the 3'-terminal base of the codon.

I should like to turn for one more moment and comment on the paradox arising from the results reported by Dr. Matthaei on the one hand and Prof. Ochoa on the other. Dr. Matthaei suggests an apparent preference for reading the 5'-terminal codeword of a message—when using the ribosomal binding system—whereas Prof. Ochoa reports that the 5'-terminal codeword is generally not read in the nucleaseless, cell-free amino acid incorporating system which he described. Our studies using the binding system, carried out first in Bethesda and extended at the Weizmann Institute in Rehovoth, indicate that certain oligonucleotide messengers permit the recognition of all possible triplet sequences, regardless of their position on the messenger. For example, the following series of oligonucleotides were prepared and characterized (2): ApApU, ApApUpU, ApApUpUpU. The oligonucleotide ApApU induces the binding of ^{14}C-Asp-NH$_2$-tRNA to ribosomes; ApApUpU, the binding of ^{14}C-Asp-NH$_2$- and ^{14}C-Ileu-tRNA; and ApApUpUpU, the binding of ^{14}C-Asp-NH$_2$-, ^{14}C-Ileu- and ^{14}C-Phe-tRNA clearly without preference for the 5'-terminal codeword. The similar series of compounds possessing a phosomonoester at the 5'-hydroxyl has been tested in the binding assay without showing clear preferences for terminal or internal codewords. Analogous results have been previously reported by Thach et al. (3). These results lead one to look either toward further chemical modifications of the 5'-terminal hydroxyl of mRNA—such as a triphosphate ester—or, in view of observations which Dr. Clark will subsequently report, toward specific initiator codons. Those corresponding to the codeword for N-formyl-met-tRNA, as already pointed out by Prof. Khorana, will be of particular interest.

REFERENCES

1. Nirenberg, M. W. and Leder, P., *Science*, **145**, 1399 (1964).
2. Leder, P., Singer, M. F. and Brimacombe, R. L. C., *Biochem.* **4**, 1561 (1965).
3. Thach, R. E., Cecere, M. A., Sundararajan, T. A. and Doty, P., *Proc. Natl. Acad. Sci.* **54**, 1167 (1965).

DISCUSSION

S. OCHOA: I wonder whether the results you mentioned ("out of frame binding") may not be due to a higher margin of error in binding than in incorporation experiments.

P. LEDER: That is entirely possible, though, in general, the binding system seems more selective in codon recognition than the peptide forming systems. I believe that Thach et al.—whose work I referred to above—used a hexanucleotide to synthesize a dipeptide which necessarily involves the reading of the 5′-terminal codeword.

B. F. C. CLARK: "Possible Codeword Signals Involved in Polypeptide Chain Initiation and Termination".

Resume: Biochemical and genetic evidence suggest that chain termination is an active process and that possible chain termination codewords are UAA and UAG. It is unlikely therefore that formyl–methionyl–sRNA is a chain terminator by passive blocking of the transmission of the polypeptide chain. We have strong evidence that the methionyl–sRNA which can be formylated, met-sRNA$_2$, is a chain initiator. This sRNA is bound to ribosomes by AUG or GUG strongly, by UUG less strongly and by CUG weakly. If the initiation signal is a triplet codeword, i.e. it could be called a commonsense codeword, the correct codeword may be NUG or AUG (N is any one of the four nucleotides). Should AUG be correct, then the other binding is strongly ambiguous. Perhaps the beginning of a messenger is sufficient for ribosomal attachment to start polypeptide chains. However, with respect to the evidence presented by Dr. Ochoa for the nonreading of a triplet at the 5′ end of the messenger, perhaps we need a leading triplet for attachment of the ribosome before the chain initiation signal is read.

It seems to me, after a consideration of the evidence for a link between the chain termination and initiation signals internal in polycistronic messengers, that a similar set of signals may be involved at the beginning of the message. For example, messengers may begin with the sequence UAAAUG and the same sequence would be involved to end and begin chains when reading off a polycistronic message. In view of published evidence that natural messenger RNAs begin with purines, perhaps a leading triplet of purines such as AAA is needed to attach the ribosome to the beginning of the message as described above.

Let me finish with a word about ribosomal sites. Our biochemical evidence suggests that there are two ribosomal sites involved in chain initiation and propagation. Site A, the aminoacyl–sRNA site, will accept aminoacyl–sRNAs not involved with chain initiation. These can be moved into site B.

the peptidyl–sRNA site, by peptide bond formation and so, in conjunction with a movement of messenger, we get chain propagation. However a chain initiator such as formyl–methionyl–sRNA, can enter site B directly, without passing through site A, thus causing chain initiation.

U. Z. Littauer: If you charge met-sRNA$_2$ with ^{14}C-methionine and another preparation with ^{3}H–formyl–methionine, which will be incorporated preferentially?

B. F. C. Clark: This experiment has not been carried out in the precise manner of the question. However we have evidence that the methionyl–sRNA capable of formylation initiates polypeptide chains at a greater rate when it is in the formylated state.

P. Leder: Does the N–formyl–methionine sRNA species bind to the 30S ribosomal subunit?

B. F. C. Clark: This has not yet been tested but we hope to do it soon.

S. Ochoa: From the statement made by Dr. Clark, that methionine from met sRNA$_2$ goes into NH$_2$-terminal positions in incorporation systems, in the presence of random poly UG, irrespective of whether there is formylation or not, it would appear that binding of met-sRNA$_2$ to the peptidyl (donor) site of the ribosome is independent of formylation and may be due to a special feature of this sRNA. This does not support the belief (Nakamoto and Kolakofsky, *Proc. Natl. Acad. Sci. U.S.* **55**, 606 (1966)) that N-formylmet-sRNA brings about chain initiation because of its analogy to peptidyl sRNA. Nakamoto's experiments strongly suggest, however, that peptidyl sRNA can also be a chain initiator.

NUCLEOTIDE SEQUENCES IN RNA

Chairman's Introductory Remarks

S. Ochoa

Yesterday's session of the symposium on Genetic Elements showed us that great strides have been made in our understanding of the molecular events involved in translation of the genetic message. It is fitting that a major portion of today's session deal with studies of the nucleotide sequence of the RNAs concerned with translation, amino acid transfer RNA in particular.

The transfer RNAs act as carriers of activated amino acids to the protein assembly line and are responsible for the correct alignment of these residues in a sequence dictated by the nucleotide sequence of messenger RNA. Base pairing between nucleotide triplets—or codons—of the messenger and complementary nucleotide triplets—or anti-codons—of the tRNA is involved in the recognition of codons by the corresponding aminoacyl ~ tRNAs. It has long been known, from the work of Paul Berg and others, that each amino acid is linked to a tRNA that is specific for it. Countercurrent distribution has been a powerful tool in resolving and purifying the individual tRNAs and has further shown the existence of two or more specific tRNAs for most of the individual amino acids. Knowledge of the primary structure, namely the nucleotide sequence of tRNA, will throw light on the nature of the sites responsible for recognition of the amino acid and for binding to the ribosomes, and should disclose the location and nucleotide sequence of the anti-codon. This would give us a clear understanding of the complex and amazingly accurate translation mechanisms.

Holley's announcement last year of the determination of the complete nucleotide sequence of yeast alanine tRNA marks a milestone of progress in this area. Other investigators are actively engaged in a similar endeavor. As we shall hear today, their efforts have led to the elucidation of the complete nucleotide sequence of two serine tRNAs and most of the sequence of one valine tRNA from yeast.

Base sequence determination of nucleic acids is still complex and time-consuming. It is encouraging, therefore, that Dr. Sanger, who was the first to successfully elucidate the amino acid sequence of a protein, and from whom we shall hear later, is now engaged in the development of simpler methods for sequence analysis.

Finally, we all wonder about the role of methylated bases in nucleic acids, in particular, in tRNA. This subject will be the topic of the last paper of this session. I am sure we can look forward to a stimulating day.

On the Primary Structure of Transfer Ribonucleic Acids*

H. G. ZACHAU, D. DÜTTING AND H. FELDMANN

Institut für Genetik der Universität Köln, Germany

In the first part of the lecture a survey of the structural work on tRNAs[†] was given, with particular emphasis on the structure of alanine tRNA, which has been elucidated by Holley and his colleagues (1). The odd nucleosides of sRNA were discussed in some detail. Figure 1 represents an attempt to summarize all odd nucleosides, which have been found in sRNA from various sources. A general discussion of the structural work will not be given here, since summaries on this subject have been published fairly recently (2, 3).

In the second part of the talk the work of our group on the structure of serine specific tRNAs from brewers yeast has been described. A brief account of their structures, which have been established recently, has been published (4). Figure 2 shows diagrammatically the nucleotide sequences and the oligonucleotides obtained by complete enzymatic digestion. Most of the experimental details of the structural work will be published elsewhere (5). In the following a few otherwise unpublished experiments will be described which were important in our work on the serine tRNAs and which illustrate some of the techniques used.

The serine tRNAs have been purified by repeated counter-current distributions in tri-n-butylamine-containing and in salt-containing solvent systems. The serine tRNAs seem to have a longer nucleotide chain and (or) may be more extended in secondary structure than most tRNAs. Therefore a partial separation can be achieved readily by simple molecular sieve chromatography on Sephadex. The separation effect can be greatly enhanced by a recycling procedure (Figure 3).

The nucleotides and nucleosides from the serine tRNAs, which were obtained by various degradation methods, were characterized by their mobility in paper chromatography and electrophoresis and by their UV-spectra. Particularly in the case of the odd nucleosides, spectra were very helpful and they will be given for future reference (Figure 4).

N(6)-γ,γ-dimethylallyl-A and N(6)-acetyl-C are of particular interest. Acetyl-C was also isolated in small amounts from pancreatic RNase digests

---

* Serine specific transfer ribonucleic acids, part IX. Part VIII = Ref. 4.
† Abbreviations: tRNA = transfer ribonucleic acid; RNase = ribonuclease; PDE = snake venom phosphodiesterase; PME = alkaline phosphomonoesterase of *E. coli*; p and - are used to represent a phosphate residue; A = adenosine; G = guanosine; C = cytidine; U = uridine; ψ = pseudouridine; UH$_2$ = 4,5-dihydrouridine; rT = ribothymidine; I = inosine; iPA = N(6)-(γ,γ-dimethylallyl adenosine or isopentenyladenosine; MeC = 5-methylcytidine; AcC = N(6)-acetylcytidine; DimeG = N(2)-dimethylguanosine; OMeG = 2'-O-methylguanosine; OMeU = 2'-O-methyluridine.

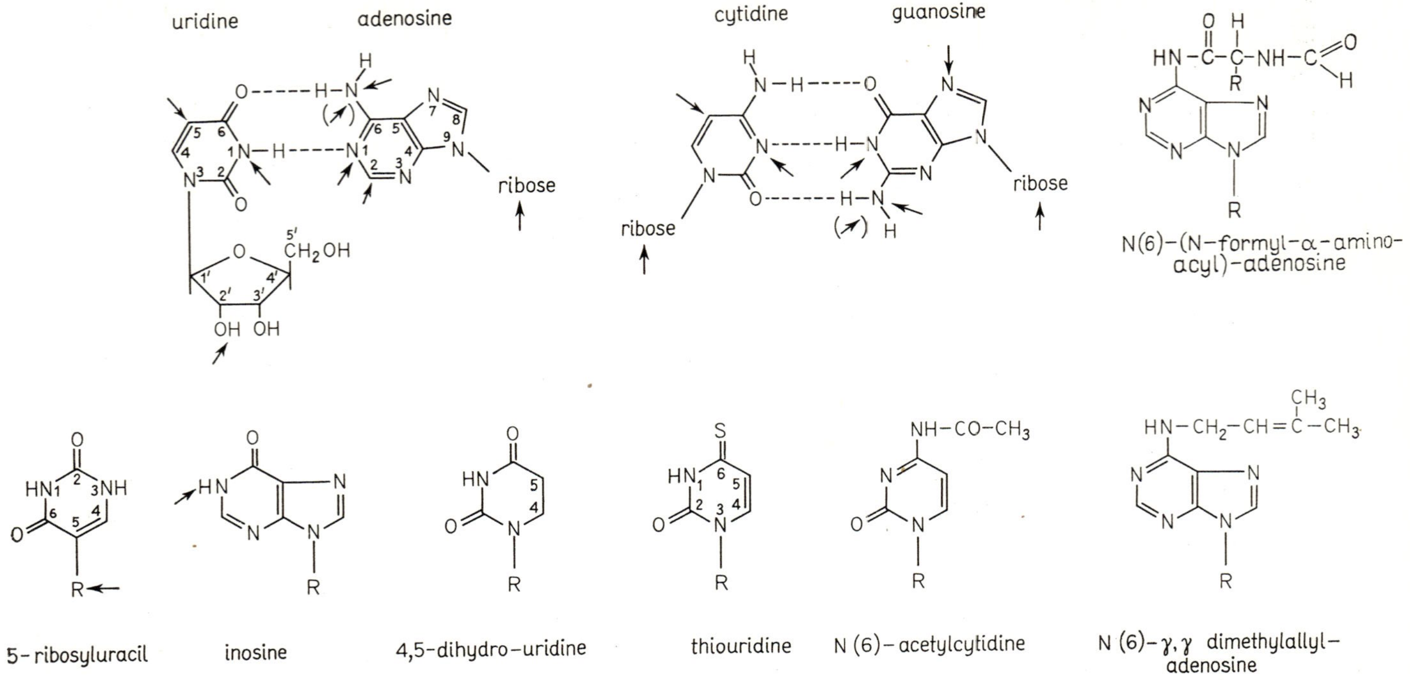

FIGURE 1. Odd nucleosides found in sRNA. Sites of methylation are indicated by arrows. Arrows in parentheses signify the finding of a dimethyl as well as a monomethyl compound.

PRIMARY STRUCTURE OF TRANSFER RIBONUCLEIC ACIDS 273

FIGURE 2. Nucleotide sequences of serine tRNA I and II. Upper line: Nucleotide sequence common to both serine tRNAs. Middle line: Continuation of the upper sequence for serine tRNA I. Lower line: Continuation of the upper sequence for serine tRNA II. Solid lines represent oligonucleotides from digestion with pancreatic (P1–P18) and T1-RNase (T1–T18). Broken lines signify overlapping sequences which could be constructed from these oligonucleotides with the aid of odd nucleotides.

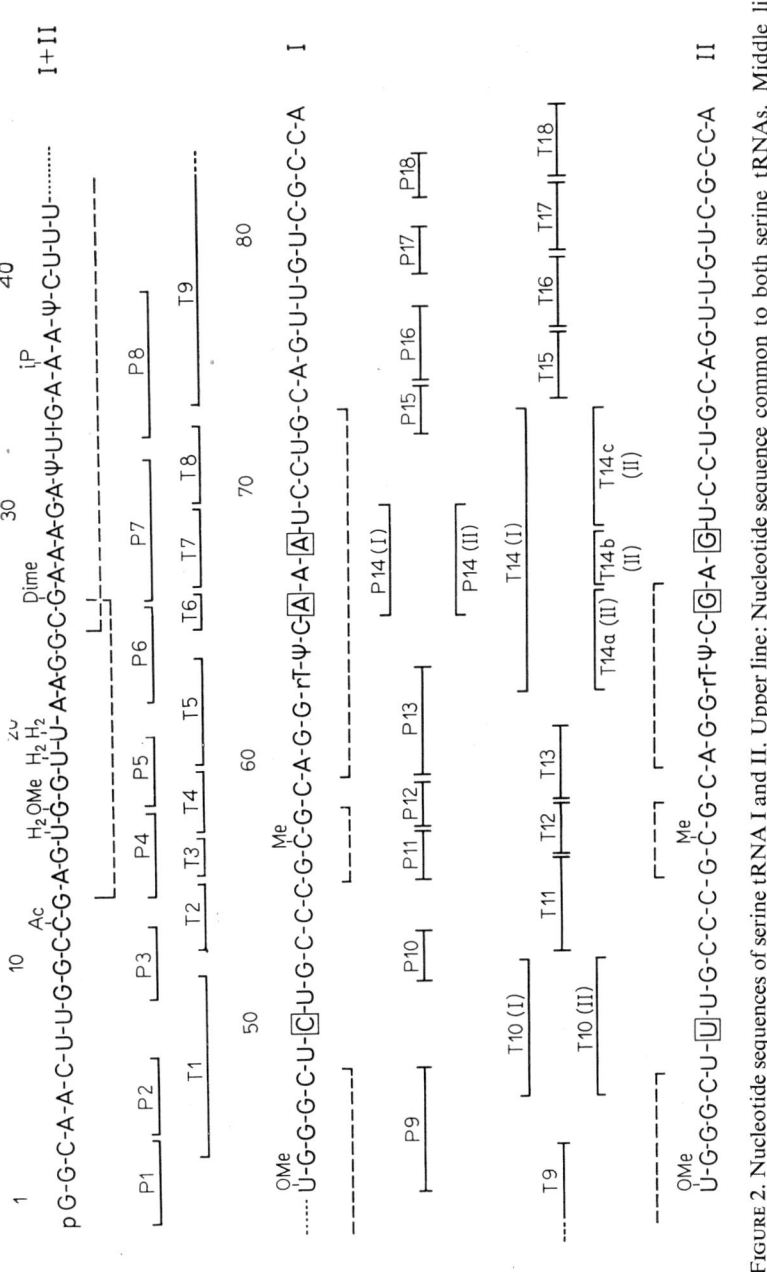

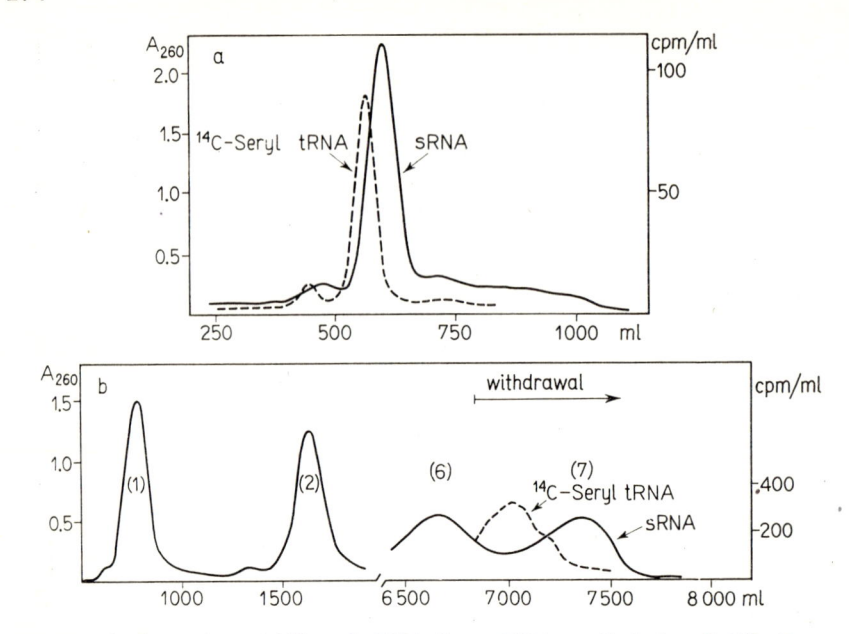

FIGURE 3. Separation of ^{14}C-seryl tRNA from sRNA on Sephadex G 100. Brewers yeast sRNA, charged with ^{14}C-serine, was chromatographed on Sephadex G 100 columns. (a) Redrawn from Ref. 6, Figure 3a, where the conditions are also described; (b) 250 A$_{260}$-units charged sRNA (2.1 nmoles serine/mg sRNA) were recycled 6 times in 0.05 M ammonium acetate, pH 4.5, on a 3.2×100 cm column with a recyclo-chromatography attachment (LKB-Produkter AB, Stockholm). Radioactivity was determined in aliquots after withdrawal. A$_{260}$ was calculated from the Uvicord recordings and the 260 mμ-extinction measured in the withdrawn fractions.

of unfractionated sRNA (Figure 5). In this way we obtained additional material to establish the structure of this nucleoside, which had not previously been found in an RNA.

The odd nucleotides were of considerable help in constructing overlapping sequences from oligonucleotides of the complete enzymatic digests. As can be seen from Figure 2, quite a number of further overlaps were required. Not even the UH$_2$-containing sequence could be linked to the DimeG-containing sequence, since in pancreatic RNase digests there are several oligonucleotides ending with -GpCp. Two approaches (7) were tried to obtain the missing overlaps: sequential degradation with an exonuclease, and partial degradation with endonucleases.

Table I illustrates an attempt to order the T1-RNase splitting products by their gradual disappearance upon exonucleolytic removal of mononucleotides from the 3'-hydroxyl end of the RNA chain with PDE. (In Table I the oligonucleotides are arranged in the order from T1 to T18, as deduced by other methods). After ca 50% degradation with PDE two T1-RNase oligonucleotides (T17 and T18) had virtually disappeared, confirming the terminal position of CpCpA and indicating UpCpG as the adjacent

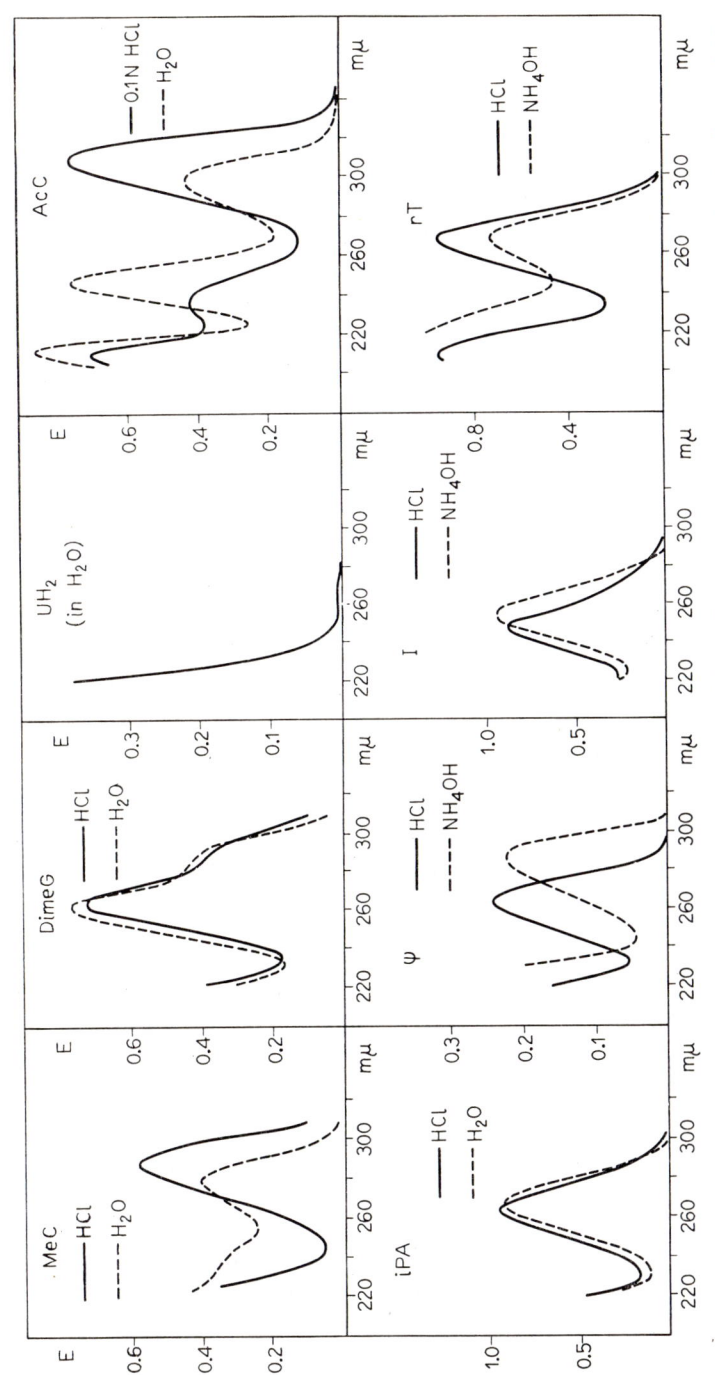

FIGURE 4. UV-spectra of the odd nucleosides found in serine tRNA. The spectra were measured against paper blanks. Water, 0.02 N HCl and 0.02 N NH₄OH were used as solvents, if not indicated otherwise.

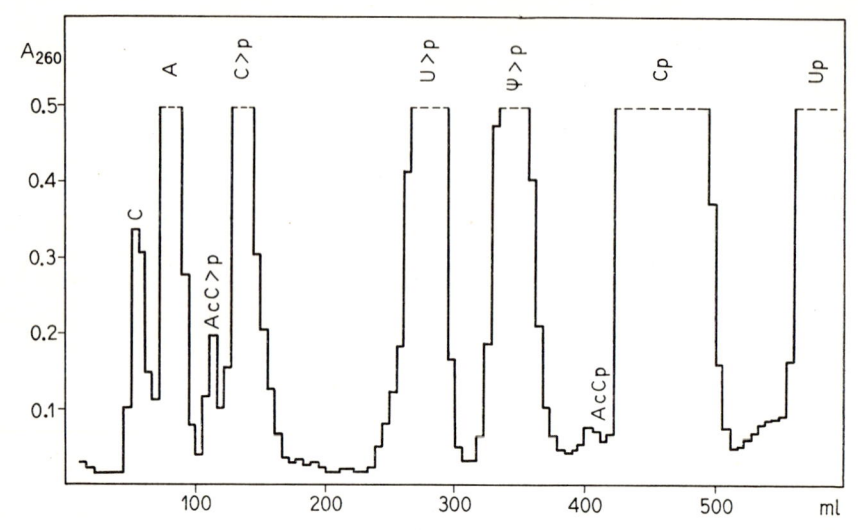

FIGURE 5. Isolation of N(6)-acetyl-Cp from sRNA. 1500 A_{260}-units of brewers yeast sRNA were incubated for 100 min at 45° and pH 7.5 (autotitrator) with 0.7 A_{280}-units of pancreatic RNase. The digest was chromatographed on a DEAE-cellulose column (0.82 meq./g, 1 × 40 cm, linear gradient with 300 ml each of 0.01 M and 0.05 M ammonium carbonate, pH 8.6). The AcC > p and AcCp containing fractions were identified by UV-spectra, pooled, freed from carbonate by flash evaporation and by drying in a desiccator and subjected to paper electrophoresis, pH 3.5, where AcC > p and AcCp are easily separated from contaminating C > p and Cp, respectively.

sequence. The oligonucleotides T14, T15, and T16 were found in small amounts, indicating that they may belong to the right half of the tRNA-chain. No numbers are given in Table I for ApG and CpApG since these sequences occur twice in the tRNAs; the amounts of these oligonucleotides cannot be distributed between the two sequences. The oligonucleotides T5, T6, and T7 occur in relatively small amounts which, in part, may be due to incomplete or non-specific splitting with T1-RNase, as was observed occasionally (5). On the other hand, an endonucleolytic splitting by the PDE in this region of the molecule cannot be ruled out, although with simple methods no endonuclease activity was detected in the purified enzyme. Considering secondary structure models of serine tRNA, it is interesting to note that oligonucleotides from the middle portion of the molecule (T9, T10, and T11) are found in large amounts. A tightly paired structure with 4 G:C-pairs is quite plausible in this region, which appears to have blocked further attack of the PDE. The resistance of this region of the molecule towards PDE was also observed in experiments with isolated fragments (T9–T11). In experiments with oligonucleotides it was found that the action of PDE is also delayed by clusters of certain odd nucleotides as rTpψ (5). The finding of a fairly high amount of MeCpG (T12 in Table 1), which should be degraded by PDE after the rTpψ-sequences, may be related to this observation. In summary it can be said that the PDE experiments helped only

TABLE I. Sequential Degradation of Serine tRNA I+II with PDE†

		nmoles
T1	CpApApCpUpUpG	90
T2	CpAcCpG+CpCpG	108
T3	ApG	—
T4	UH$_2$pOMeGpG	137
T5	UH$_2$pUH$_2$pApApG	99
T6	CpDimeG>p	66
T7	ApApApG	75
T8	ApψpUpI	140
T9	ApiPApApψpCpUpUpUpOMeUpG	148
T10(I)	CpUpCpUpG	~50
T10(II)	CpUpUpUpG	~70
T11	CpCpCpG	156
T12	MeCpG	112
T13	CpApG	—
T14(I)	rTpψpCpApApApUpCpCpUpG	26.5
T14(II)		
a	rTpψpCpG	37
b	ApG	—
c	UpCpCpUpG	~30
T15	CpApG	—
T16	UpUpG	30
T17	UpCpG	<10
T18	CpCpA	<10

† Amounts of oligonucleotides are given which were obtained after total splitting with T1-RNase and PME of the sequentially degraded tRNA. 320 A$_{260}$-units serine tRNA I+II, previously dialyzed against EDTA and water, were incubated at 48° in a total volume of 2 ml 0.005 M MgCl$_2$ with 1600 units of PDE (purified as in Ref. 9); pH 9.0 was maintained with an autotitrator. The reaction was stopped by addition of phenol after half of the NaOH, required for total degradation (determined separately), had been taken up (45 min). After 4 extractions with phenol and 4 extractions with ether the partially degraded tRNA (115 A$_{260}$-units) was separated from the mononucleotides (86 A$_{260}$) by precipitation with 1.5 ml N HCl at 0° and 2 washings with 0.07 N HCl. The precipitate was dissolved in water by addition of NH$_4$OH to pH 7.5. Degradation with T1-RNase/PME, separation of the oligonucleotides on a DEAE-cellulose column and their characterization and determination were carried out as described in Ref. 9.

little in the ordering of oligonucleotides, but contributed to the understanding of secondary structure.

The approach which finally allowed reconstruction of the complete sequence was the partial degradation of the tRNAs with T1-RNase and pancreatic RNase and the isolation of large oligonucleotide fragments. A few examples for the separation of these large fragments and for the determination of their structure will be given in the following.

The fractionation of a partial T1-RNase digest under three different conditions is shown in Figure 6c–e. The digest was prepared in such a way as to give a splitting of the tRNA molecules into large fragments only. In most of the fractions a definite residual acceptor activity for serine was found. This may be due to remaining undegraded serine tRNA or to an acceptor activity of the large fragments themselves, which, because of the presence of Mg^{++}, were not completely dissociated from each other. It is unlikely that the remaining acceptor activity originates from tRNA dimers(10) or aggregates, since the serine tRNA had been freed from them by passing through a Sephadex column (Figure 6a) prior to the T1-RNase treatment.

It is somewhat interesting that the aggregate fractions (first peak in Figure 6a, and second peak in Figure 6b) exhibit some acceptor activity, which may be due to a facile disaggregation of the aggregates under our conditions of isolation (flash evaporation and dialysis against water). This is supported by two findings:

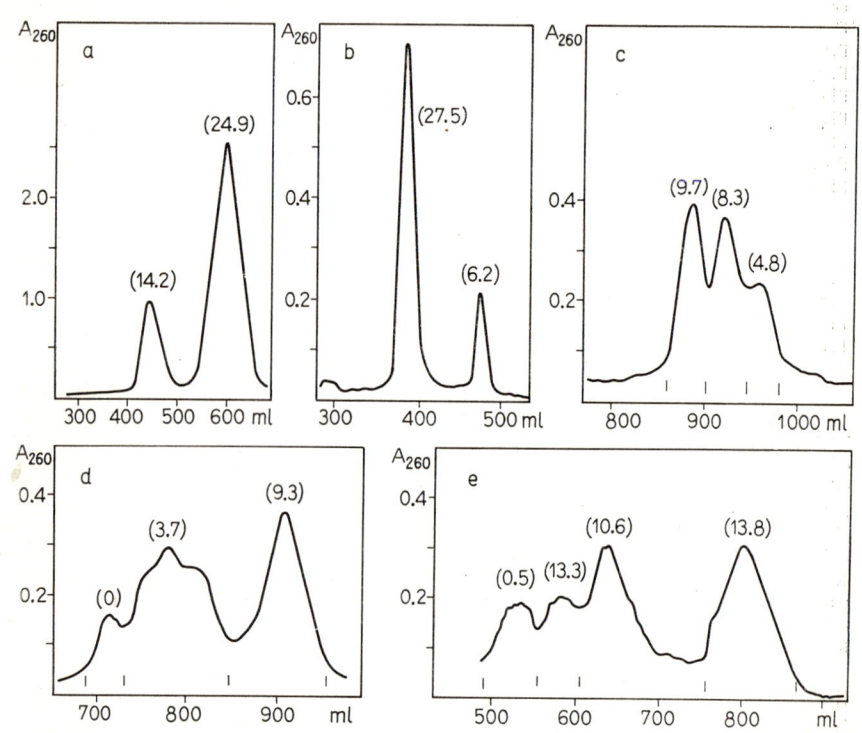

FIGURE 6. Comparison of column types for the separation of serine tRNA, tRNA aggregates, and large oligonucleotide fragments. (a) 240 A_{260}-units serine tRNA I+II (exp. E in Figure 1, Ref. 11) were chromatographed on a Sephadex G 100-column (4.4 × 100 cm) in 0.025 M ammonium acetate, pH 5.5, as described in Ref. 6; (b) 30 A_{260}-units of the same tRNA were chromatographed on a DEAE-Sephadex A 50-column (0.4 × 210 cm) with a linear gradient of 500 ml each of 0.3 M and 0.55 M NaCl in 7 M urea, 0.02 M Tris HCl, pH 7.5. (c-e) 170 A_{260}-units aggregate-free serine tRNA (peak 2 of panel a of this Figure) were digested for 4 min at 0° with 135 units T1-RNase (Sankyo) in 2 ml 0.02 M MgCl₂, 0.02 M Tris·HCl, pH 7.5. After extractions with phenol and ether 138 A_{260}-units of digest were obtained. 0.4 × 210 cm-columns with DEAE-Sephadex A 25 (c, d) and A 50 (e) were used, applying linear salt gradients in 7 M urea. The fractions were isolated after dialysis and tested for ¹⁴C-serine incorporation. The results are given in nmoles serine/mg tRNA next to each peak; (c) 40 A_{260}-units digest, 600 ml each of 0.1 M and 0.5 M NaCl, 0.02 M Tris·HCl, pH 8.0; (d) as (c), without Tris·HCl, elution fluid adjusted to pH 3.0 with HCl; (e) 50 A_{260}-units digest, 550 ml each of 0.13 M and 0.5 M NaCl, pH 3.0.

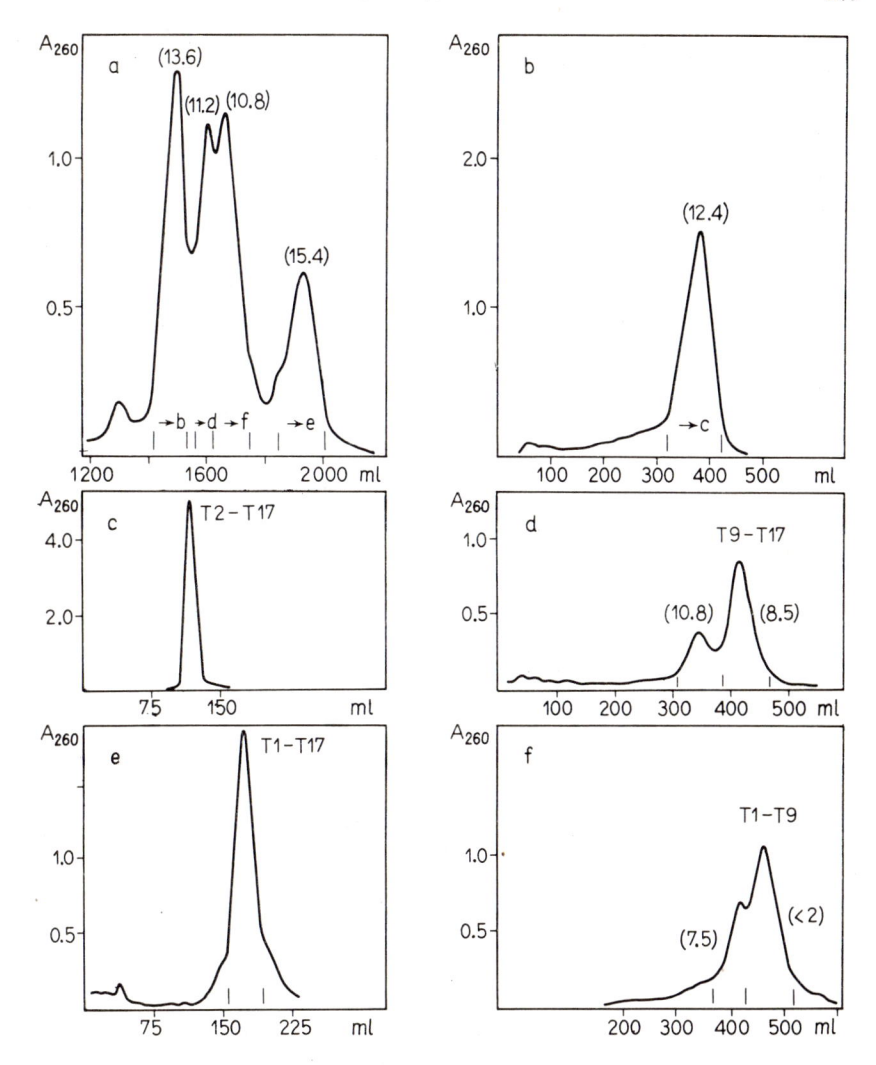

FIGURE 7. Separation of large tRNA fragments from serine tRNA I+II. 500 A$_{260}$-units tRNA were digested for 4 min at 0° with 400 units Tl-RNase in 3 ml 0.02 M MgCl$_2$, 0.02 M Tris·HCl pH 7.5. After 6 extractions with phenol and 5 extractions with ether the digest (412 A$_{260}$-units) was chromatographed on DEAE-Sephadex A 50 (1.26×220 cm, linear gradient with 1500 ml each of 0.18 M and 0.55 M NaCl, in 7 M urea, adjusted to pH 3.0 with HCl) (Figure 7a). The peaks were rechromatographed, as indicated, on DEAE-Sephadex A 50 (0.68×210 cm, linear gradients with 800 ml each of 0.25 M and 0.5 M NaCl in 7 M urea, pH 3.0) (Figure 7b, d, f) and on DEAE-cellulose (0.68 meq./g) at 55° (0.4× ×210 cm, linear gradients with 150 ml each of 0.25 M and 0.55 M NaCl in 7 M urea, 0.02 M Tris·HCl, pH 7.0) (Figure 7c, e). The fractions were isolated after dialysis, tested for ^{14}C-serine incorporation (nmoles serine/mg RNA are given in parentheses) and degraded for analysis.

(1) Rechromatography of peak 1 (Figure 6a) under identical conditions yields only serine tRNA in the position of peak 2 (Figure 6a).

(2) Rechromatography of peak 2 (Figure 6b) under the conditions of Figure 6a shows both aggregates and disaggregated tRNA.

The conditions, determined in the experiments of Figure 6, were used in an experiment with larger amounts of tRNA (Figure 7). Fractions from the first column were rechromatographed, until a reasonable degree of homogeneity was achieved. The series of rechromatographies (Figure 7b, d, f) shows further separation and demonstrates the reproducibility of the acidic DEAE-Sephadex columns. By complete digestion with T1-RNase the composition of the fragments was determined. The designations T9–T17 etc. refer to fragments extending from oligonucleotide T9 to oligonucleotide T17 (cf. Figure 2).

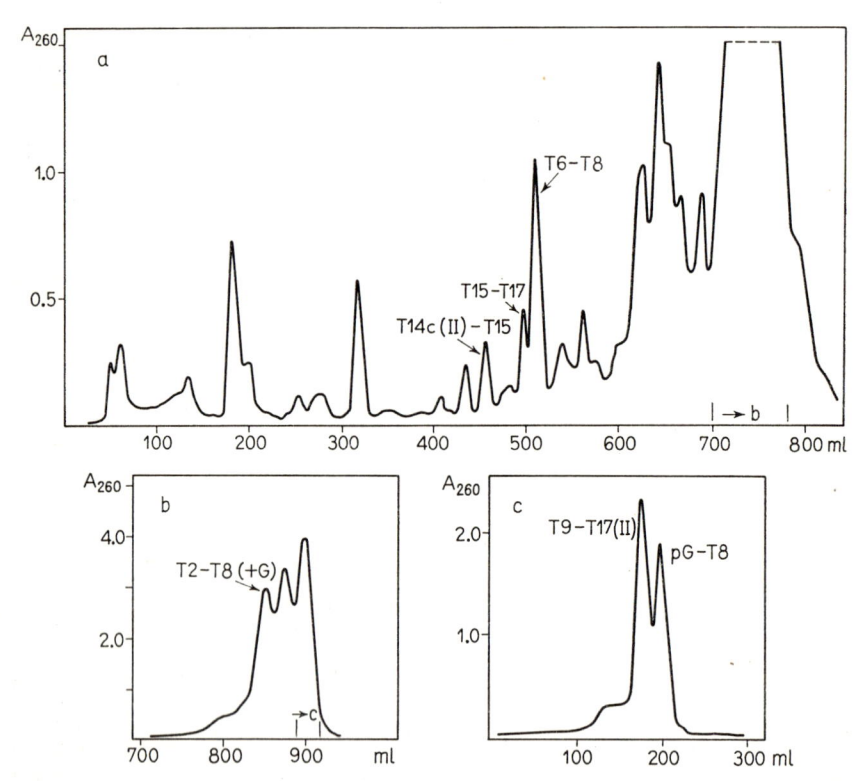

FIGURE 8. Isolation of "half molecules" from serine tRNA II. 550 A_{260}-units serine tRNA I+II were digested for 30 min at 0° with 5000 units T1-RNase in 3 ml 0.2 M Tris · HCl, pH 7.5, and chromatographed on a DEAE-cellulose column (0.4×210 cm) in 7 M urea, 0.02 M Tris, pH 7.5, at 45° with 700 ml each of 0.045 M and 0.45 M NaCl (Figure 8a). Rechromatography of peaks was as indicated on DEAE-Sephadex A 25 columns (0.4×210 cm) in 7 M urea, pH 3, with gradients of 0.05–0.45 M NaCl (800 ml each) (Figure 8b) and 0.25–0.45 M NaCl (400 ml each) (Figure 8c).

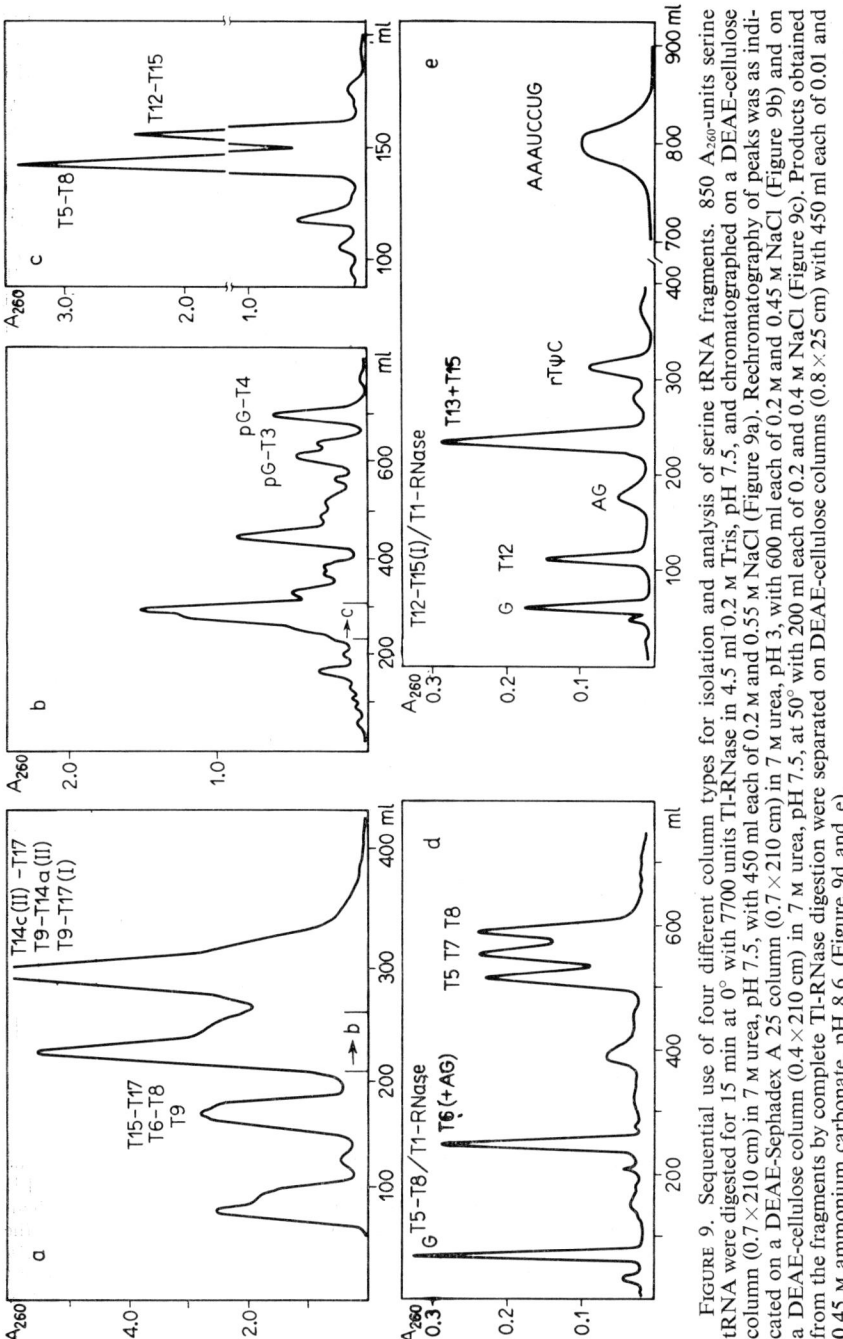

PRIMARY STRUCTURE OF TRANSFER RIBONUCLEIC ACIDS 281

FIGURE 9. Sequential use of four different column types for isolation and analysis of serine tRNA fragments. 850 A$_{260}$-units serine tRNA were digested for 15 min at 0° with 7700 units T1-RNase in 4.5 ml 0.2 M Tris, pH 7.5, and chromatographed on a DEAE-cellulose column (0.7×210 cm) in 7 M urea, pH 7.5, with 450 ml each of 0.2 M and 0.55 M NaCl (Figure 9a). Rechromatography of peaks was as indicated on a DEAE-Sephadex A 25 column (0.7×210 cm) in 7 M urea, pH 3, with 600 ml each of 0.2 M and 0.45 M NaCl (Figure 9b) and on a DEAE-cellulose column (0.4×210 cm) in 7 M urea, pH 7.5, at 50° with 200 ml each of 0.2 and 0.4 M NaCl (Figure 9c). Products obtained from the fragments by complete T1-RNase digestion were separated on DEAE-cellulose columns (0.8×25 cm) with 450 ml each of 0.01 and 0.45 M ammonium carbonate, pH 8.6 (Figure 9d and e).

Partial T1-RNase degradation of serine tRNA under stronger conditions gave a number of shorter fragments and, in addition, a small amount of the tRNA-"halves", the purification of which is shown in Figure 8. It should be noted that on the acidic column (Figure 8c) the 47 nucleotide-long fragment T9–T17, derived from serine tRNA II, is eluted before the shorter fragment pG–T8 (34 nucleotides).

Figure 9 gives an example of how certain tRNA fragments can be purified only by the successive use of three different types of columns (pH 7.5, pH 3, and pH 7.5 at 50°). The two fragments obtained this way were analysed after complete degradation with T1-RNase and PME. As in most of our analyses, the oligonucleotides from these degradations were separated on DEAE-cellulose columns with ammonium carbonate gradients (Figure 9d

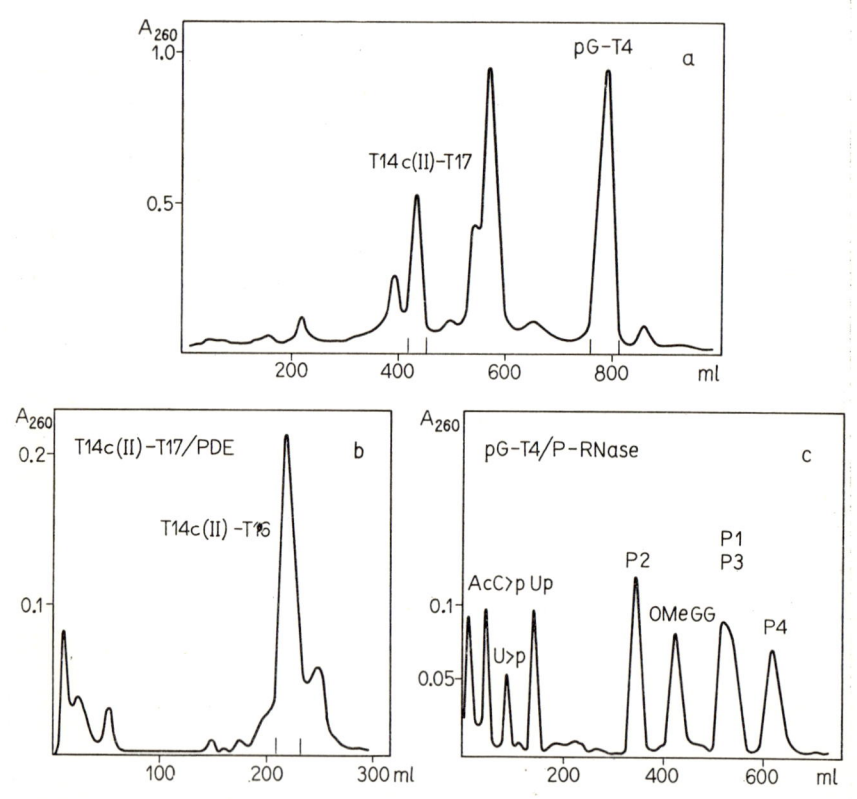

FIGURE 10. Additional methods for the characterization of certain tRNA fragments. The last but one peak from a chromatography of a 45 min digest of serine tRNA I+II was rechromatographed on a DEAE-Sephadex column A 25 (0.7×210 cm, linear gradient with 600 ml each of 0.2 M and 0.4 M NaCl, in 7 M urea, pH 3.0) (Figure 10a). Part of the peak T14c (II)–T17 was partially digested with PDE and chromatographed on a DEAE-cellulose column (1×8 cm, linear gradient with 220 ml each of 0.0 M and 0.4 M NaCl in 7 M urea, pH 7.5) (Figure 10b). The peak pG–T4 was degraded with pancreatic RNase and the digest chromatographed as described in Figure 9d and e (Figure 10c).

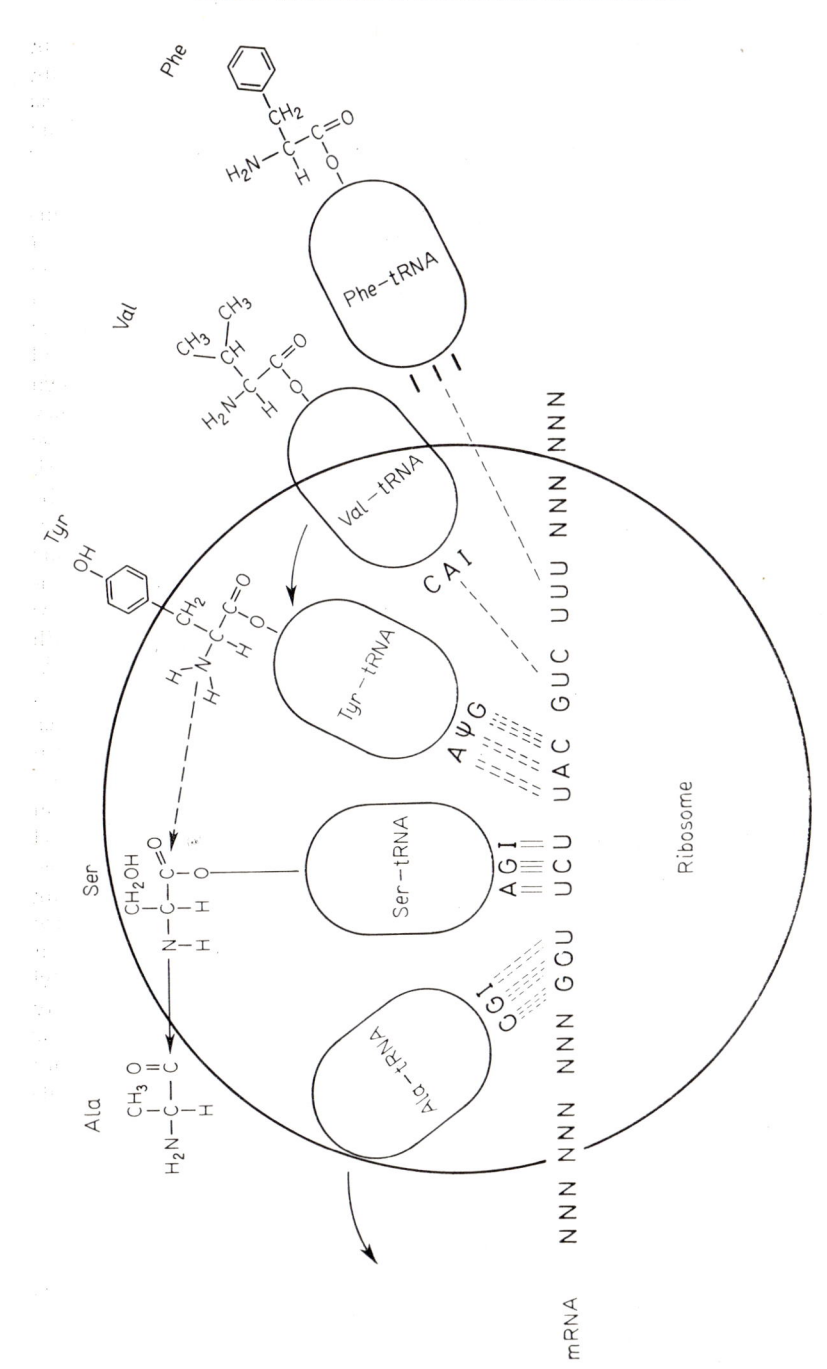

FIGURE 11. A schematic presentation of the adaptor hypothesis.

and e) and identified by paper electrophoresis, UV-spectra and splitting to mononucleotides. The first peak (T5–T8) gave the above mentioned missing overlap between the UH$_2$-region and the DimeG-sequence. The analysis of the second peak (T12–T15 from serine tRNA I) shows an example of the occasionally observed (12) unorthodox splitting by T1-RNase between C and A.

In partial digests no fragments were found which permitted of the assignment of the relative position of the two trinucleotides UpUpG and UpCpG (T16 and T17). The problem was solved by partial PDE-degradation of the fragments T14c (II)–T17 (Figure 10a). PDE very preferentially removed the 3'-terminal nucleotides. Thus, on rechromatography (Figure 10b), the fragment T14c (II)–T16 could be isolated, from which UpCpG was removed but still contained UpUpG (5). The analysis of pG–T4 (Figure 10a) is shown in Figure 10c as an example of a total digestion with pancreatic RNase.

In this paper some experiments have been described which may serve as examples for the approaches and techniques used in the structural work on the serine tRNAs. For full documentation these experiments have to be considered together with the other experimental evidence published elsewhere (5). With respect to the general features of the serine tRNAs just one point should be brought up, which is particularly pertinent to this symposium on "Properties and Function of Genetic Elements": The sequence IpGpA (position 34–36 in Figure 2) is most probably the anticodon of both serine tRNAs. This is suggested by comparison with alanine and valine tRNA, which also contain one I per molecule in sequences which in an antiparallel fashion are complementary to the proposed codons for these amino acids. In secondary structure models, which are based on the principle of maximal base pairing, IpGpA is located in a loop, as would be required for the functioning of the tRNA in the adaptor complex. A schematic presentation of the adaptor hypothesis is given in Figure 11: The tRNAs are arranged with their anticodons along the codons of the messenger RNA and discharge their amino acids on to the growing peptide chain. The IpGpA-containing loop in the secondary structure models of serine tRNA has the same size as the corresponding loop in alanine tRNA. In both tRNAs there is an unusual nucleoside immediately adjacent to the anticodon, dimethylallyl adenosine in serine tRNA and 1-methylinosine in alanine tRNA. It may be speculated that these odd nucleosides play a particular role in the adaptor complex. Since work on a number of further tRNAs is in progress, a more detailed understanding of their common structural features should be available soon.

REFERENCES

1. Holley, R. W., Apgar, J., Everett, G. A., Madison, J. T., Marquisee, M., Merrill, S. H., Penswick, J. R. and Zamir, A., *Science*, Wash. **147**, 1462 (1965).
2. *Cold Spring Harbor Symp. Quant. Biol.* **28**, (1963); **31** (1966) Articles on Transfer RNA.
3. Rajßhandery, U. L. and Stuart, A., *Ann. Rev. Biochemistry* **35**, 759 (1966).
4. Zachau, H. G., Dütting, D. and Feldmann, H., *Angew. Chem.* **78**, 392 (1966).

5. Manuscripts in preparation Z. physiol. Chem. **247** (in press).
6. Zachau, H. G., *Biochim. Biophys. Acta* **108**, 355 (1965).
7. Holley, R. W., Apgar, J., Everett, G. A., Madison, J. T., Merrill, S. H. and Zamir, A., *Cold Spring Harbor Symp. Quant. Biol.* **28**, 117 (1963).
8. Razzell, W. E., in Colowick, S. P. and Kaplan, N. O. (Eds.), *Methods in Enzymology*, Acad. Press, N. Y. **VI**, 236 (1963).
9. Melchers, F., Dütting, D. and Zachau, H. G., *Biochim. Biophys. Acta* **108**, 182 (1965).
10. Schleich, T. and Goldstein, J., *Proc. Nat. Acad, Sci.*, Wash. **52**, 744 (1964).
11. Zachau, H. G., Dütting, D., Melchers, F., Feldmann, H. and Thiebe, R., Ribonucleic Acid—Structure and Function, Symp. FEBS, Vienna 1965, Pergamon Press, Oxford, 1966, p. 21.
12. Dütting, D. and Zachau, H. G., *Biochim. Biophys. Acta* **91**, 573 (1964).

DISCUSSION

(Chairman: S. OCHOA)

U. Z. LITTAUER: For some time we have been engaged in developing a general chemical method for the isolation of amino acid specific tRNA. Initially any desired amino acid is enzymatically coupled to those tRNA chains specific for it; the resulting mixture of tRNA and aminoacyl tRNA is then reacted with β-benzyl-N-carboxy-L-aspartate, leading to the formation of a polypeptidyl tRNA derivative. When the reaction is carried out in a mixture of dioxane-water, a preferential percipitation of the polypeptidyl-RNA from tRNA occurs. However, we were unable to recover biologically active tRNA from the polypeptidyl-RNA (*Biochim. Biophys. Acta* **80**, 169, 1964). More recently we found that the inactivation of the purified RNA is due to the presence of trace amounts of ribonuclease in the tRNA preparations. When the bound RNAse is removed from the tRNA, biologically active tRNA can be recovered and the approximate purity of the amino acid specific tRNA obtained is about 60% (E. Katchalsky, S. Yanotsky, A. Novogrodsky, Y. Galenter and U. Z. Littauer, *Biochim. Biophys. Acta* (In press)).

D.G. KNORRE: Not only counter-current distribution, but also some chemical methods have been elaborated for sRNA fractionation, such as the use of polyacryloylhydrazide in agar gel in the Zamecnik procedure (D.G. Knorre, S. D. Mysina and L. S. Sandakhchiev, *Izvestija Sibirskogo Otdelenija Akad. Nauk* SSSR; Seria Chimitscheskich nauk **11**, 135, 1964) and the use of polyacrylic anhydride for the specific binding of aminoacyl-sRNA.

The partial degradation by phosphodiesterase, with subsequent action of guanylyl–RNase as mentioned by Dr. Zachau, has also been used in Novosibirsk to determine the oligonucleotide sequences in valine-specific sRNA molecules.

H.G. ZACHAU: I should like to repeat the remark made earlier in my talk: If the separation methods involving chemical modification could be made to work, and to give good yields of biologically active material, they really would be preferable to the tedious (but efficient) counter-current distribution techniques.

Primary Structure of the Valine Transfer RNA. Partial Reconstruction of the Molecule

A. A. BAYEV, T. V. VENKSTERN, A. D. MIRZABEKOV,
A. I. KRUTILINA, V. A. AXELROD, L. LI AND V. A. ENGELHARDT

Institute of Molecular Biology of the USSR Academy of Sciences, Moscow

Even a few years ago the determination of the nucleotide sequence of a tRNA could be considered only as a theoretical possibility. Today it has become a reality, mainly due to the brilliant works of Holley and his collaborators (1–5) on the tRNA^{Ala}. The study of other transfer RNA's has not yet reached the same level, but recent works on the structure of serine tRNA (6, 7), valine tRNA (8–12), etc., indicate a continuous advance in this field.

We are reporting here on the study of the structure of valine tRNA which we began about 3 years ago. The investigation has now reached a stage which may definitely present general interest.

Valine transfer RNA was obtained from bakers' yeast. The total tRNA was isolated by phenol treatment followed by precipitation with ethanol, cetavlon (cetyltrimethylammonium bromide) and NaCl (13).

As described earlier (14), the initial step of the individual tRNA^{Val} isolation involved counter-current distribution (CCD) in the slightly modified system of Apgar *et al.* (15). This was performed in an apparatus with 102 units of "Quickfit and Quartz" production. The distribution pattern of total tRNA after 230–250 transfers is shown in Figure 1. The most slowly moving tRNA^{Ala} and the most rapid tRNA^{Tyr} are the reference points of the distribution. The tRNA^{Val} occurs as two incompletely resolved fractions, designated as tRNA$_1^{Val}$ (the slowly moving species) and tRNA$_2^{Val}$ (the more rapid moving one). Other investigators have also detected by end group analysis two valine-acceptor tRNA's (16–21). The same number of tRNA^{Val} species has been found by Goldstein *et al.* (22) in the tRNA of *E. coli* by means of the CCD method. On the other hand, Thiebe and Zachau (23, 24) have found in brewer's yeast three valine specific tRNA's. In the system used we observed a valine specific tRNA fraction remaining in the first tube of the CCD apparatus. This species has the same (or a similar) acceptor end as tRNA$_1^{Val}$. This fraction is being examined, but at present it is not yet certain whether it represents another individual species (tRNA$_3^{Val}$).

After 230–250 transfers the purity of tRNA$_1^{Val}$ reaches 20–30% and the yield is 60 %. The resolution of tRNA$_1^{Val}$ and tRNA$_2^{Val}$ is rather good, the admixture of the latter in the tRNA$_1^{Val}$ preparation not exceeding 10–15%

One can get a much better resolution with 1000 transfers in a steady state distribution system, when the upper and the lower phases move past each other in a true counter-current fashion.

A separation of tRNA$_1^{Val}$ and tRNA$_2^{Val}$ (without significant increase in total degree of purity) can be achieved by chromatography of the crude fraction from CCD on a DEAE-cellulose column, using a combined temperature- and NaCl-gradient (25).

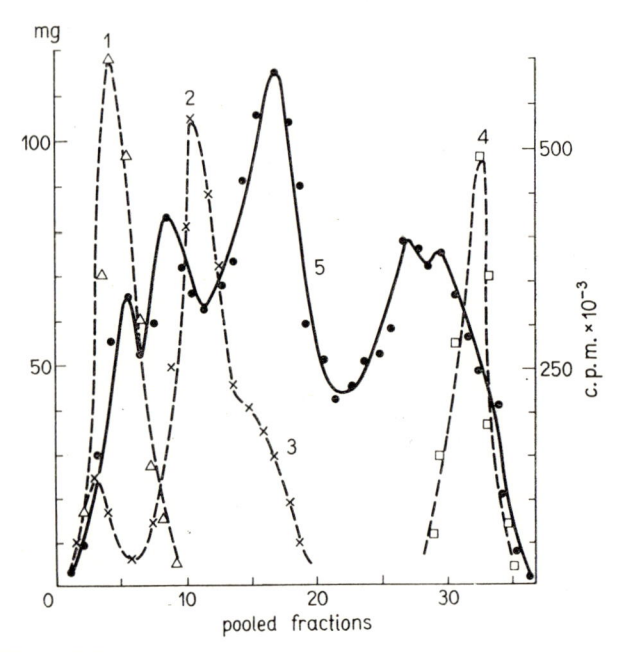

FIGURE 1. Counter-current distribution of total tRNA from baker's yeast, 1—tRNA^{Ala}; 2—tRNA$_1^{Val}$; 3—tRNA$_2^{Val}$; 4—tRNA^{Tyr}; 5—mg of tRNA.

The next step was periodate oxidation after aminoacylation by valine, according to Zamecnik et al. (26, 27), followed by the removal of the oxidized products on a polyacrylic acid hydrazide column; instead of the soluble form (28, 29), it is much more convenient to use the hydrazide incorporated in an agar gel, according to the technique proposed by Knorre et al. (30). This form has no disadvantages peculiar to the soluble polyacrylic acid and can be employed in the form of a column. At the same time it does not prevent the oxidized tRNA's from reacting with active groups of the polyacrylic acid hydrazide.

The most tedious part of the isolation of individual tRNA is the enzymatic aminoacylation of tRNA's after CCD. It is associated with technical difficulties: the isolation of aminoacyl-tRNA-ligase as well as the obligatory exhaustive aminoacylation of tRNA$_1^{Val}$ are time-consuming procedures.

As a result, we usually obtain tRNA$_1^{Val}$ of 85–95% purity with a yield of about 50%. The admixtures appear to include tRNA$_2^{Val}$ as the main component. The purity of the preparation was established on the basis of an analysis of oligonucleotides resolved on fingerprints and by column chromatography. This is necessary because the determination of the acceptor activity does not discriminate between tRNA$_1^{Val}$ and tRNA$_2^{Val}$ and cannot be used as a basis for the purity calculation.

Most stages of the isolation procedure were greatly facilitated by the use of an experimental technique devised during the course of this investigation, and which may be designated as "flotation", by analogy with the well-known method used in ore-dressing (31). The suspension of the cetavlon-RNA complex, containing a suitable amount of salt (e.g. 0.1–0.6 M phosphate) is shaken with ethyl ether under moderate heating. Within a few minutes the cetavlon-tRNA is accumulated at the interphase between the ether and the aqueous layers and can be easily removed. In this way very small quantities of tRNA can be isolated from large volumes of highly diluted solutions (down to 1 μg/ml).

The study of the primary structure was begun in the usual manner with the exhaustive enzymatic hydrolysis of the tRNA^{Val} preparations by two kinds of RNases, namely pancreatic pyrimidylo-ribonuclease (EC 2.7.7.16) and guanylo-ribonuclease (EC 2.7.7.26) from Actinomycetes. The latter is similar to the T$_1$-RNase from taka diastase (32). It has been isolated in highly purified form from the culture-medium of *Actinomyces aureoverticillatus* Kras et Di Shen, strain 1306 (33, 34). In contrast to the T$_1$RNase, this enzyme splits not only the bonds adjacent to inosinic and guanylic acids, but also all bonds adjacent to methylated derivatives of guanylic acid. Other enzymes, used in later stages, were phosphomonoesterase (prostatic and from *E. coli*) (E. C. 3.1.3.1, 3.1.3.2) and phosphodiesterase (E.C. 3.1.4.1) from the venom of *Vipera lebetina*.

For the separation, isolation and identification of the products of hydrolysis column chromatography (35) as well as the fingerprint method were used (36, 37).

The ion exchange separation of a pyrimidylo-ribonuclease digest is shown in Figure 2. The isopliths obtained on DEAE-cellulose in 7 M urea (38) were transferred (without drying, desalting or removal of urea) to a Dowex 1×4 column (HCOO$^-$) and separated by a linear gradient of HCOOH from 0.0 to 4.0 M and then in a gradient of NH$_4$COOH (0.0 to 0.6 M) at a constant 4 M concentration of HCOOH. The greater part of oligonucleotides thus obtained was homogeneous; there were only two exceptions, namely, the peaks II$_3$ adn III$_6$. In these two cases rechromatography was carried out in the same system, but at a constant eluant concentration.

The guanyloribonuclease digest was separated in the same ion exchange system and also yielded homogeneous oligonucleotides. Paper chromatograms were developed in the first direction with isobutyric acid—0.5 M NH$_4$OH (10:6), pH 3.7 and in the second with tert. butanol-HCOOH (1:1), pH 4.8 [34, 35].

The separation of the guanyloribonuclease digest by the fingerprint method is shown in Fig. 3. The resolution of 1–2 mg tRNA$_1^{Val}$ digest on What-

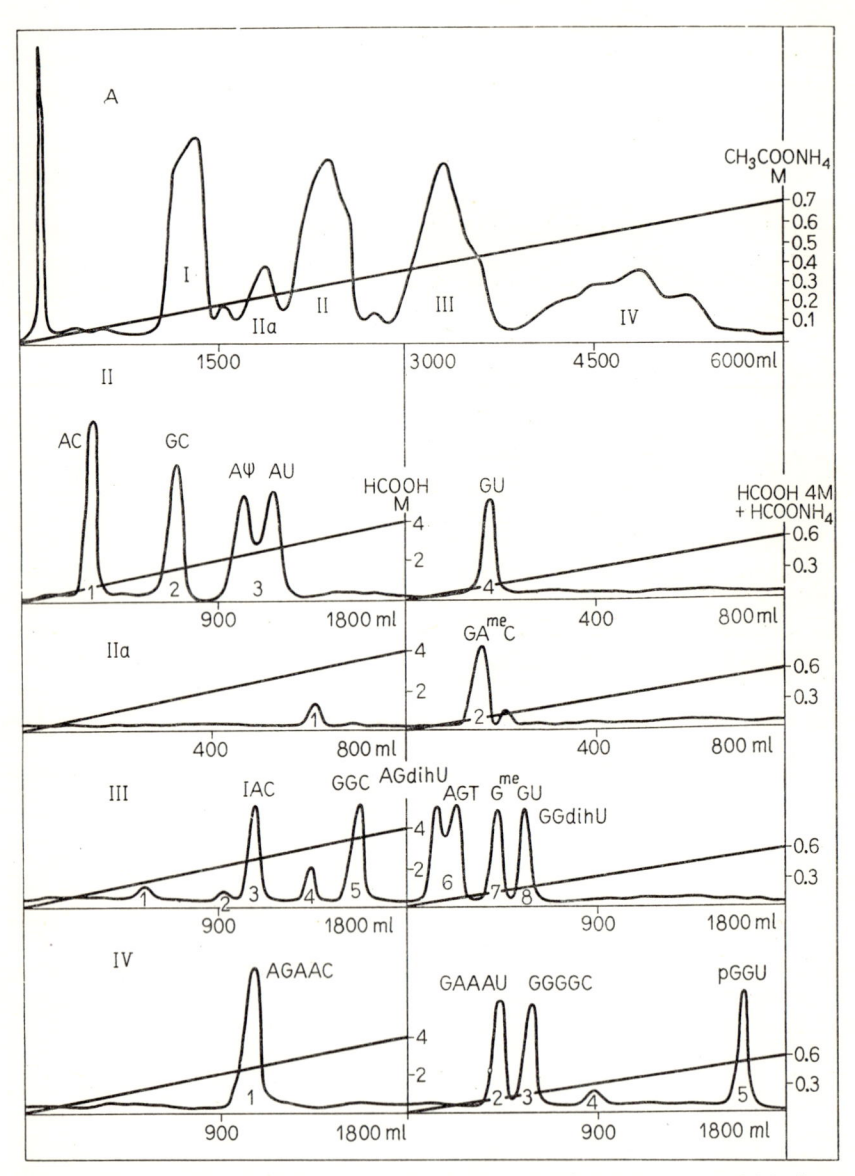

FIGURE 2. Ion-exchange chromatography of oligonucleotides from a pyrimidyloribonuclease digest of baker's yeast tRNA$_1^{Val}$. A—Separation on the DEAE-cellulose column with ammonium acetate gradient in 7 M urea, pH 7.5. II, IIa, III and IV—rechromatography of corresponding fractions of A run on the Dowex 1×4 (HCOO⁻) column with formic acid and ammonium formiate.

man 3 is usually satisfactory; overloading results in merging of the spots, mainly in the trinucleotide area.

The nucleotide composition of the separate oligonucleotides was established by the usual methods of alkaline hydrolysis and paper chromatography. A useful tool for the direct spectrophotometric identification of di- and trinucleotides was provided by the "set" of spectra collected during the last two years and partly published in the form of an atlas (39).

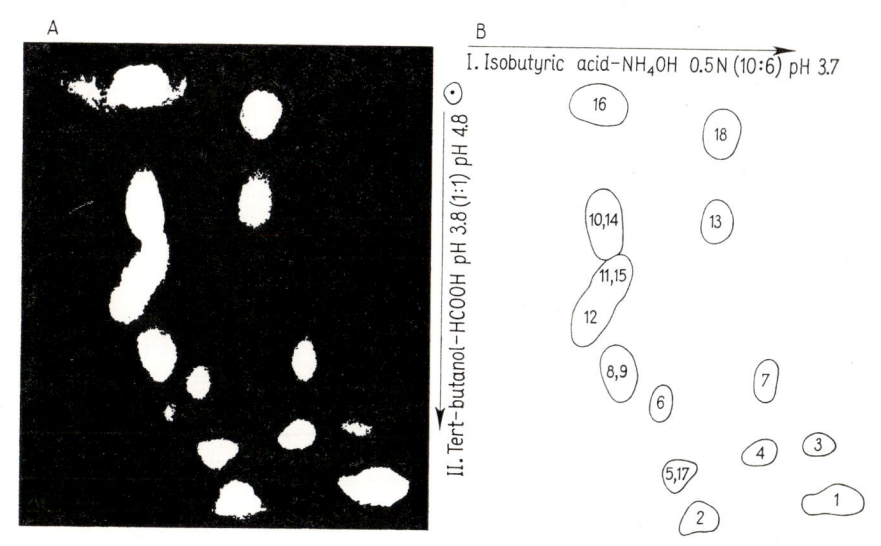

FIGURE 3. Two-dimensional paper chromatography of oligonucleotides from a guanyloribonuclease digest of baker's yeast tRNA$_1^{Val}$. (A)—photo- (B)—key to the photo.

The nucleotide composition of tRNA$_1^{Val}$ is shown in Table I. These data are compiled from the results of the alkaline digest analysis and those of the oligonucleotides of the guanylo- and pyrimidyloribonuclease digests. For a long time we could not detect 5-methylcytidylic acid (12), but later it was found in the mononucleotide fraction of pyrimidyloribonuclease digest and in the oligonucleotide AAC(C$_3$, diHU, 5 MeC) AGp of the guanyloribonuclease digest.

The comparison of our data with the results of Holley et al. (1) and Ingram and Sjöquist (8) can be made only with caution since their papers were published three years ago. But it is difficult to explain the absence of IMeGp from Holley's list and the occurrence of an additional 2MeA. With this exception the agreement of the data can be considered as satisfactory.

The mono- and oligonucleotides of the pyrimidylo- and guanylo RNase digests of tRNA$_1^{Val}$ are listed in Table II. The parentheses indicate that the nucleotide sequence has not yet been established. This concerns some oligonucleotides of the guanylo-RNase digest with three or more adjacent pyrimidine units. The only method of sequence determination in such cases

Table I. Nucleotide Composition of Baker's Yeast tRNA$_1^{Val}$

Nucleotides	Our data	Moles/mole tRNA Holley et al. (1)	Armstrong et al. (11)
Ap	14	14*	13*
Gp	21	23	18
Up	12	15	14
Cp	21	21	25–26
ψp	4	4	3
1MeAp	1	1†	1
lMeGp	1	—	1
5MeCp	1	1	1–2
RiboTp	1	1	1
Ip	1	1	1
diHUp	4	—††	—††
2MeAp	—	1	—
	81	81	77–79

* The terminal Ap should be added.
† N$_{(6)}$MeAp.
†† Dihydrouridylic acid was unknown in 1963.

Table II. Oligonucleotides of Exhaustive Pyrimidylo- and Guanyloribonuclease Digests of tRNA$_1^{Val}$

	Pyrimidyloribonuclease digest			Guanyloribonuclease digest	
NN	Oligo (mono)-nucleotides	Moles/mole tRNA	NN	Oligo (mono)-nucleotides	Moles/mole tRNA
1	Cp	12	1	Gp	7
2	Up	5	2	C–Gp	2
3	diHUp	2	3	A–Gp	—
4	ψp	3	4	U–IMeGp	1
5	5MeCp	1	5	U–Gp	1
6	A–Cp	2	6	diHU–C–Gp	1
7	G–Cp	3	7	C–A–Gp	1
8	A–ψp	1	8	T–ψ–C–Gp	1
9	A–Up	1	9	C–ψ–U–Ip	1
10	G–Up	2	10	diHU–A–U–Gp	1
11	G–IMeA–Up	1	11	U–ψ–C–A–Gp	1
12	I–A–Cp	1	12	U–U–U–C–Cp	1
13	G–G–Cp	1	13	A–C–A–C–Gp	1
14	A–G–diHUp	1	14	C–A–ψ–U–C–G	1
15	A–G–Tp	1	15	IMeA–U(C, C, U, diHU) Gp	1
16	IMeG–G–Up	1	16	A–A–C (C, C, C, diHU, 5 MeC)A–Gp	1
17	G–G–diHUp	1	17	pGp	1
18	A–G–A–A–Cp	1	18	A–A–A–U(C, A) C–C–A$_{OH}$	1
19	G–A–A–A–Up	1			81
20	G–G–G–G–Cp	1			
21	pG–G–Up	1			
22	A	1			
		81			

is incomplete digestion with snake venom phosphodiesterase (1,40) or pancreatic RNase.

The list of oligonucleotides of both digests completely agree with each other, i.e., there is no discrepancy in the number of nucleotides and in the nucleotide sequence of the fragments. There remains some uncertainty about the dinucleotide AGp; its amount varies in different experiments and therefore we suppose that it may originate not from RNA_1^{Val} but from impurities or it is very likely a breakdown product of some labile oligonucleotides.

We encountered various difficulties in analysing the oligonucleotides, due to a large number of monomeric units in many of the fragments. This becomes obvious when examining the oligonucleotides listed in Table II. There were other difficulties which are less obvious, but nevertheless very important from the practical point of view: e.g. the degradation of longer oligonucleotides during the analytical procedure as a result of secondary effects of RNases; on the other hand, we must take into consideration the lability of certain internucleotide bonds. Spontaneous breakdown of oligonucleotides has also been noticed by other investigators (5,6). A high degree of instability is observed, for instance, in the case of the decanucleotide No. 16 of the guanyloribonuclease digest. The lability is not always dependent on the chain length but can result from the presence of dihydrouridylic acid (41). The degradation of the larger oligonucleotides not only leads to the accumulation of breakdown products, but also reduces the amount of available material. This seriously complicates the analysis because the quantity of oligonucleotides is as a rule already small due to unavoidable losses during the multistage analytical procedure. The dihydrouridylic acid introduces special difficulties owing to its lability and spectral inperceptibility in 260 mμ region.

When comparing the data in Table II with those of other authors we notice similarities as well as some discrepancies. There are only three common oligonucleotides in our list and in that of Holley (1), namely A-G-Tp, I-A-Cp and G-G-Cp (we do not take into consideration the dinucleotides). There is better agreement between our data and those of Ingram (8, 11); we find seven common oligonucleotides: A-ψp, I-A-Cp, G-G-Cp, A-G-Tp, pG-G-Up (terminal), A-G-T-(C, ψ)-Gp, U-lMeG-G-Up. There are considerable differences between our data and those of Staehelin (42), but his preparation of $tRNA^{Val}$ was only of moderate purity.

More striking is the resemblance between the oligonucleotides of $tRNA_1^{Val}$ and $tRNA^{Ala}$. A comparison of Holley's data with ours shows the following oligonucleotides to be common for both tRNA's: A-G-diHUp, G-G-diHUp, U-lMeGp, diHU-C-Gp, T-ψ-C-Gp; at least four other oligonucleotides differ in only one nucleotide. It is difficult to decide whether this similarity reflects some structural features in common which may be accounted for by the fact that alanine and valine both belong to the aliphatic nonpolar amino acids. It should be noticed that the above mentioned common oligonucleotides contain minor components (diHU, 1MeG, T, ψ). This fact underlines the peculiar nature of this resemblance, since the position of the minor components is highly specific, though their function is not yet clear.

The determination of the primary structure of a nucleic acid as a whole begins, strictly speaking, from the stage which we have just described, namely, after the establishment of the sequences within the oligonucleotides of the exhaustive RNase digests. We fully agree with Holley when he regards this preliminary part of the work to be the most tedious and time consuming (2).

The oligonucleotides of the two RNase digests present very few opportunities for overlapping, and therefore give us scanty information about the structure of the whole molecule. This is especially due to the small length of the oligonucleotides obtained by pancreatic RNase digestion. The data derived from the two exhaustive enzymatic digests permitted us to reconstruct only a limited number of sequences which are represented in Table III.

TABLE III. Reconstruction of tRNA$_1^{Val}$ Fragments Based on the Oligonucleotide Sets from Exhaustive RNase Digests

The lines above and below the symbols designate the oligonucleotides of the pyrimidilo- and guanyloribonuclease digests, respectively

1. ... G–C–ψ–U–I–A–C–A–C–G

2. ... G–U–IMeG–G–U

3. ... pyr A–G–T–ψ–C–G

4. ... pyr G–IMeA–U(C, C, U, diHU) G

5. ... pyr G–A–A–A–U(A, C) C–C–A$_{OH}$ acceptor 3'-end

6. ... pyr A–G–A–A–C (C, C, C, diHU, 5MeC) A–G

A–G of 5'-end of the sequence "6" is overlapped by C–A–G or U–ψ–C–A–G (from a guanyloribonuclease digest). A–G of the 3'-end may be overlapped by oligonucleotide A–G–diHU or by the sequence "3".

7. pG–G–U 5'-end

The reconstructed fragments contain a total of 57 nucleotides from among the 80 nucleotides of the whole tRNA$_1^{Val}$ chain; this includes both terminal sequences—the acceptor as well as the non-acceptor 5'-end.

Although of considerable length, these reconstructed fragments do not permit of any further conclusions. In principle several procedures could provide further necessary information. One of these would be to find a nuclease which splits the chain at sites of adenylic residues (adenylo-ribonuclease). R. Tatarskaja in our laboratory undertook an extensive search for this purpose, mainly among a large variety of Actinomycetes, but the efforts have as yet been unsuccessful. Meanwhile other authors have reported that this enzyme is unspecific and splits other nucleotidic linkages besides those formed by adenylic residues. Another approach would be to modify chemically certain nucleotides, thus changing the conditions for the action of the two ribonucleases used (43). This may be an efficient method, but we lack at present the necessary means to produce the modifications fully adapted for the sequence analysis. Finally, it is known that different nucleotide linkages, even of the same general kind (i.e. pyrimidylic or guanylic) differ in their susceptibility to the action of the corresponding ribonucleases, depending on the immediate neighbour or on the broader environment. First of all, it is well known that the rate of phosphodiester bond hydrolysis by both kinds of RNases depends on the nature of the base of the right-hand neighbour of the nucleotide of which the 3'-bond is split (34, 44, 54). Another important factor is the secondary structure in which the nucleotides may be involved. Hydrogen bonding considerably increases the resistance to enzymatic hydrolysis whereas the coplanar stacking of the bases seems to have no noticeable effect. Cousin (46), using different natural and synthetic polyribonucleotides, has shown that the rate of splitting by pancreatic ribonuclease depends to a large extent on the secondary structure of the substrate molecule. Van Holde *et al.* (47) have clearly shown the important role of the helical two-stranded structure of polynucleotides for their resistance to RNase action. According to the data of Cantoni *et al.*, tRNA is relatively resistant to the nuclease action, the property also being dependent on its secondary structure (48, 49).

These data explain the effect of Mg^{2+} ions on the rate of hydrolysis by RNase described by Nishimura and Novelli (50).

The different stability of diester bonds of tRNA does not manifest itself under the optimal conditions of exhaustive hydrolysis, but may become apparent under less favourable conditions of pH, temperature etc. Maintaining proper conditions sufficiently constant would help to obtain products of partial hydrolysis, necessary for the purpose of primary structure analysis.

We have applied this principle in our studies on the tRNA structure in relation to its function, with encouraging results.

The idea of partial enzymatic hydrolysis for the investigation of the primary structure of RNA was formulated by Holley in 1963 (1), and was followed by his fundamental studies on the structure of alanine tRNA (3, 4), in which the method of incomplete enzymatic digestion was effectively employed.

We have performed the partial hydrolysis of RNA by guanylo-RNase and can report here on the data obtained concerning the sequences of some large $tRNA_I^{Val}$ fragments. The conditions of digestion were chosen on the basis of our experience on partial tRNA digestion which we obtained in our preliminary studies. The tRNA was digested with a minute amount of enzyme at 0° in the presence of 0.01 M $MgCl_2$ for 4 hours. Under these conditions only one phosphodiester bond appears to be split and the molecule breaks into two nearly equal parts. The larger corresponds to the acceptor sequence (3'-end), the shorter fragment represents the non-acceptor 5'-end. The two fragments of $tRNA_I^{Val}$ were completely hydrolyzed further with guanyloribonuclease. In this way we have obtained two sets of the oligonucleotides of the guanyloribonuclease digest of $tRNA_I^{Val}$ (Table IV). This analysis allowed us to distribute between the two halves of $tRNA_I^{Val}$ nearly all oligonucleotides identified in the exhaustive digest of the original $tRNA_I^{Val}$ (Table III).

TABLE IV. Oligonucleotides of 5'-and 3'-Fragments of tRNA$_1^{Val}$ Partially Digested with Guanyloribonuclease

No. of Table III	5'-fragment	No. of Table III	3'-fragment (acceptor)
4	U–IMeGp	2	C–Gp
5	U–Gp	7	C–A–Gp
6	diHU–C–Gp	8	T–ψ–C–Gp
9	C–ψ–U–I	13	A–C–A–C–G
10	diHU–A–U–Gp	15	1MeA–U(C, C, U, diHU) Gp
11	U–ψ–C–A–G	16	A–A–C(C, C, C, diHU, 5MeC)A–G
12	U–U–U–C–Gp	18	A–A–A–U(A, C) C–C–A$_{OH}$
14	C–A–ψ–U–C–G		
17	Gp		

The analytical data thus collected resulted in an increased number of reconstructed sequences as compared to those listed in Table III. In particular, we could discriminate between the alternative oligonucleotides adjacent to 5'- and 3'-ends of the reconstructed oligonucleotide No. 6 (see Table III). In both cases the concurrent oligonucleotides distributed into different halves of the molecule, the oligonucleotide C–A–G remaining as the only possible left sequence, and A–G–T–ψ–C–G as the right one, of the oligonucleotide No. 6, which belonged to the 3'(acceptor) half of the tRNA$_1^{Val}$ molecule.

It now becomes possible to propose a nucleotide sequence for the tRNA$_1^{Val}$ molecule, although the available information required for the reconstruction of its primary structure is incomplete. It may be assumed, however, that the location of some characteristic sequences, which are identical in tRNA$_1^{Val}$ and tRNA^{Ala}, is similar in both cases. Taking into account this assumption, we can suggest the probable primary structure of tRNA$_1^{Val}$ as shown in Table V.

TABLE V. Probable Primary Structure of tRNA$_1^{Val}$

```
                Me           2H        2H
                 |            |         |
p–G–G–U–U–U–C–G–U–G–G–U–$\psi$–C–A–G–U–C–G–G–U–A–U–G–G–G–C–A–$\psi$–

                                              2H Me
                                               |  |
–U–C–G–U–G–C–$\psi$–U–I–A–C–A–C–G–C–A–G–A–A–C (C, C, C, U, C) A–G–T–$\psi$–C–

     Me          2H
      |           |
–G–A–U (C, C, U, U)–G–C–G–G–C–G–A–A–A–U (C, A) C–C–A$_{OH}$
```

It is possible to draw several conclusions concerning general features of the tRNA structure. First, we see that individual tRNA's from the same source (yeast), namely tRNA^{Ala} (Holley et al., Ref. 2), tRNA^{Ser} (Zachau et al., Ref. 6) and tRNA$_1^{Val}$ in our experiments show considerable differences in their gross nucleotide composition. This is not astonishing, in view of their very dissimilar behaviour in various analytical procedures, such as CCD, chromatography etc. From the data presented here we see that both the acceptor 3′-terminal sequence (excluding the C–C–A link) and the non-acceptor, 5′-sequence of the tRNA$_1^{Val}$ differ markedly from those in the alanine and the serine-tRNA's (2, 6).

The first cleavage by guanylo-RNase in the course of partial hydrolysis occurs, in the case of valine tRNA, between the inosinic and the adenylic residues localized within the sequence G–C–ψ–U–I–A–C–A–C–G. According to Penswick and Holley (3), the first bond to be split in the alanine-tRNA is between a guanylic and cytidylic residue. Alanine-tRNA is split in two parts of equal length; in the case of valine-tRNA the two parts differ slightly in length.

The localization of the ubiquitous sequence T–ψ–C–G– is analogous in valine and alanine tRNA's, since it is observed in the acceptor half of the molecule. Two fragments containing dihydrouridylic acid, namely diHU–A–U–G and diHU–C–G, are found in the digest of the 5′-end of tRNA$_1^{Val}$. Valine tRNA contains the sequence A–G–diHU–C–G–G–diHU–A–U–C, which resembles closely the sequence A–G–diHU–C–G–G–diHU–A–G present in alanine tRNA; in both cases they are localized in the 5′-OH half of the molecule.

These data allow us to conclude that, despite the obvious individuality of the primary structure of each acceptor specific tRNA, these molecules are designed according to a general pattern. Further studies will show whether this viewpoint is correct.

According to generally accepted views, the coding triplet of tRNA is not involved in the secondary structure, thus possessing a greater chemical activity and being free to enter in complementary interaction with its partner in the template RNA. On the other hand, it is known that its participation in the secondary structure, as a result of hydrogen bonding, makes the corresponding parts of the polynucleotide molecule more resistant to the action of hydrolytic enzymes (see above). It seems reasonable to expect the first attack by RNase to be directed towards a point of least resistance; as such, the site of the coding triplet should be given primary consideration.

Before discussing how far this assumption can be applied to the two tRNA's at present available, viz. the alanine- and the valine tRNA's, some preliminary consideration are necessary. The pairing of nucleotide chains, be it of the Watson–Crick double-stranded type, or intramolecular pairing by loop formation, involves not only complementarity of the pairing bases, but also their antiparallel arrangement. There are grounds to believe that the second principle is also valid in the interaction between codon and anticodon.

Evidence is accumulating in favour of the assumption that the information carried by mRNA is read from left to right (5′ → 3′). This view is supported by many recent findings (51, 52, 53) in contrast to the previously widespreaded contrary opinion.

Taking into consideration the data mentioned above, we must attribute to the anticodons a complementary composition and an inverted order. Therefore, if the codon is, for example, 5'– ... C–G–A– ... 3', the anticodon will be, not 5'– ...G–C–U– ... 3'–OH, but 3'– ... G–C–U– ...–5', or, written in the conventional way, 5'– ... U–C–G– ... 3'–OH.

Now we have for the codons ascribed to valine (GUU, GUC, GUA, GUG) (54–56) the following anticodons (inverted): AAC, GAC, UAC, CAC. Of these one, namely GAC, is functionally equivalent to IAC, and it is exactly this sequence which in our experiments is split first during limited hydrolysis by guanyloribonuclease. In the case of tRNA^{Ala}, in the experiments of Holley, the situation is quite similar: the nucleotide chain is halved at the sequence I–G–C, equivalent to G–G–C–, which is the (inverted) anticodon to one of the alanine codons, G–C–C–. Thus, in both cases, at least with one of the known codons for each, the previously advanced assumption that the coding triplet represents the point of first attack by RNase, appears justified. Further experiments will demonstrate how far this can be regarded as a general rule. Should this be the case, it would greatly simplify the exact localization of the coding triplets within the complete sequence of the intact molecule of a tRNA. Without this clue, it will be very questionable to ascribe the coding function to a definite triplet. We must consider the difficulties introduced, on the one hand, by the degeneracy of the code and also by the great variety of trinucleotide combinations which may be composed from any given primary structure.

The final goal in the elucidation of the complete nucleotide sequence of tRNA is obviously the establishment of correlations between the chemical structure and biological function. The role of the coding triplet is self-evident, as is also that of the acceptor segment. But nothing is yet known about the remainder of the large molecule of tRNA. In small molecules all their properties are determined by the presence of specific chemical groupings. In macromolecules this is complicated by the part played by definite assemblies of structural units. In enzymes we have to deal often with very intricately composed active sites, and modern research reveals the decisive significance of the tridimensional structure, as evidenced by the recent brilliant studies on heme proteins or lysozyme. There is every reason to expect that the same holds for the nucleic acids and the transfer RNA's in particular.

The coding triplet contributes the specificity concerning the spacial localization of individual tRNA on the template mRNA. The same grouping together with other parts of the molecules may participate in the specific interaction of tRNA's with the corresponding coding enzymes. But alongside these highly specialized properties, the tRNA's fulfil other functions which lack the high specificity of those mentioned, and may be of a practically universal character, being common to all the individual tRNA's to the same degree. This would represent, on a limited scale, a case of "unity among diversity" which manifests itself so often in the field of biology. Thus arises the problem of "functional topology" regarding the structure of tRNA.

Such a generalized function lies in the interaction with the complex "messenger RNA-ribosome". It may be regarded as well established that tRNA somehow reacts with the ribosome, probably entering the functional site where the formation of the peptide bond actually takes place. Perhaps the tRNA also participates in the process of the gliding of the mRNA over this active site of the ribosome. Whichever it may be, here we deal with functions that bear no signs of specificity, being common to all the individual tRNA's. It may be assumed that definite structural features, common to each individual tRNA, are responsible for these functions. We known one sequence which appears to be ubiquitous for all tRNA's, i.e. the sequence T–ψ–C–G, discovered by Holley et al. It would be only logical to attribute a common function to structural assemblies of this type. Each success on this difficult but fascinating pathway will be highly rewarding for the often tedious, and sometimes exasperating, efforts which have to be made to achieve progress.

SUMMARY

tRNA$_1^{Val}$ was isolated from baker's yeast total tRNA. The oligonucleotide composition of its exhaustive pyrimidylo- and guanyloribonuclease digests was established. Two large slightly unequal fragments of tRNA$_1^{Val}$, obtained by its partial digestion with guanylo-RNase, were separated and their oligonucleotide composition determined. The smaller fragment belonged to the 5'-end of the tRNA$_1^{Val}$ molecules and the larger one was derived from the 5'-(acceptor) end. Thus, preliminary formulation of the primary structure of tRNA$_1^{Val}$ became possible.

REFERENCES

1. Holley, R. W., Apgar, J., Everett, G. A., Madison, J. T., Merrill, S. H. and Zamir, A., *Cold Spring Harbor Symp.* **28**, 117 (1963).
2. Holley, R. W., Apgar, J., Everett, G. A., Madison, J. T., Marquisee, M. and Merrill, S. H., *Science* **147**, 1465 (1965).
3. Penswick, J. R. and Holley, R. W., *Proc. National Acad. Sci. USA* **53**, 543 (1965).
4. Apgar, J., Everett, G. A. and Holley, R. W. *Proc. National Acad. Sci.* **53**, 546 (1965).
5. Holley, R. W., Everett, G. A. and Madison, J. T., Zamir, A., *J. Biol. Chem.* **240**, 2122 (1965).
6. Dütting, D., Karau, W., Melchers, F. and Zachau, H. G., *Biochim. Biophys. Acta* **108**, 194 (1965).
7. Bergquist, P. L. and Robertson, J. M., *Biochim. Biophys. Acta* **95**, 357 (1965).
8. Ingram, V. M. and Sjöquist, J. A., *Cold Spring Harbor Symp.*, **28**, 113 (1963).
9. Staehelin, M., Schweiger, M. and Zachau, H. G. *Biochim. Biophys. Acta* **68**, 129 (1963).
10. Holley, R. W., Apgar, J., Merrill, S. H. and Zubkoff, P. L., *J. Amer. Chem. Soc.* **83**, 4861 (1961).
11. Armstrong, A., Hagopian, H., Ingram, V. M., Sjöquist, I. and Sjöquist, J., *Biochemistry* **3**, 1194 (1964).
12. Bayev, A. A., Venkstern, T. V., Mirzabekov, A. D., Krutilina, A. I., Li, L. and Axelrod, V. D., *Biochim. Biophys. Acta* **108**, 162 (1965).
13. Mirzabekov, A. D., Venkstern, T. V. and Bayev, A. A., *Biokhimia* **30**, 825 (1965).
14. Mirzabekov, A. D., Krutilina, A. I., Reshetov, P. D., Sandakhchiev, L. S., Knorre, D. G., Khokhlov, A. S. and Bayev, A. A., *Dokl. Akad. Nauk SSSR* **160**, 1200 (1965).

15. Apgar, J., Holley, R. W. and Merrill, S. H., *J. Biol. Chem.* **237**, 796 (1962).
16. Smith, C. J. and Herbert, E., *Feder. Proc.* **22**, No. 2, p. 1, 230 (1963).
17. Herbert, E., Smith, C. J. and Wilson, C. W., *J. Mol. Biol.* **9**, 376 (1964).
18. Smith, C. J., Herbert, E. and Wilson, C. W., *Biochim. Biophys. Acta* **87**, 341 (1964).
19. Herbert, E., Smith, C. J. and Wong, J. T., *Abstr.* VIth ICB, 1–79 (1964).
20. Ishida, T. and Miura, K. I., *J. Mol. Biol.* **11**, 341 (1965).
21. Grachev, M. A., Budowsky, E. I., Mirzabekov, A. D., Krutilina, A. I. and Sandakhchiev, L. S., *Biochim. Biophys. Acta* **108**, 506 (1965).
22. Goldstein, J., Bennett, T. P. and Graig, L. G., *Proc. National Acad. Sci. USA* **51**, 119 (1964).
23. Thiebe, R. and Zachau, H. G., *Biochim. Biophys. Acta Previews* **4**, No. 10 (1965).
24. Thiebe, R. and Zachau, H. G., *Biochim. Biophys. Acta* **103**, 568 (1965).
25. Mirzabekov, A. D., Krutilina, A. I. and Bayev, A. A., *Dokl. Akad. Nauk SSSR* (In press).
26. Zamecnik, P., Stephenson, M. L. and Scott, J. F., *Proc. National Acad. Sci. USA* **46**, 811 (1960).
27. Stephenson, M. L. and Zamecnik, P. C., *Biochem. Biophys. Res. Comm.* **7**, 91 (1962).
28. Portatius, V. H., Doty, P. and Stephenson, M. L., *J. Amer. Chem. Soc.* **83**, 3351 (1961).
29. Zachau, H. G., Tada, M., Lawson, W. B. and Schweiger, M., *Biochim. Biophys. Acta* **53**, 221 (1961).
30. Knorre, D. G., Misina, S. D. and Sandakhchiev, L. S., *Izv. Sibirsk. Otd. Akad. Nauk SSSR*, No. 11, 143 (1964).
31. Mirzabekov, A. D., Krutilina, A. I., Gorshkova, V. I. and Bayev, A. A., *Biokhimia* **29**, 1158 (1964).
32. Egami, F., Takahashi, K. and Uchida, T., in "Progress in Nucleic Acid Research", N.Y. and L., vol. 3, p. 59 (1964).
33. Tatarskaya, R. I., Abrossimova-Amelyanchik, N. M., Axelrod, V. D., Korenyako, A. T., Venkstern, T. V., Mirzabekov, A. D. and Bayev, A. A., *Dokl. Akad. Nauk SSSR* **157**, 725 (1964).
34. Abrossimova-Amelyanchik, N. M., Tatarskaya, R. I., Venkstern, T. V., Axelrod, V. D. and Bayev, A. A., *Biokhimia* **30**, 1269 (1965).
35. Axelrod, V. D., Venkstern, T. V. and Bayev, A. A., *Biokhimia* **30**, 999 (1965).
36. Krutilina, A. I., Venkstern, T. V. and Bayev, A. A., *Biokhimia* **29**, 333 (1964).
37. Li L., Venkstern, T. V., Mirzabekov, A. D., Krutilina, A. I. and Bayev, A. A., *Biokhimia* **31**, 117 (1965).
38. Tomlinson, R. V. and Tener, G. M., *Biochim. Biophys. Res. Comm.* **10**, 304 (1963).
39. Venkstern, T. V. and Bayev, A. A. Absorption Spectra of Minor Bases. "Nauka", Moscow (1965).
40. Holley, R. W., Madison, J. T. and Zamir, A., *Biochem. Biophys. Res. Comm.* **17**, 389 (1964).
41. Madison, J. T. and Holley, R. W., *Biochem. Biophys. Res. Comm.* **18**, 153 (1965).
42. Staehelin, M., Schweiger, M. and Zachau, H. G., *Biochim. Biophys. Acta* **68**, 129 (1963).
43. Kochetkov, N. K. and Budowsky, E. I., in "Molecular Biology. Problems and Perspectives", Moscow, "Nauka", 1964, p. 139.
44. Witzel, H. and Barnard, E. A., *Biochem. Biophys. Res. Comm.* **7**, 289 and 295 (1962).
45. Whitfeld, P. R. and Witzel, H., *Biochim. Biophys. Acta* **72**, 338 (1963).
46. Cousin, M., *Bull. Soc. Chimie Biol.* **45**, 1363 (1963).
47. Van Holde, K. E., Brahms, J. and Michelson, A. M., *J. Mol. Biol.* **12**, 762 (1965).

48. Singer, M. F., Luborsky, S. W., Morrison, R. A. and Cantoni, G. L., *Biochim. Biophys. Acta* **38**, 568 (1960).
49. Cantoni, G. L., Gelboin, H. V., Luborsky, S. W., Richards, H. H. and Singer, M. F., *Biochim. Biophys. Acta* **61**, 354 (1962).
50. Nishimura, S. and Novelli, G. D., *Biochim. Biophys. Res. Comm.* **11**, 161 (1963).
51. Ochoa, S., Abstr. Comm. 2nd Meeting FEBS, Vienna, p. 273 (1965).
52. Matthaei, J. H., Heller, G., Voigt, H. P., Kleinhauf, H., Küntzel H., Vogt, M. and Matthaei, H., *Naturwiss.* **52**, 653 (1965).
53. Thach, R. E., Cecere, M. A., Sundararajan, T. A. and Doty, P., *Proc. Nat. Acad. Sci.* **54**, 1167 (1965).
54. Nirenberg, M. and Leder, P., *Science* **145**, 1399 (1964).
55. Brimacombe, R., Trupin, J., Nirenberg, M., Leder, P., Bernfield, M. and Jaouni, T., *Proc. Nat. Acad. Sci. USA* **54**, 954 (1965).
56. Söll, D., Ohtsuka, E., Jones, D. S., Lohrmann, R., Hayatsu, H., Nishimura, S. and Khorana, H. G., *Proc. Nat. Acad. Sci. USA* **54**, 1378 (1965).

ADDENDUM (added in proof): The tentative primary structure of $tRNA_I^{Val}$ was confirmed experimentally. By means of partial hydrolysis with guanyloribonuclease many large oligonucleotides were obtained separately from 3′-and 5′-fragments of the molecule. The resulting overlapping was sufficient for a complete reconstruction of the $tRNA_I^{Val}$ molecule. This structure differs from the tentative one only in localization of two dinucleotides (UG and CG), underlined in the following formula:

$$\text{me} \qquad\qquad\qquad 2H \qquad\qquad 2H$$
$$pG-G-U-U-U-C-G-U-G-U-\dot{G}-G-U-\psi-C-A-G-\dot{U}-C-G-G-\dot{U}-A-U-G-$$
$$-G-G-G-C-A-\psi-U-C-G-C-\psi-U-I-A-C-A-C-G-C-A-G-A-A-C(C, C,$$

$$2H\text{ me} \qquad\qquad\qquad \text{me} \qquad\qquad 2H$$
$$C, \dot{U}, C)A-G-T-\psi-C-G-\dot{C}-G-\dot{A}-U(C, C, U, \dot{U})G-G-C-G-A-A-A-U-A-$$
$$-C-C-C-A_{OH}$$

The determination of the sequences of two oligonucleotides (T 15 and T 16, Table II) is now under way. The tentative $tRNA_I^{Val}$ primary structure was formulated on the basis of the idea, that the primary structures of $tRNA_I^{Val}$ and of $tRNA^{Ala}$ are similar, at least in their main features. This idea has apparently a more wide significance since the position of some specific sequences (those containing dihydrouridylic acid, the anticodon and the "ubiquitous" oligonucleotide) proved to be similar also in $tRNA^{Ser}$ (H. Zachau et al., this Symposium, p. 271) and $tRNA^{Tyr}$ (J.T. Madison et al., Science, **153**, 531, 1966).

Fractionation of Radioactive Nucleotides

F. SANGER AND G. G. BROWNLEE

Medical Research Council Laboratory of Molecular Biology, Cambridge, England

During studies on the sequences of amino acids in proteins the main limiting factor to progress has been in the development of methods of fractionation of the polypeptides produced on partial degradation, and it seems probable that the same will be true for nucleic acids and that progress will depend very largely on the development of fractionation techniques for oligonucleotides. The most generally used method is ion exchange chromatography on DEAE-cellulose columns and this has been used with particular success by Holley and his colleagues (1) in their determination of the sequence of alanine transfer-RNA. In our own studies of amino acid sequences we have frequently preferred to use, where possible, fractionation techniques on paper rather than column chromatography, since such methods are relatively rapid and simple to perform, and it is also possible to employ the two-dimensional approach which gives a very high degree of resolution. The standard methods of fractionation on paper have not led to very good resolution with nucleotides since the larger ones, and particularly those containing G residues, tend to streak rather badly.

We have recently developed a two-dimensional fingerprinting technique for separating radioactive nucleotides from RNA which employs two modified celluloses and gives better resolution than other reported techniques (2). Since only small amounts of material can be used for the method it has only been possible to apply it to highly radioactive RNA. This can readily be prepared in biological labelling experiments with ^{32}P. The nucleotides are then detected by radioautography and can be estimated by counting techniques. Digests are prepared by the action of ribonuclease T_1 or pancreatic ribonuclease. The former is the most useful since it is more specific, splitting only at G residues, and has been studied in more detail. One of the dimensions used in the fingerprinting technique makes use of DEAE–paper and is somewhat analogous to the DEAE–cellulose column methods. Fractionation is however carried out by ionophoresis on this material, and the resulting separation is due partly to ionophoresis and partly to ion-exchange chromatography. The two effects complement one another and prove rather effective, particularly in the separation of isomers.

The first dimension of the fingerprinting technique is carried out by ionophoresis, usually at pH 3.5 on cellulose acetate membrane. Nucleotides move as much sharper spots on this material than on paper, and with most of those studied we have not found any serious streaking. Figure 1 shows the

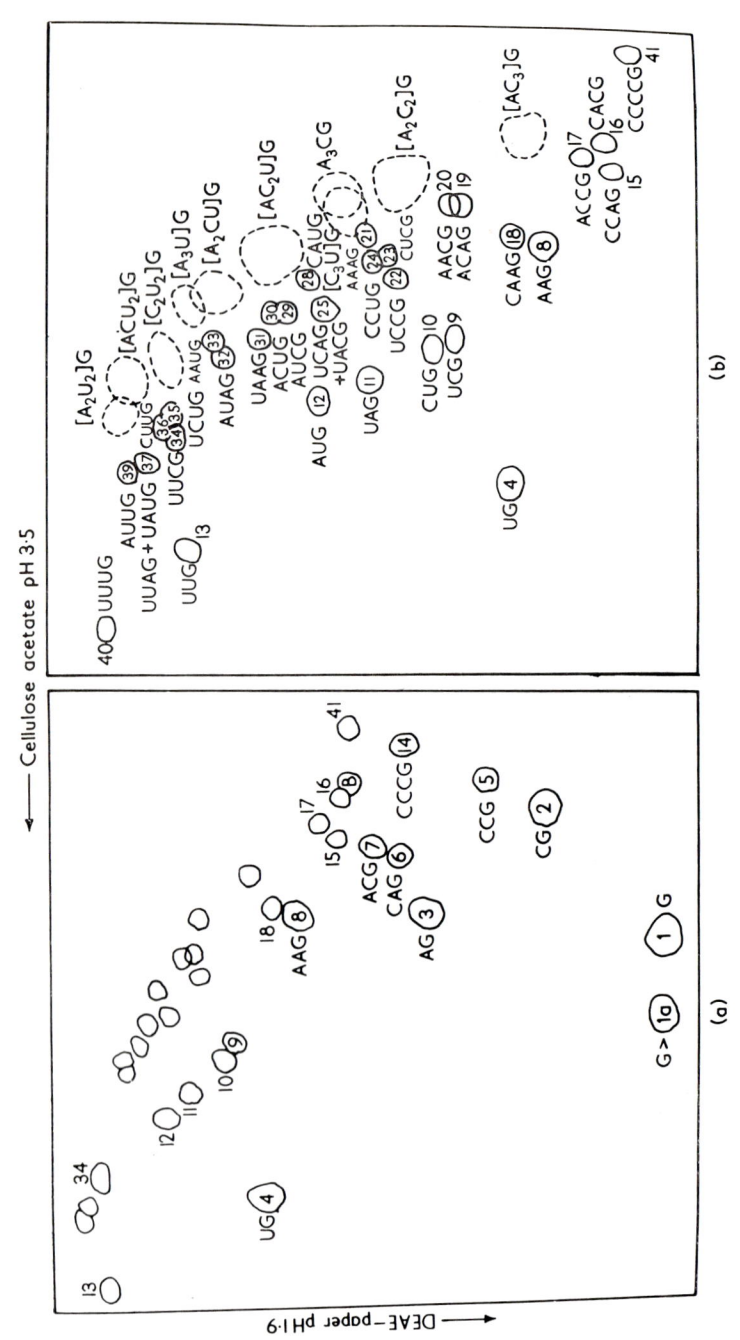

FIGURE 1. Diagram showing the distribution of nucleotides on the two-dimensional system using a mixture at pH 1.9 for the fractionation on DEAE-paper. (a)—the fractionation has been run for a relatively short time; (b)—run for a longer time.

distribution of the various oligonucleotides obtained from a ribonuclease T₁ digest of ribosomal RNA, using a mixture at pH 1.9 for the fractionation on DEAE-paper. Since the enzyme splits at the G residues, all of the nucleotides contain G in the 3'-terminal position. The three dinucleotides and nine trinucleotides are all completely separated from one another on the system. About half of the 27 possible tetranucleotides are resolved as single pure spots, the remainder occurring as mixtures of isomers, most of which can be partially resolved. There are 81 possible pentanucleotides and most of these are present in spots which are mixtures of isomers. The areas occupied by these different isomers are indicated in Figure 1b by dotted lines.

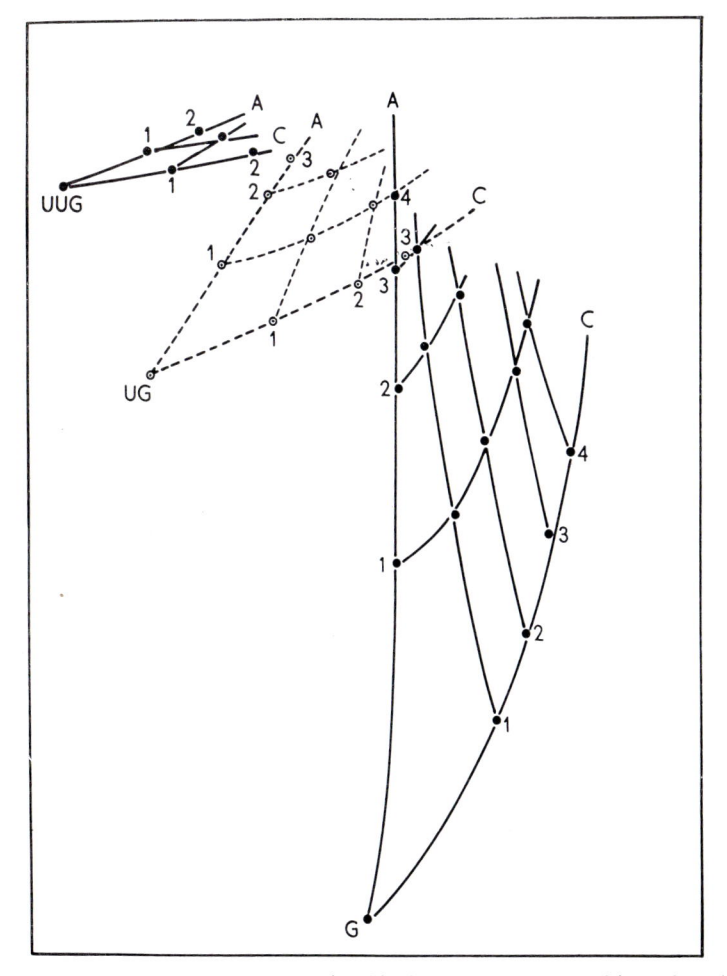

FIGURE 2. Diagram showing the relationship between the composition of nucleotides from a ribonuclease T₁ digest and their position on the two-dimensional system using ionophoresis at pH 1.9 for the DEAE-paper dimension.

Figure 2 is a diagram showing the relationship between the composition of the different nucleotides and their position on the two-dimensional fractionation procedure. It was prepared as follows. The "centre of gravity" of each set of isomers was first marked with a point. Lines were then drawn to link up the points representing nucleotides that differed only in the number of C residues they contained. Thus, for instance, a line was drawn joining UG, (CU) G, (C$_2$U)G and C$_3$U)G, etc. Similarly, lines were drawn connecting

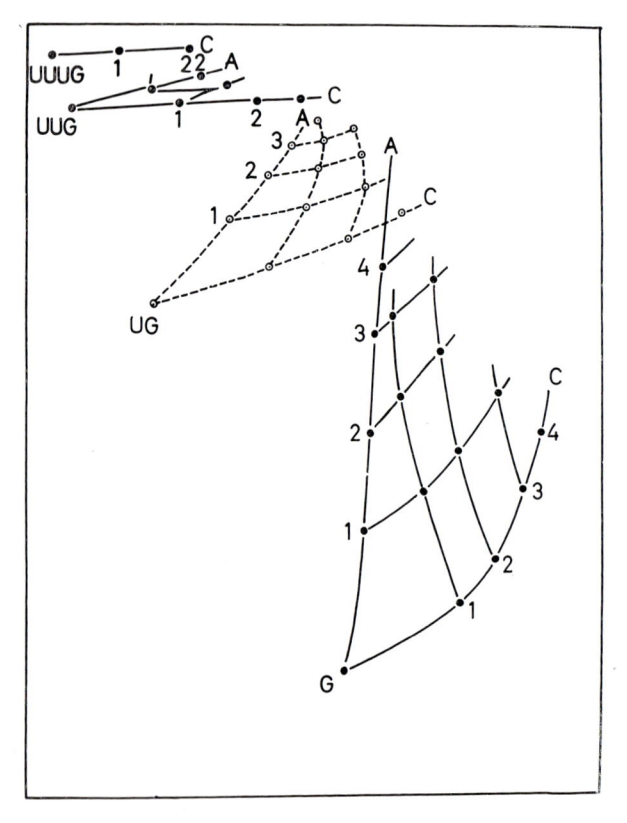

FIGURE 3. Diagram as in Figure 2 but using 7% formic acid for the DEAE-paper dimension.

nucleotides that differed only in A residues. It can be seen that the fingerprint is divided up into three graticules which vary in the number of U residues present since U is the mononucleotide which has the greatest effect on the mobility in the DEAE dimension. From a diagram of this sort it is possible to determine the composition of a particular nucleotide from its position on the fingerprint. However, in certain areas there is overlapping between different graticules and no unequivocal conclusion about the composition can be deduced for a spot coming in such areas. For instance, (A₃C) G and (C₃U)G would come in the same area. It was found that if a higher concentration of formic acid (7-10%) was used for the DEAE-paper dimension there was better separation between the different graticules and a diagram as in Figure 3 was obtained. Resolution of some fo the faster moving nucleotides is not as good on this system, but the slower ones move faster and

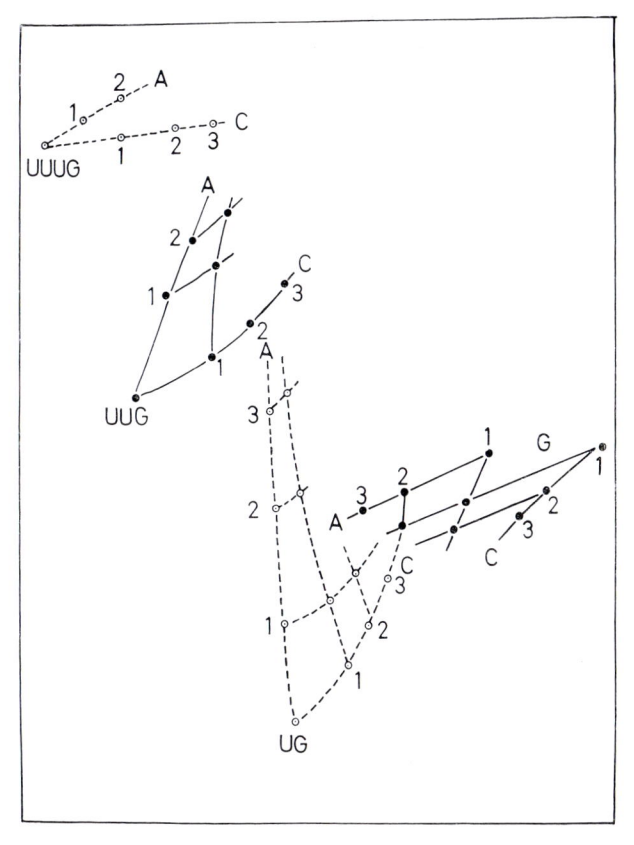

FIGURE 4. Diagram showing the relationship between the composition and position of nucleotides from a ribonuclease T₁ and phosphatase digest using 7% formic acid for the DEAE-paper dimension.

the separation of the graticules makes identification easier. The method is efficient for fractionation of nucleotides having up to two U residues per molecule, but ones containing more U residues do not move very far away from the origin on the DEAE-paper. Some improvement of the fractionation of slower moving nucleotides can be obtained by treating the digest with phosphatase which splits off the 3'-terminal phosphate groups and thus reduces the net negative charge. Figure 4 shows a diagram of the fractionation obtained using 7% formic acid on a digest that had been prepared by the action of ribonuclease T_1 and phosphatase. In this case four graticules are obtained representing nucleotides with 0, 1, 2 and 3 U residues respectively. There is somewhat less resolution of the nucleotides containing no U residues on this system and it will be noted that they in fact move more slowly in the DEAE dimension than the more acidic nucleotides containing one U residue. This is probably dependent on the complicated nature of the fractionation and suggests that in the case of the nucleotides containing no U residues the ionophoresis is becoming more important than the ion-exchange chromatography, which appears to be the predominant factor in the fractionation of the other nucleotides.

The systems hitherto described have made use of acidic conditions for fractionation on the DEAE-cellulose. Fractionation can equally well be carried out in alkaline solutions and Figure 5 shows a diagram of a fractionation which was carried out using triethylamine carbonate buffer. In this system there is no marked effect due to the number of U residues as there is on the acid side. This has the disadvantage that it is not possible to draw separate graticules from which one can identify nucleotides from their positions. On the other hand, it has the advantage that nucleotides containing large numbers of U residues will move faster on this system and can sometimes be resolved. It also seems probable that the separation of isomers is frequently better at the alkaline pH. Thus, for instance, the two isomers AACG and ACAG are readily separated in triethylamine carbonate, whereas they come in a single spot using the acidic systems.

The main use of the technique is likely to be for the determination of sequences in small RNAs. However it can also be used as a fingerprinting method for the identification and characterization of nucleic acids and for detecting small differences such as may be expected between corresponding nucleic acids from different mutants or species. Characteristic and different patterns were obtained from the two large ribosomal RNA components, indicating that the resolution is sufficient to obtain a specific fingerprint from an RNA of a few thousand nucleotides long.

In order to investigate the potentialities of the method for determining sequences, we have used it in a study of the 5S ribosomal component (3). This is composed of about 115 residues. All of the products from a ribonuclease T_1 digest and from a pancreatic ribonuclease digest can be separated on the two-dimensional system and the determination of their structure is described in a separate communication to this conference (Brownlee and Sanger). This information is insufficient to deduce a complete sequence and the use of partial digestion methods to obtain longer fragments is now under investigation.

Preliminary studies have also been carried out on transfer-RNAs. Plate I shows a "fingerprint" of a ribonuclease T_1 digest of mixed transfer-RNA from *E. coli*. The pattern is rather different from that obtained with ribosomal RNA and this is mainly due to the content of "minor" nucleotides in transfer-RNA. In confirmation of the work of others, it was seen that the minor nucleotides are frequently present in unexpectedly strong spots, suggesting that they are found in sequences which are common to more than one transfer-RNA. The strongest spot on the fingerprint is usually

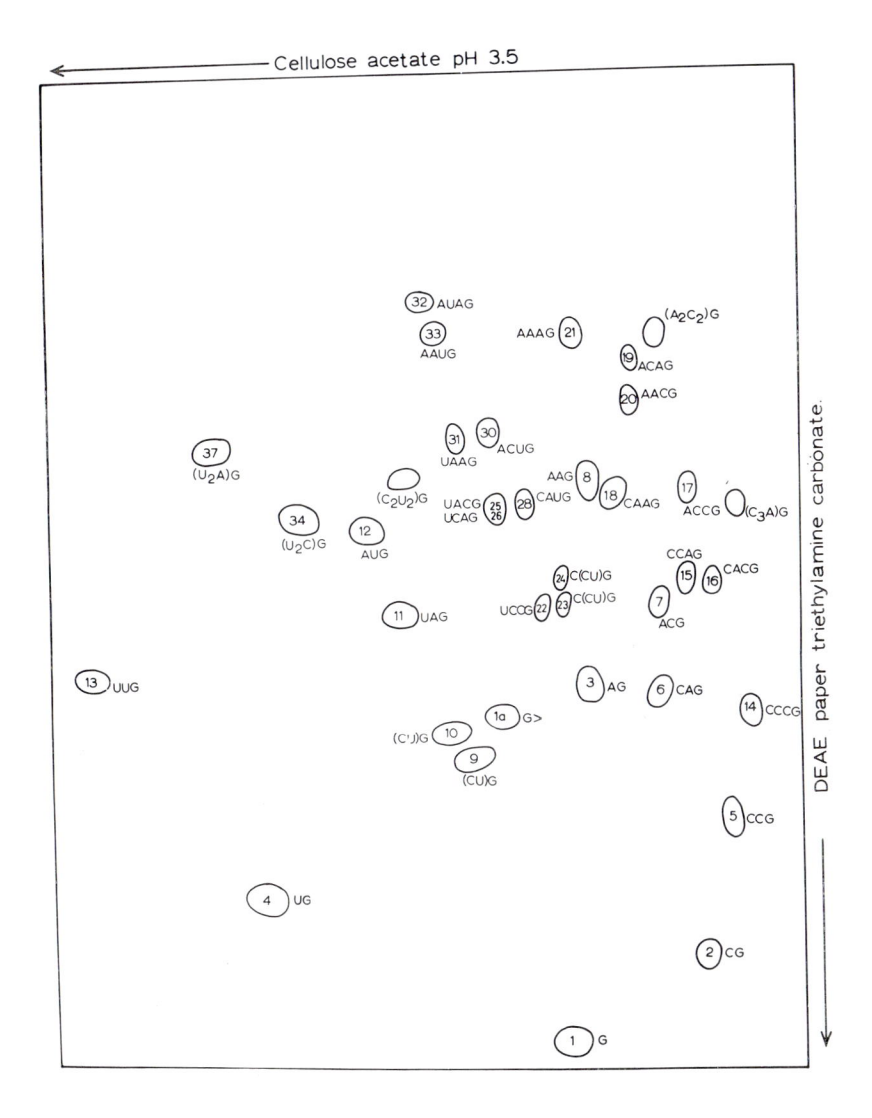

FIGURE 5. Diagram showing the distribution of some nucleotides on the two-dimensional system using triethylamine carbonate buffer for the fractionation on DEAE–paper. 7.5% triethylamine saturated with CO_2 was used as buffer, and 2% triethylamine was added to the coolant (white spirit 100).

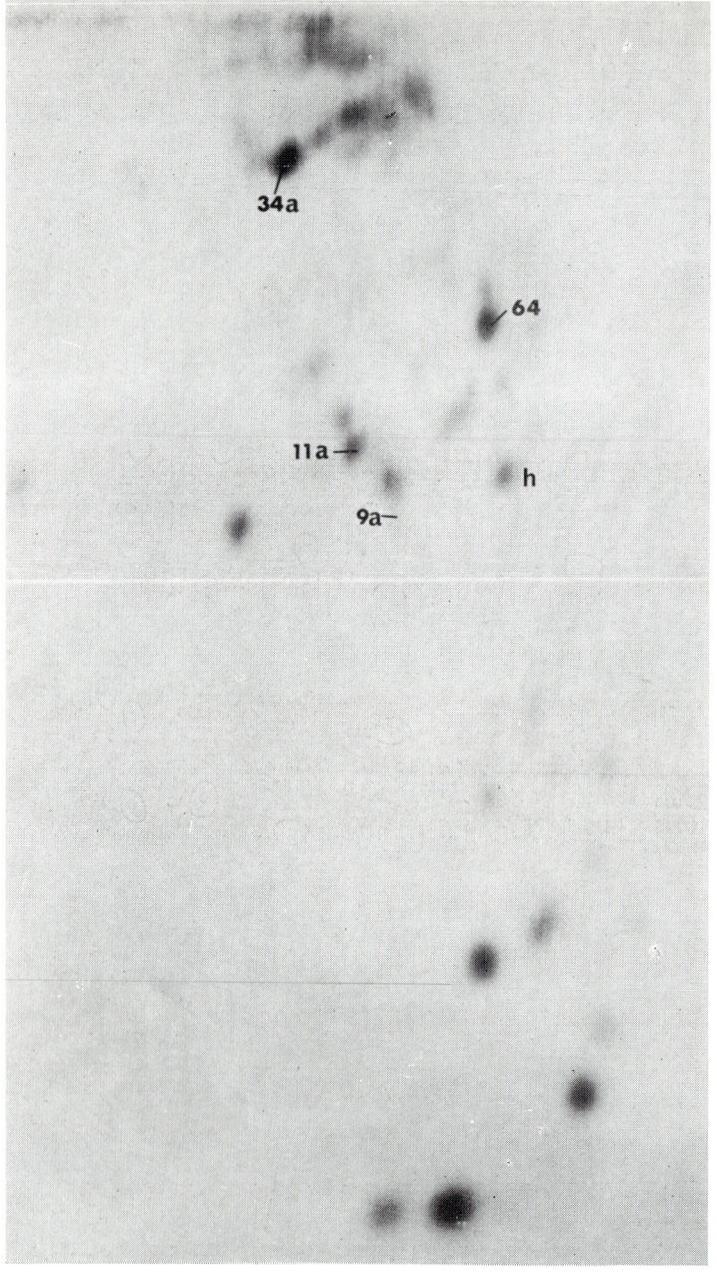

PLATE I. Radioautograph of two-dimensional fractionation of a ribonuclease T₁ digest of mixed transfer-RNA from *E. coli*.

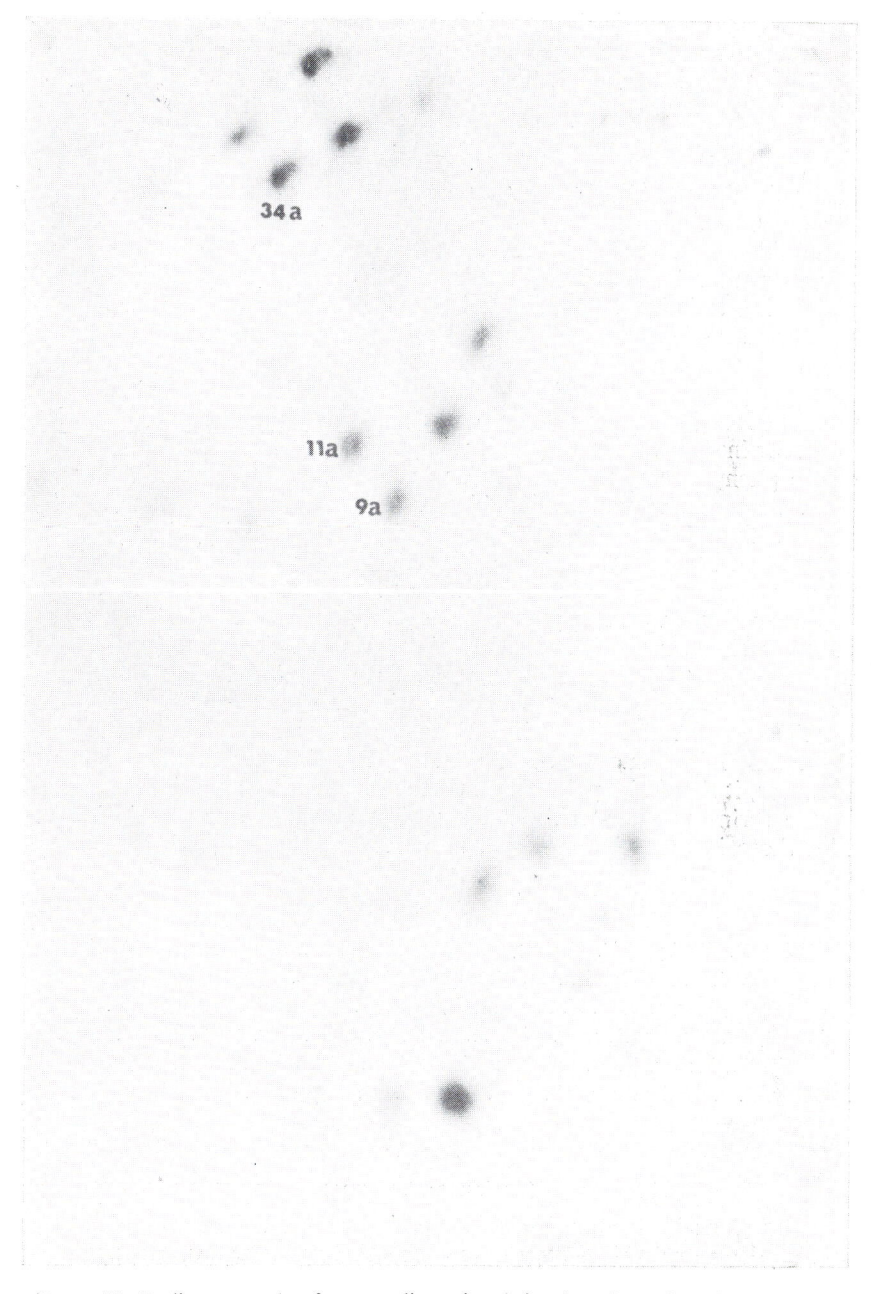

PLATE II. Radioautograph of a two-dimensional fractionation of a ribonuclease T₁ digest of purified phenylalanyl-transfer-RNA.

spot 34a (TψCG*), which is probably present in most of the transfer-RNAs (4). Other strong spots are those containing dihydrouridylic acid, particularly DCG (9a) and DAG (11a). Spot h is 7mGUCG and contains a high proportion of the 7mG which is present in transfer-RNA. There are also certain spots which do not contain minor bases but are present in high concentrations, and this clearly is a further effect of the homologies which must be present in different transfer RNAs. One of the strongest of these is in spot 64, which is CUCAG.

Considerable difficulty has been experienced in the purification of individual transfer RNAs on the small scale required for the radioactive fingerprinting technique. One method that has been used with some success and may be of considerable use since it should be of general application to all the transfer-RNAs was the use of the specific binding properties of synthetic polynucleotides in the presence of ribosomes, as has been used for the deduction of the genetic code by Nirenberg and his colleagues (5). Thus, for instance, using poly U it was possible to obtain a selective binding of the phenylalanine transfer-RNA to ribosomes and this could be eluted at low magnesium concentration. The procedure can be repeated and Plate II shows a fingerprint of material purified by two successive bindings in this way; it can be seen that it is considerably simpler than that obtained from the mixed transfer-RNA, and from the ratio of the yields of the different monucleotides it would appear to be at least 50 per cent pure. The structure of the various nucleotides is under investigation. A comparison of these with those obtained by Holley et al. from alanine transfer-RNA from yeast suggests the presence of the following two sequences in both RNAs: GGTψCG and AGDCGGDAG. These homologies between the two transfer-RNAs might suggest some common function for the sites involved. However, it should be pointed out that although there are related sequences in the serine transfer-RNAs neither of the above sequences is identical with those in one of the serine transfer-RNAs (Zachau, this symposium).

REFERENCES

1. Holley, R. W., Apgar, J., Everett, G. A., Madison, J. T., Marquisee, M., Merrill, S. H., Penswick, J. R. and Zamir, A., *Science* **147**, 1462 (1965).
2. Sanger, F., Brownlee, G. G. and Barrell, B. G., *J. Mol. Biol.* **13**, 373 (1965).
3. Rosset, R., Monier, R. and Julien, J., *Bull. Soc. Chim. Biol.* **46**, 87 (1964).
4. Zamir, A., Holley, R. W. and Marquisee, M., *J. Biol. Chem.* **240**, 1267 (1965).
5. Leder, P. and Nirenberg, M., *Proc. Nat. Sci.* **52**, 420 (1964).

DISCUSSION

(Chairman: S. OCHOA)

F. J. BOLLUM: I have considerable experience with DEAE chromatography of oligodeoxynucleotides sequences on DEAE paper sheets, most of which

---

* Abbreviations used for minor nucleotides: T—ribothymidylic acid; ψ—pseudouridylic acid; D—dihydrouridylic acid; 7mG—N^7-methyl guanylic acid.

was published 3–5 years ago by myself, and by Setlow and myself. Jacobson and his coworkers have published the two-dimensional chromatography at pH 8.0 and 3.5 separately as a fingerprinting method. I would like to ask Dr. Sanger if he has experienced any difficulty with the low wet strength of the DEAE paper.

F. SANGER: Initially, yes, but these difficulties were subsequently overcome by new developments. The DEAE paper sheet is supported on a rack which can be lifted in and out of the electrophoresis tanks.

Properties and Function of Methyl-deficient Phenylalanine Transfer RNA*

U. Z. LITTAUER AND M. REVEL

Biochemistry Section, Weizmann Institute of Science, Rehovoth, Israel

THE OCCURENCE OF METHYLATED BASES IN RNA

Studies in several laboratories have shown that RNA from plant, animal and bacterial sources contain trace amounts of methylated purines and pyrimidines at levels ranging from 0.005 to 2 mole % (1, 2, 3, 4). It was subsequently discovered that there is a substantially higher proportion of methylated bases in sRNA (soluble RNA) than in the rRNA (ribosomal RNA) (4–14).

More than ten different methylated purines and pyrimidines have been identified in sRNA of various sources among which are: thymine, 2-methyladenine, 6-methylaminopurine, 6-dimethylaminopurine, 1-methylguanine, 6-hydroxy-2-methylaminopurine, 6-hydroxy-2-dimethylaminopurine (3, 4, 6, 8, 11, 12, 15), 5-methylcytosine (16, 17), 1-methyladenine, 7-methylguanine (8, 18, 19, 20), 3-methyluracil, 3-methylcytosine (21), and 1-methylhypoxanthine (22, 23).

Tansfer RNA (tRNA) species isolated from different sources differ in the number and type of methylated bases they contain. Moreover it appears that each of the amino acid-specific tRNA chains thus far studied has a specific content of minor nucleotides which occupy definite positions along the RNA chain (i.e., they are not uniformly distributed) (24–28). Yeast alanine tRNA contains methylated guanylic acids but not methylated adenylic acids, while yeast valine tRNA contains methylated adenylic acids but not methylated yeast guanylic acids. The tyrosine tRNA contains a total of nine methylated nucleotides, accounting for approximately 10% of the total RNA molecule (24). In addition, ribothymidilic acid seems to occupy a unique position along the polynucleotide chain. The pentanucleotide sequence GpTpΨpCpGp was detected in alanine, valine and tyrosine yeast tRNAs (29), as well as in yeast serine tRNA II (30). It was suggested (29) that this pentanucleotide sequence is a common structural feature of all tRNA molecules. However, in the case of yeast serine tRNA I, this sequence is replaced by GpTpΨpCpAp (30).

---

* This research was supported, in part, by United States Public Health Service Grant RG-5217.

The RNA Methylases

In vivo experiments with methyl-labeled methionine have shown that the methyl group of this amino acid serves as a precursor of the methylated bases in sRNA from bacteria (31, 32) and ascites tumors (33), in nucleolar RNA (34), in ribosomal RNA (35–38) and also in bacterial DNA (39).

It was observed by Mandel and Borek (40, 41) that methyl-deficient sRNA is synthesized by a "relaxed" methionine-requiring mutant of *E. coli* ($K_{12}W6$) grown in the absence of methionine. Using such a methyl-deficient sRNA preparation, it has been shown that crude extracts of *E. coli* are capable of catalyzing the methylation of sRNA at the polynucleotide level (42, 36, 35) with S-adenosyl-methionine serving as the actual methyl donor (43, 44). Hurwitz *et al.* (45, 46) purified six different tRNA methylases from extracts of *E. coli*. Each enzyme fraction has a definite base specificity: two different fractions methylate guanine at N_1 position, one methylates guanine at N_7, one converts uracil to thymine, one methylates cytosine at C_5, and one enzyme fraction produces 2-methyladenine, 6-methylaminopurine and 6-dimethylaminopurine (this fraction may contain more than one enzyme). The enzymes do not methylate ribosomal RNA, f2 phage RNA, synthetic polymers or their homologous sRNA *in vitro* but do methylate methyl-deficient sRNA or heterologous sRNA (47). The sRNA methylases have been detected in a large number of organisms. These enzymes were shown to be species-specific, and consequently sRNA, while fully saturated with respect to its homologous enzymes, offers new sites for methylation to a heterologous sRNA methylase (36, 48–51). Furthermore, different amino acid-specific tRNA chains are methylated to different degrees with the same heterologous enzyme (51).

Amino Acid Acceptance by "Methionine-Starved" sRNA

The study of the role of methylated bases in tRNA has been made possible by the discovery of Mandel and Borek (40, 41) that methyl-deficient sRNA is synthesized by a "relaxed" methionine-requiring mutant of *E. coli* ($K_{12}W6$) during methionine starvation. sRNA isolated from such a "methionine-starved" *E. coli* $K_{12}W6$ culture is approximately a one-to-one mixture of methylated and methyl-deficient sRNA.

The capacity of "methionine-starved" sRNA to accept several amino acids was tested in several laboratories and shown to be of the same order as that of normal sRNA (36, 52–56). These results may indicate that the methylated bases do not participate in this function. However, the sensitivity and precision of these experiments does not rule out subtle differences in the ability of the two RNA species to accept amino acids (cf. 57). In fact, recent experiments show that small differences may exists between the two RNA species. It has been observed (58) that while *E. coli* leucyl RNA synthetase can aminoacylate both normal and "methionine-starved" tRNA, the yeast amino acid activating enzyme shows a reduced ability to attach leucine to "starved" tRNA. These results indicate that the leucine-activating enzyme of yeast has a much more stringent requirement for tRNA methyl groups than does the *E. coli* enzyme. In addition we have found (59) that the two RNA species differ in their Mg^{++} requirement with respect to the phenylalanine acceptor function. The methyl-deficient phenylalanine tRNA requires less Mg^{++} than does the normal species.

Amino Acid Transfer to Ribosomes by "Methionine-Starved" sRNA

The participation of "normal" and "methionine-starved" tRNA in the transfer of phenylalanine (54–56) or leucine (55) to ribosomes was studied in polyuridylate-directed synthesis of polyphenylalanine. The results obtained indicate that methyl-deficient tRNA can function in the transfer reaction. However at high concentrations of tRNA a small preference for "normal" methylated tRNA over methyl-deficient tRNA was observed. Further studies (57) revealed that the small differences found for the two RNA species are probably not relevant to the presence or absence of methyl groups in tRNA, as similar differences were encountered with tRNA prepared from cells starved for amino acids other than methionine, in which the tRNA is not deficient in methyl groups. The lowered transfer capacity of tRNA prepared from amino acid-starved auxotrophs was shown to be a general phenomenon and due to an inhibitor formed during the starvation period. The formation of this inhibitor does not require RNA synthesis since it takes place in starved cultures of both RC "relaxed" and "stringent" mutants. Our experiments indicate that the inhibitor seems to compete with "normal" tRNA at the ribosomal level, inhibiting thereby the overall transfer reaction. We may conclude that methyl-deficient tRNA can participate in the transfer reaction of phenylalanine and leucine in the presence of poly U*.

The Isolation of Methyl-Deficient Phenylalanyl Transfer RNA

Regarding the function of methylated bases in tRNA, it should be recalled that "methionine-starved" RNA preparations consist of a mixture of equal amounts of methyl-deficient and normal tRNA. Therefore, small differences in the activities of these RNA species *in vitro*, which could be very critical *in vivo*, might not have been detected with such crude RNA preparations. These considerations prompted us to develop a method for the isolation of methyl-deficient tRNA. We therefore tried chromatography

---

* Many other possibilities have been considered for the role of methylated bases in tRNA (60, 57). One possible function is the involvement of methylation in some growth regulating mechanism. As a model system, we chose to compare the pattern of methylated bases of normal and regenerating liver sRNA. No significant differences were found. We have also compared a partially purified sRNA methylase from rat liver with that obtained from regenerating liver. The enzyme preparations were assayed for their ability to methylate "normal" *E. coli* sRNA. The data demonstrate that the specific activities, extent and pattern of methylation are almost identical for the two enzyme preparations (R. Rodeh and U.Z. Littauer, unpublished results).

on methylated albumin kieselguhr (MAK) columns. Chromatography of "methionine-starved" tRNA which was labeled with ^{32}P-orthophosphate during the starvation period shows that the ^{32}P-label follows very closely the optical density pattern (the label is in the methyl-deficient tRNA). It follows that this column cannot separate all the different methyl-deficient RNA chains from the normal species. However, when a single species of tRNA was followed (61, cf. 57) namely phenylalanine tRNA (phe-tRNA), a double peak appeared (Figure 1). This can be more clearly seen in the next

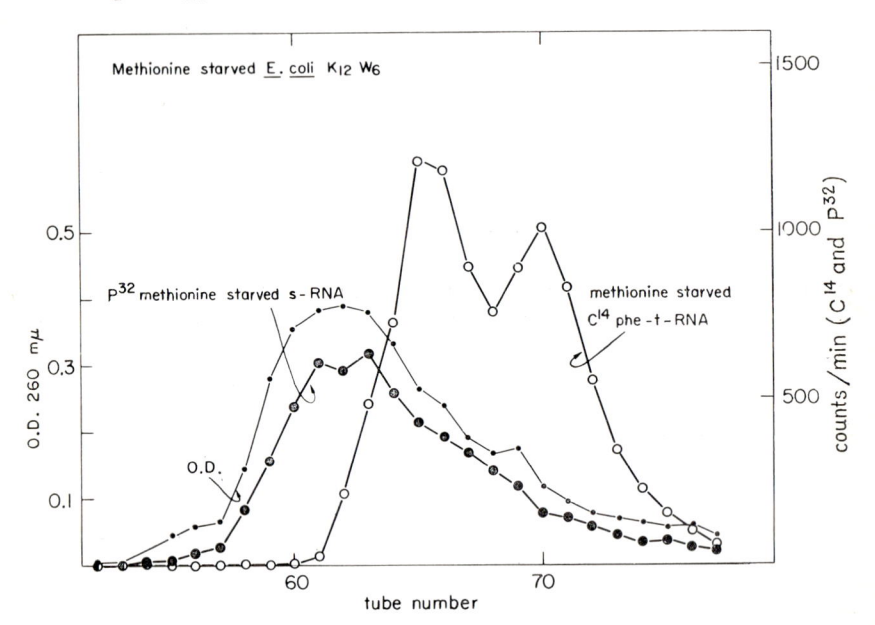

FIGURE 1. MAK column chromatography of "methionine-starved" sRNA. *E. coli* K$_{12}$W6 RC^{rel}Met$^-$ cells were grown in minimal medium supplemented with 5 μg per ml L-methionine. Cells were harvested in the middle of the logarithmic growth phase, washed twice and suspended in the same medium containing (^{32}P) orthophosphate but no methionine was added. The cell suspension was incubated for 3 hours with vigorous shaking at 37°C, harvested and sRNA isolated. The (^{32}P)-labeled sRNA was then charged with ^{14}C-phenylalanine and submitted to MAK column chromatography.

Figure (Figure 2) where the tRNA obtained from the methionine-starved cultures was charged with ^{14}C-phenylalanine and mixed with normal tRNA charged with ^{3}H-phenylalanine. In addition to the normal peaks of phe-tRNA, the RNA after methionine starvation contains a new fraction of phe-tRNA. The new component is eluted at a salt concentration lower than the normal phe-tRNA and amounts to about 60% of the total phe-tRNA, in agreement with the amount of total methyl-deficient tRNA accumulating during methionine starvation.

In order to establish that the appearance of the new phe-tRNA is specific to methionine starvation, an auxotroph of *E. coli* G-15 (RC^{rel}Met$^-$His$^-$

Biotin⁻, obtained from Dr. G. Stent) was used. sRNA was isolated from methionine-starved cultures of from histidine-starved cultures and charged with ^{14}C-phenylalanine or ^{3}H-phenylalanine respectively. The two labeled RNA preparations were then subjected to MAK column chromatography

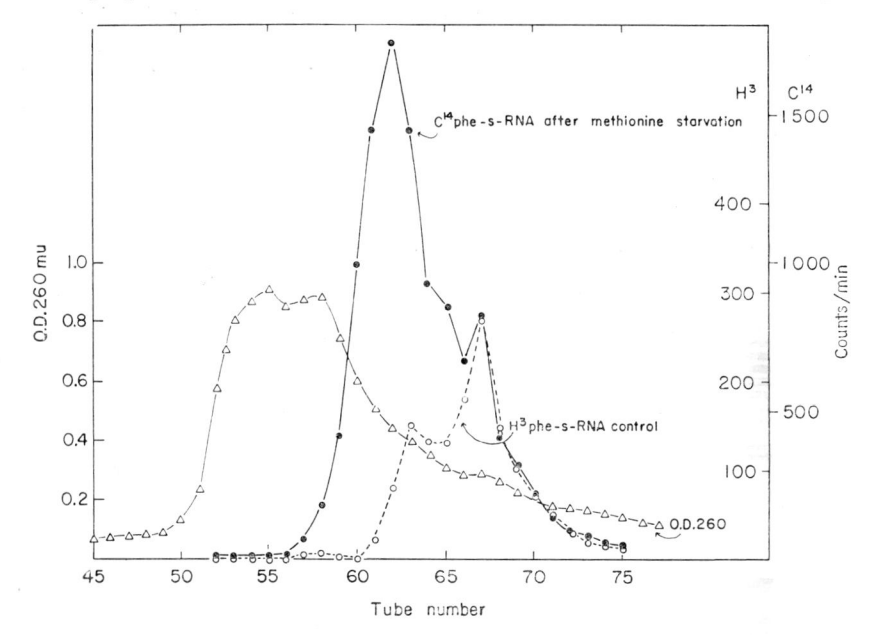

FIGURE 2. MAK column chromatography of phenylalanyl-sRNA from *E. coli* G-15 after methionine starvation (61).

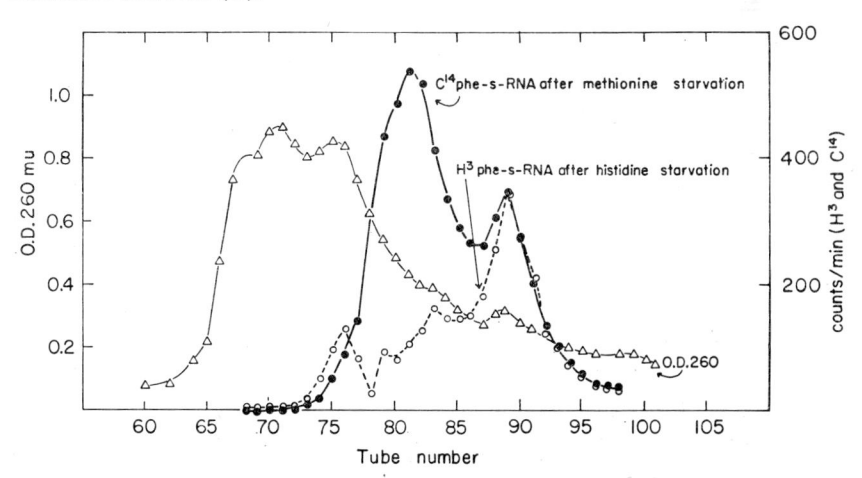

FIGURE 3. MAK column chromatography of phenylalanyl-sRNA from *E. coli* G-15 after histidine starvation (61).

(Figure 3). The new phe-tRNA peak was only observed with "methionine-starved" sRNA but does not appear when "histidine-starved" sRNA was used. The methyl content of each sRNA preparation was measured. With a crude *E. coli* RNA methylase, methionine-starved sRNA accepted 19.7 mμmoles of methyl residues per mg RNA. The corresponding value for both normal and histidine-starved RNA was 0.1 mμmole per mg RNA. From these studies it can be concluded that in methionine-starved cells, a new phe-tRNA appears which might correspond to methyl-deficient phe-tRNA. Similar results were obtained by Lazzarini and Peterkofsky (62) who detected a new leucyl-tRNA peak after MAK column chromatography of "methionine-starved" sRNA.*

The Identification of Methyl-deficient phe-tRNA

Is the phe-tRNA produced during methionine starvation methyl-deficient phe-tRNA? To answer this question, the amount of methyl groups per phe-tRNA chain present in this fraction was compared to that in normal phe-tRNA. Since in the fractions obtained from the MAK column chromatography, phe-tRNA represents less than 10% of the chains, additional purification of this particular tRNA was required. This purification was achieved by using the property of phe-tRNA to attach specifically to poly U ribosome complexes, which can be separated from other free tRNA chains (63). In a first series of experiments (method A, Table I), tRNA from methionine-starved cultures was charged with ^{3}H-phenylalanine and submitted to chromatography on the MAK column. The fractions of phe-tRNA were pooled as indicated in Figure 4, treated with pronase to remove proteinaceous materials, † and then methylated *in vitro* with ^{14}C–CH$_3$–S-adenosyl methionine and a crude *E. coli* enzyme fraction (streptomycin supernatant). Under these conditions, only the methyl-deficient tRNA would be expected to receive (^{14}C)-labeled methyl groups. Since most of the radioactive phenylalanine had been lost from the tRNA at this stage, the charging with ^{3}H-phenylalanine was repeated. The doubly labeled RNA was adsorbed on ribosomes previously supplemented with poly U and the complex was isolated by filtration on Millipore membranes (64).

Table II shows that both fractions were methylated to the extent of about 100 methyl groups per μmole phe-tRNA. Therefore, in order to see a difference, both fractions had to be further purified on poly U ribosomes. The results indicate that per chain of phe-tRNA adsorbed, fraction I which

---

* The separation of normal and methyl-deficient phe-tRNA was dependent upon amino acylation of the tRNA. On a column in which unacylated tRNA was chromatographed and phenylalanine acceptance assayed after fractionation, no separation of methyl-deficient from normal tRNA could be observed. The change in chromatographic behavior was found to be a general phenomenon, in that the position of the column of normal as well as methyl-deficient species of phe-tRNA differs markedly from the profile of unacylated chains (R. Stern and U.Z. Littauer, unpublished results).

† The removal of this material by pronase was found essential for the recovery of the biological activity of tRNA.

TABLE I. Identification of Methyl-deficient phe-tRNA

Method A Non labeled methionine-starved RNA	Charge with ³H-phe →	MAK column chromatography ↗↘	Fraction III normal RNA → Fraction I methyl-deficient RNA →	Methylate with ¹⁴C–CH₃–SAM → →	Charge with ³H-phe → →	Adsorb to poly U ribosomes and filter on Millipore → →	Count
Method B (¹⁴C)-methyl methionine-starved RNA	Charge with ³H-phe →	MAK column chromatography ↗↘	Fraction III (¹⁴C)-methyl normal RNA → Fraction I methyl-deficient →		Charge with ³H-phe → →	Adsorb to poly U ribosomes and surcrose gradient → →	Count

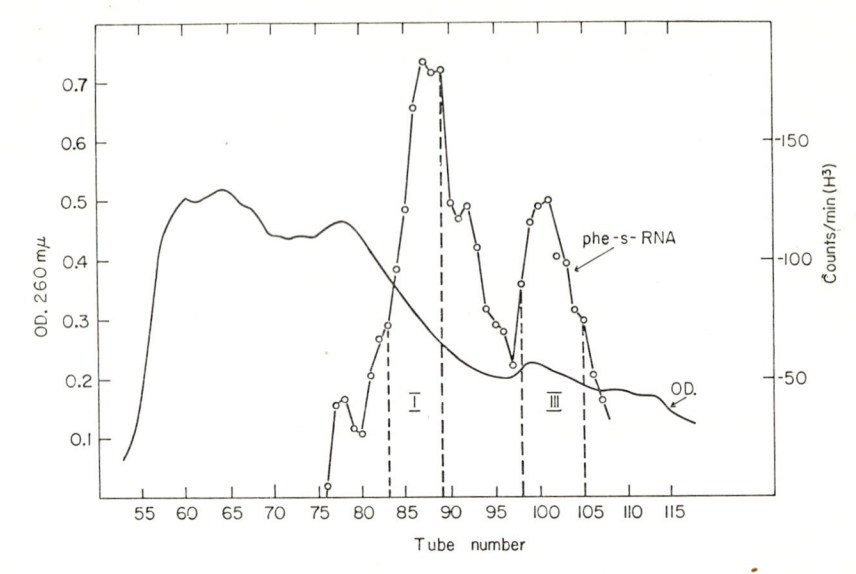

FIGURE 4. Separation by MAK column chromatography of normal and methyl-deficient phe-tRNA. 1.7 mg of "methionine-starved" sRNA from *E. coli* G-15 was charged with ³H-phenylalanine, and used for chromatography. The tubes between the vertical lines were pooled, dialyzed against water and lyophilized. The tRNA dissolved in 0.1 M NaCl was precipitated with 2.5 vol of ethanol, incubated in 1.0 ml of 0.01 M Tris, pH 7.4 at 37°C for 5 hours with 1 mg pronase. The RNA was then treated with phenol and recovered by ethanol precipitation. To eliminate possible traces of RNase the pronase solution was pre-incubated for 90 minutes in Tris-HCl, pH 7.4, 0.01 M at 37°C (61).

corresponds to the RNA appearing after methionine starvation is much richer in radioactive methyl groups than fraction III which contains normal phe-tRNA (expressed in Table II as a relative ratio). These data, therefore, support the idea that fraction I is methyl-deficient phe-tRNA while fraction III is fully methylated.

TABLE II. Ratio of ¹⁴C-methyl to ³H-phenylalanine in the two tRNA Fractions Isolated from the MAK Column and Methylated *in vitro* (61)

RNA fraction*	RNA from MAK column		RNA absorbed on poly U ribosomes	
	³H-phe	¹⁴C-methyl	³H-phe	¹⁴C-methyl
	μμmole		μμmole	
Fraction I	1.21	112	0.68	4.15
Fraction III	1.51	91	0.71	0.53
I/III†			1.00	8.20

* The numbers refer to the fractions indicated in Figure 4.
† Relative ratio.
(Method A, Table I).

The drawback of this method is that a significant amount of *non specific* binding of tRNA takes place in the absence of poly U and has to be subtracted from the values obtained. To reduce this background, we isolated the poly U polyribosomes by sucrose gradient centrifugation. In these experiments (method B, Table I) we used *in vivo* labeled ^{14}C-methyl RNA. The bacteria were first grown in the presence of ^{14}C-methyl-methionine and then starved for methionine during 3.5 hours. The normal phe-tRNA should now be labeled with ^{14}C-methyl while the methyl-deficient species should not contain any radioactivity. The tRNA prepared from such a culture was charged with ^{3}H-phenylalanine and fraction I and III were isolated after MAK column chromatography. The chromatographic pattern indicated that the ^{14}C-CH$_3$ label followed the distribution of the optical density. The ratio of (^{14}C) to (^{3}H) in fraction I was much higher than in fraction III, due to the higher amount of other tRNA chains contaminating fraction I. To purify phe-tRNA, the doubly labeled RNA fractions were incubated with ribosomes and a limiting amount of poly U and the complex formed was examined by sucrose gradient centrifugation. The ^{14}C/^{3}H ratio was measured in the polyribosome region where most of the ribosomes have combined with the poly U, and where a minimal binding of other tRNA chains would be expected. The results of this experiment are presented in Table III. With both RNA fractions, about 25% of the phe-tRNA is

TABLE III. Ratio of ^{14}C-methyl to ^{3}H-phenylalanine in the two tRNA Fractions Isolated from The MAK Column after *in vivo* Labelling with ^{14}C-CH$_3$-methionine (61)

RNA fraction*	Distribution of label in the RNA from various portions of sucrose gradient					
	> 100 *S*		70–100 *S*		< 70 *S*	
	^{3}H-phe	^{14}C-methyl	^{3}H-phe	^{14}C-methyl	^{3}H-phe	^{14}C-methyl
		per cent of total label				
I	24	1.4	55	11	21	88
III	25	5.0	40	17	35	79
III/I†	1.0	3.5	1.0	2.2	1.0	0.55

* The numbers refer to the fractions indicated on Figure 4.
† Relative ratio.
Method B, Table I.

bound to ribosomal aggregates heavier than 100*S*. However, the amount of ^{14}C-CH$_3$ in this region is 3.5 times higher in the case of RNA III than with RNA I (expressed in Table III as a relative ratio). A two-fold difference is still observed in the 70 to 100*S* region of the gradient where about one half of the total phenylalanine input is bound. These results are again in agreement with the conclusion that fraction I contains a methyl-deficient phe-tRNA.

Coding Properties

The availability of purified methyl-deficient phenylalanine tRNA made possible a more comprehensive study of this RNA species. It was shown by Bernfield and Nirenberg (65) that normal *E. coli* phe-tRNA is coded by UUU and UUC triplets. These two codewords for phenylalanine may be synonymous in that one tRNA species can recognize more than one codon. To investigate whether the absence of methylated bases affects the normal phenylalanine code, we first compared the activity of a purified

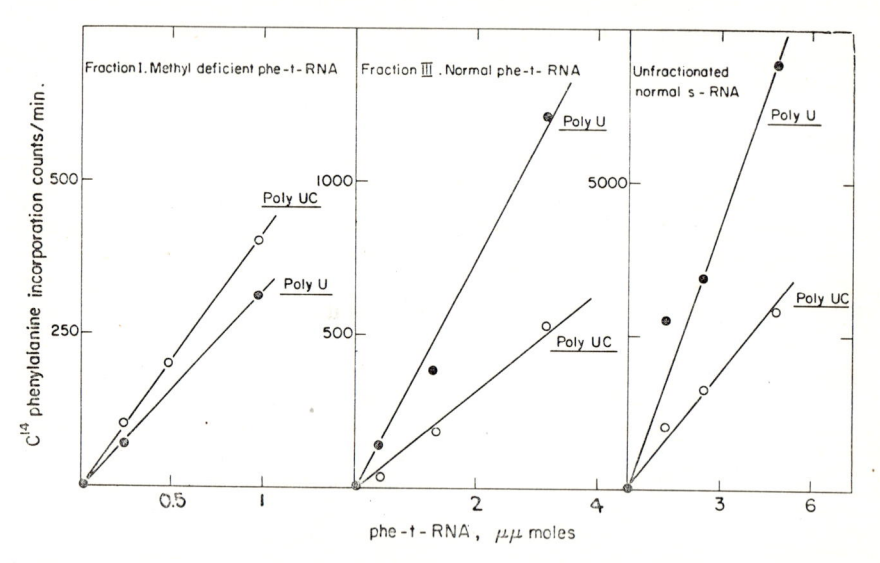

FIGURE 5. Coding properties of normal and methyl-deficient phenylalanine transfer RNA. The assay system consisted of puromycin-treated ribosomes supplemented with poly U or poly UC (1:1). Streptomycin-treated supernatant fraction from *E. coli* B served as enzyme source; this system was dependent on the addition of tRNA.

fraction of normal and methyl-deficient phe-tRNA in a messenger RNA–tRNA dependent amino acid incorporating system supplemented with poly U or poly UC (1:1) (59). Comparing the two fractions derived from "methionine-starved" tRNA, it appears that the methyl-deficient and normal species of phe-tRNA respond equally well to poly U, as had been previously noted (54). In contrast, when poly UC is used as template, methyl-deficient phe-tRNA is more than twice as active as the normal fraction (Figure 5). Poly UC is less active than poly U when normal RNA or fraction III is used, but is more effective than poly U when fraction I is used. It should be noted that the RNA recovered after MAK column chromatography shows a lowered transfer activity. The decrease, however, affects both the normal and the methyl-deficient components. As a result of fractionation, the different fractions of phe-tRNA used have a variable component of *non* phe-tRNA chains. For example, some of them may code for leucine or serine in response to poly UC. The similarity of the poly UC/poly U ratio of RNA from peak III and unfractionated normal tRNA indicates that the increased response to poly UC observed with methyl-deficient phe-tRNA (peak I) is not due to differences in *non* phe-tRNA chains present in these preparations. To further examine this finding, the coding properties of normal and methyl-deficient phe-tRNA were reexamined by measuring the ability of each species to bind to ribosomes in the presence of various synthetic polynucleotides. Binding of ^{14}C-phenylalanine tRNA to ribosomes

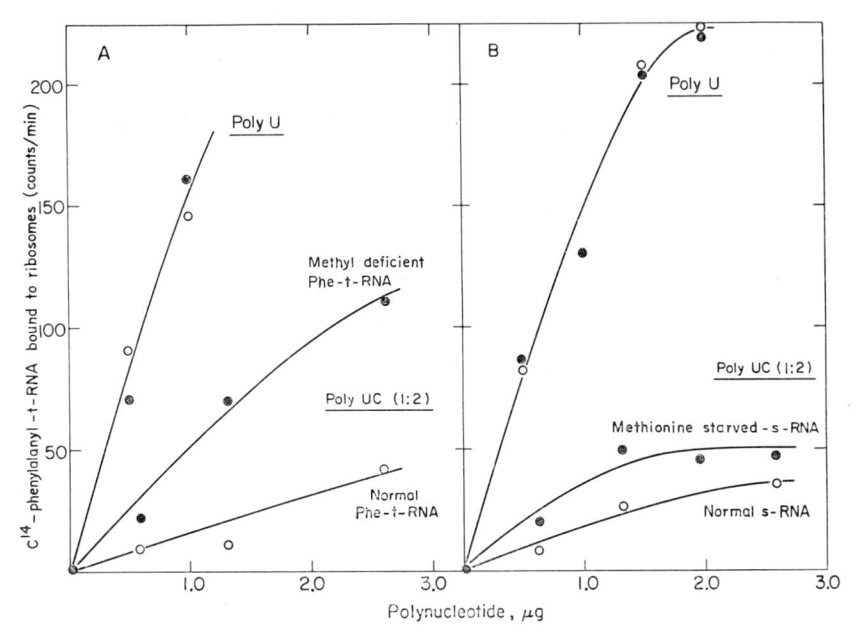

FIGURE 6. Binding of normal and methyl-deficient phenylalanyl tRNA to ribosomes by poly U and poly UC (1:2). (a) Comparison of normal (○) and methyl-deficient (●) phenylalanine tRNA fractions (I and III) from "methionine-starved" tRNA. (b) Comparison of normal (○) and "methionine-starved" unfractionated tRNA (●) (59).

requires the presence of UUU or UUC triplets (65). In order to decrease the response of normal phe-tRNA to poly UC, a copolymer with a C/U ratio 2/1 was used and binding was measured under conditions in which the polynucleotide was limiting. A typical experiment is shown in Figure 6a in which the two fractions obtained from "methionine-starved" tRNA are compared. The relative binding by poly UC over that by poly U (ambiguity ratio) was 10% for the normal phe-tRNA, but 32% for the methyl-deficient species. The binding assay was sufficiently sensitive to detect differences between unfractionated tRNA preparations (Figure 6b). Thus, while the ambiguity ratio of normal tRNA was 12%, that of unfractionated "methionine-starved" tRNA was 22% as expected from a one-to-one mixture of normal and methyl-deficient phe-tRNA. In another control experiment, tRNA isolated from histidine-starved cells was found to have an ambiguity ratio of 11%. Since the combined calculated frequencies of UUU and UUC in poly UC (1:2) is about 11%, the fact that methyl-deficient phe-tRNA binds over 30% to poly UC, as compared to poly U, indicates that this tRNA species must recognize codons other than UUU or UUC.

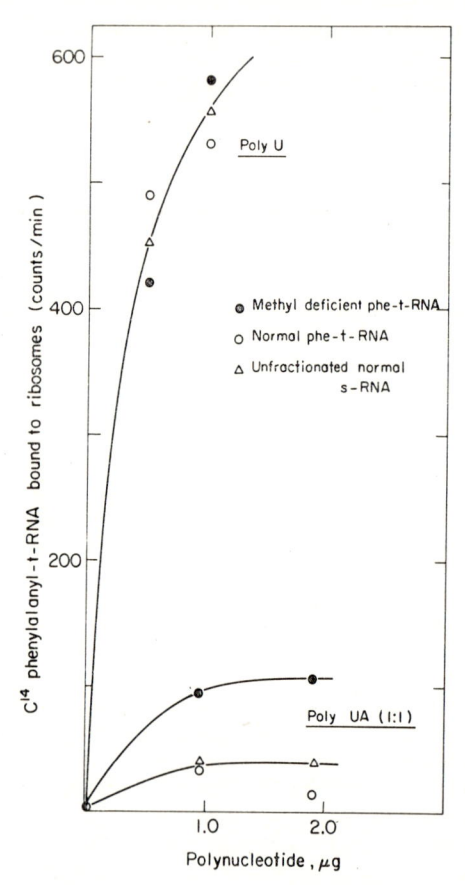

FIGURE 7. Binding of normal and methyl-deficient phenylalanyl-tRNA to ribosome by poly U and poly UA (1:1) (59).

In view of the existence of the UUU-UUC ambiguity in normal phe-tRNA, it was of interest to know if the "miscoding" exhibited by the methyl-deficient species could be demonstrated also with other copolymers. Figure 7 shows that poly UA (1:1) binds three-times more methyl-deficient phe-tRNA than normal tRNA.

The experiments presented above have established that the absence of methyl groups in phe-tRNA increased the level of error in the reading of UC or UA containing triplets. If this type of miscoding is similar to those which have been described as the poly U directed incorporation of leucine or isoleucine (66, 67), it should be very sensitive to Mg^{++} concentration. Therefore, we undertook a study of the effect of Mg^{++} on the transfer of phenylalanine in the presence of poly U and poly UC (1:1). Table 4 shows that the ambiguity ratio of normal phe-tRNA is more than doubled when Mg^{++} concentration is raised from 17 to 31 mM. On the other hand, methyl-deficient phe-tRNA miscodes even at low Mg^{++} concentrations. This is in agreement with the concept that the absence of methylated bases allows the tRNA to read erroneous triplets under conditions where very little miscoding is observed with normal tRNA.

TABLE IV. The Effect of Magnesium Ions on the Coding Properties of Normal and Methyl-deficient phe-tRNA (59)

Fraction from "methionine-starved" sRNA	Methyl-deficient phe-tRNA (I*)			Normal phe-tRNA (III*)		
Mg^{++} concentration	17 mM	21 mM	31 mM	17 mM	21 mM	31 mM
	Phenylalanine ^{14}C incorporation, cpm					
Poly U	1,111	1,260	852	814	732	376
Poly UC	273	567	259	105	130	118
	per cent					
$\frac{\text{Poly UC}}{\text{Poly U}} \times 100$	25	45	30	12	18	32

* Fractions obtained from MAK column chromatography.

Methylation may produce significant changes in the secondary structure and complexing ability of synthetic polynucleotides. It was observed by Shugar and Szer (68) that, in contrast to poly U which exhibits a low degree of secondary structure below 8°C, polyribothymidilic acid possesses a highly ordered configuration at room temperature with a T_m of 36°C. Moreover, polyribothymidilic acid is more extensively bound to poly A than poly U. On the other hand, poly-N(3)-methyluridylic acid exhibits no secondary structure under any condition and does not form a complex with poly A. It was also shown by Bollack, Keith and Ebel (69) that chemical methylation alters significantly the secondary structure of sRNA. Thus, the introduction of methyl groups may strengthen or weaken the secondary structure of tRNA. This change depends upon the nature of the base, the position of the attached methyl group and possibly the nucleotide sequence.

Our lack of knowledge concerning the nature and localization of methyl groups in *E. coli* phe-tRNA makes it difficult to understand the mechanism by which they influence coding properties. It is possible that one of the nucleotides in the anticodon is methylated. On the other hand, the presence of methyl groups in another section of the tRNA molecules might sufficiently modify the configuration of the RNA chain to change its coding properties. The altered chromatographic behavior of methyl-deficient phe-tRNA and the differential effect of Mg^{++} concentration on coding properties of normal and methyl-deficient phe-tRNA (Table IV) suggest that the latter might have a modified secondary structure.

Experiments are now in progress to determine the identity of the methylated bases that are responsible for maintaining the specificity of the translation process in phe-tRNA. It is hoped that these will lead us to a better understanding of the factors that govern the biological activity of tRNA.

REFERENCES

1. Adler, M., Weissmann, B. and Gutman, A. B., *J. Biol. Chem.* **230**, 717 (1958).
2. Kemp, J. W. and Allen, F. W., *Biochim. et Biophys. Acta* **28**, 51 (1958).
3. Littlefield, J. W. and Dunn, D. B., *Biochem. J.* **70**, 642 (1958).
4. Smith, J. D. and Dunn, D. B., *Biochem. J.* **72**, 294 (1959).
5. Davis, F. F., Carlucci, A. F. and Roubein, J. F., *J. Biol. Chem.* **234**, 1525 (1959).
6. Dunn, D. B., *Biochim. et Biophys. Acta* **34**, 286 (1959).
7. Bergquist, P. L. and Matthews, R. E. F., *Biochim. et Biophys. Acta* **34**, 567 (1959).
8. Dunn, D. B., Smith, J. D. and Spahr, P. F., *J. Molecular Biol.* **2**, 113 (1960).
9. Monier, R., Stephenson, M. L. and Zamecnik, P. C., *Biochim. et Biophys. Acta* **43**, 1 (1960).
10. Lane, B. G. and Allen, F. W., *Biochim. et Biophys. Acta* **47**, 36 (1961).
11. Sluyser, M. and Bosch, L., *Biochim. et Biophys. Acta* **55**, 479 (1962).
12. Bergquist, P. L. and Matthews, R. E. F., *Biochem. J.* **85**, 305 (1962).
13. Brawerman, G., Hufnagel, D. A. and Chargaff, E., *Biochim. et Biophys. Acta* **61**, 340 (1962).
14. Cantoni, G. L., Gelboin, H. V., Luborsky, S. W., Richards, H. H. and Singer, M. F., *Biochim. et Biophys. Acta* **61**, 354 (1962).
15. Price, T. D., Hinds, H. A. and Brown, R. S., *J. Biol. Chem.* **238**, 311 (1963).
16. Amos, H. and Korn, M., *Biochim. et Biophys. Acta* **29**, 444 (1958).
17. Dunn, D. B., *Biochim. et Biophys. Acta* **38**, 176 (1960).
18. Dunn, D. B., *Biochim. et Biophys. Acta* **46**, 198 (1961).
19. Dunn, D. B., *Biochem. J.* **86**, 14P (1963).
20. Dunn, D. B., Hitchbarn, J. H. and Trim, A. R., *Biochem. J.* **88**, 34P (1963).
21. Hall, R. H., *Biochem. Biophys. Research Communs.* **12**, 361 (1963).
22. Hall, R. H., *Biochem. Biophys. Research Communs.* **13**, 394 (1963).
23. Holley, R. W., Everett, G. A., Madison, J. T. and Zamir, A., *J. Biol. Chem.* **240**, 2122 (1965).
24. Holley, R. W., Apgar, J., Everett, G. A., Madison, J. T., Merrill, S. H. and Zamir, A., *Cold Spring Harbor Sympos. Quant. Biol.* **28**, 117 (1963).
25. Cantoni, G. L., Ishikura, H., Richards, H. H. and Tanaka, K., *Cold Spring Harbor Sympos. Quant. Biol.* **28**, 123 (1963).
26. Ingram, V. M. and Sjoquist, J. A., *Cold Spring Harbor Sympos. Quant. Biol.* **28**, 133 (1963).
27. Bank, A., Gee, S., Mehler, A. and Peterkofsky, A., *Biochemistry* **3**, 1406 (1964).
28. Dütting, D. and Zachau, H. G., *Biochim. et Biophys. Acta* **91**, 573 (1964).

29. Zamir, A., Holley, R. W. and Marquisee, M., *J. Biol. Chem.* **240**, 1267 (1965).
30. Zachau, H. G., Dütting, D. and Feldmann, H., Personal communication.
31. Mandel, L. R. and Borek, E., *Biochem. Biophys. Research Communs* **6**, 138 (1961).
32. Mandel, L. R. and Borek, E., *Biochemistry* **2**, 555 (1963).
33. Biswas, B. B., Edmonds, M. and Abrams, R., *Biochem. Biophys. Research Communs.* **6**, 146 (1961).
34. Sirlin, J. L., Jacob, J. and Tandler, C. J., *Biochem. J.* **89**, 447 (1963).
35. Starr, J. L., *Biochem. Biophys. Research Communs.* **10**, 428 (1963).
36. Svensson, I., Boman, H. G., Eriksson, K. G. and Kjellin, K., *J. Molecular Biol.* **7**, 254 (1963).
37. Starr, J. L. and Fefferman, R., *J. Biol. Chem.* **239**, 3457 (1964).
38. Brown, G. M. and Attardi, G., *Federation Proc.* **24**, 292 (1965).
39. Theil, E. C. and Zamenhof, S., *J. Biol. Chem.* **238**, 3058 (1963).
40. Mandel, L. R. and Borek, E., *Biochem. Biophys. Research Communs.* **4**, 14 (1961).
41. Mandel, L. R. and Borek, E., *Biochemistry* **2**, 560 (1963).
42. Fleissner, E. and Borek, E., *Proc. Natl. Acad. Sci. U.S.* **48**, 1199 (1962).
43. Gold, M., Hurwitz, J. and Anders, M., *Biochem. Biophys. Research Communs.* **11**, 107 (1963).
44. Fleissner, E. and Borek, E., *Biochemistry* **2**, 1093 (1963).
45. Hurwitz, J., Gold, M. and Anders, M., *J. Biol. Chem.* **239**, 3462 (1964).
46. Hurwitz, J., Gold, M. and Anders, M., *J. Biol. Chem.* **239**, 3474 (1964).
47. Gold, M. and Hurwitz, J., *Cold Spring Harbor Sympos. Quant. Biol.* **28**, 149 (1963).
48. Srinivasan, P. R. and Borek, E., *Proc. Natl. Acad. Sci. U.S.* **49**, 529 (1963).
49. Gold, M., Hurwitz, J. and Anders, M., *Proc. Natl. Scad. Sci. U.S.* **50**, 164 (1963).
50. Birnstiel, M. L., Fleissner, E. and Borek, E., *Science* **142**, 1577 (1963).
51. Srinivasan, P. R. and Borek, E., *Biochemistry* **3**, 616 (1964).
52. Neidhardt, F. C. and Eidlic, L., *Biochim. et Biophys. Acta* **68**, 380 (1963).
53. Starr, J. L., *Biochem. Biophys. Research Communs.* **10**, 181 (1963).
54. Littauer, U.Z., Muench, K., Berg. P., Gilbert, W. and Spahr, P.F., *Cold Spring Harbor Sympos. Quant. Biol.* **28**, 157 (1963).
55. Littauer, U.Z., *Abst. Intern. Congr. Biochem. 6th, New York* **1**, 11 (1964).
56. Peterkofsky, A., Jesensky, C., Bank, A. and Mehler, A. H., *J. Biol. Chem.* **239**, 2918 (1964).
57. Littauer, U.Z. and Milbauer, R., *Fed. European Biochem. Soc.* Pergamon Press, London (In press) 1965.
58. Peterkofsky, A., *Proc. Natl. Acad. Sci. U.S.* **52**, 1233 (1964).
59. Revel, M. and Littauer, U.Z., *J. Molecular Biol.* **15**, 389 (1966).
60. Srinivasan, P. R. and Borek, E., *Science* **145**, 548 (1964).
61. Revel, M. and Littauer, U. Z., *Biochem. Biophys. Research Communs.* **20**, 187(1965).
62. Lazzarini, R. A. and Peterkofsky, A., *Proc. Natl. Acad. Sci. U.S.* **53**, 549 (1965).
63. Kaji, A. and Kaji, H., *Biochem. Biophys. Research Communs.* **13**, 186 (1963).
64. Nirenberg, M. W. and Leder, P., *Science* **145**, 1399 (1964).
65. Bernfield, H. R. and Nirenberg, M. W., *Science* **147**, 479 (1965).
66. Szer, W. and Ochoa, S., *J. Molecular Biol.* **8**, 823 (1964).
67. Davies, J., Gilbert, W. and Gorini, L., *Proc. Natl. Acad. Sci. U.S.* **51**, 833 (1964).
68. Shugar, D. and Szer, W., *Abstr. Intern. Congr. Biochem. 6th, New York* **1**, 5 (1964).
69. Bollack, C., Keith, G. and Ebel, J. P., *Bull. Soc. Chim. Biol.* **47**, 765 (1965).

DISCUSSION

W. Szer: A number of 5'-diphosphates of base analogues has been synthesized in recent years and polymerized, taking advantage of the low specificity of polynucleotide phosphorylase. Some of the bases investigated are indentical or analogous to the minor components discussed here. It appears from these model studies that the naturally occurring minor components may be divided into several groups according to their complexing ability and, hence, contribution to modification of secondary structure. Firstly, there are nucleosides, e.g. pseudouridine, 5-methyluridine and 5-methyl-cytidine, with an unchanged complementarity but with enhanced affinity to the Watson–Crick complementary base. Secondly, methylamino derivatives of adenosine, cytidine and, presumably guanosine (the latter model has not yet been prepared at the polymer level), with unchanged base-pairing properties, form much weaker complexes with the complementary base. The third group consists of at least three nucleotides: N-methyluridine, N-methylcytidine and 5,6-dihydrouridine, not capable of interacting with the complementary nucleoside. Another group contains the 2'-O-methyl derivatives. Obviously the presence of this methyl group inhibits phosphodiesterase bond hydrolysis by RNase, but some structural modifications induced by the methyl group can also be expected. I think that conclusions from these model investigations should be taken into account when s-RNA structure proposals are advanced, or the role of minor components in it is considered.

D. Fan: Does methylation on recovery from methionine starvation proceed slowly enough to allow for the study of suppression *in vivo* of mistakes in enzymes?

U. Z. Littauer: We do not know whether protein synthesis resumes before all the methyl-deficient tRNA has been methylated.

Author Index

Note: A number in heavy type is the page on which reference is made to a given author.

Where an author is not cited by name, the number in brackets is the reference number of the article of which he is author or co-author.

A number in italics refers to the page on which the full reference is presented.

Abrams, R., **316** (33), *329*
Abrossimova-Amelyanchik, N. M., **289** (33, 34), **295** (34), *300*
Acs, G., **180** (16), *199*, **185** (62), *201*
Adams, J. M., **60** (44), *78*, **109** (157), *115*
Adelberg, E. A., **170**, *177*
Adler, M., **293** (1), **295** (1), **315** (1), *328*
Afzelius, B. A., **85** (41), **95** (41), *112*
Alberts, B. M., **52**, *54*, **57** (12), *78*
Aliyev, K. A., **104** (115), **105** (115), *114*
Allard, C., **95** (61), *113*
Allen, D. W., **189** (74), *201*, **203** (97), *202*
Allen, E. M., **185** (61), *201*
Allen, F. W., **315** (2, 10), *328*
Allende, J. E., **182** (46), *200*, **189** (76), *201*
Amelunxen, F., **61** (51), *79*, **181** (35), **187** (35), **203** (35), *200*, **234** (12), **237** (12), **242** (12), **243** (12), **245** (12, 17), **246** (12, 17), **250** (17), *250*
Ames, B., **178**
Amos, H., **315** (16), *328*
Ammann, J., **120** (44), **129** (44), **134** (44), **140** (44), **141** (44), *152*
Anders, M., **209** (6), *231*, **316** (43, 45, 46, 49) *329*
Andre, J., **105** (133), *115*
Apgar, J., **57** (2), *77*, **234** (6), *250*, **271** (1), *284*, **274** (7), *285*, **287** (1, 2, 4, 10), *299*, **287** (15), *300*, **297** (2), *299*, **303** (1), *312*, **315** (24), *328*
Aposhian, H. V., **3** (5), *14*
App, A. A., **104** (123), *114*
Arca, M., **59**, *78*
Arcamore, F., **34** (37), **34** (38), *38*

Arlinghaus, R., **181** (42), **186** (42, 68), **187** (42), *200,* **189** (77), *201*
Armstrong, A., **287** (11), **292**, *299,* **293** (11), *299*
Arnott, S., **23**, **25**, **26**, *37*
Asano, K., **62**, *79*
Astier-Manifacier, S., **118** (25), *152*
Attardi, G., **121** (51), *152,* **316** (38), *329*
August, J. T., **118** (28), **122** (28), **125** (28), *152,* **125** (68), *153*
Augusti-Tocco, G., **64**, *79*
Aurisicchio, A., **131** (89), *153*
Avers, C. J., **98** (80), **98** (81), *113,* **107** (148), *115*
Axelrod, V. D., **234** (1), *249,* **287** (12), *299,* **289** (33, 34, 35), *300,* **291** (12), *299,* **295** (34), *300*

Bader, J. P., **118** (23), *152*
Bahr, G. F., **95** (60), *113*
Baladin, I. G., **118** (17), *151*
Baldwin, R. L., **132** (95), *154*
Baltimore, D., **117** (8), **118** (8, 14, 15), *151,* **118** (21), **119** (21, 29), **120** (34), *152,* **121** (57), *153,* **123** (21), *152,* **123** (8), *151,* **127** (34), *152,* **141** (21, 34), *152,* **147** (121), **148**, *154,* **149** (29), *152*
Bank, A., **315** (27), *328,* **316** (56), **317** (57), *329*
Barbieri, W., **34** (37), *38*
Barclay, R. K., **20** (12), *37*
Barnard, E. A., **295** (44), *300*
Barnett, L., **36** (46), *38,* **180** (11), *199*
Barnett, W. E., **175** (22), *177*

AUTHOR INDEX

Barrell, B. G., **180** (26), *200*, **303** (2), *312*
Bartley, W., **107** (147), *115*
Basilio, C., **73** (62), *79*, **246** (26), **247** (26), *250*
Bassel, A., **103** (107), *114*
Bassett, D. W., **117** (5), *151*
Bayer, A. A., **234** (1), *249*, **287** (12, 14), *299*, **288** (25), **289** (31, 33, 34, 35, 36, 37), **291** (39), *300*, **291** (12), *299*, **285** (34), *300*
Beaudreau, G., **125** (69), **130** (69), *153*
Becker, Y., **110** (160), *116*, **118** (19), **119** (30), **120** (30, 34), **121** (19), **127** (34), **141** (34), **149** (30), *152*
Beers, R. F., **45** (10), *54*
Befort, N., **58** (23), *78*
Bell, P. R., **107** (150), *115*
Bellett, A. J. D., **119** (31), *152*
Bennett, T. P., **287** (22), *300*
Benoît, H., **43** (3), *54*
Benzer, S., **180** (14), *199*, **180** (21), *200*
Berg, P., **132** (93, 95), *154*, **180** (17), *199*, **185** (63), *201*, **209** (7), **209**, **231**, **230**, **242**, **269**, **316** (54), **317** (54), **324** (54), *329*
Berger, S., **104** (113), **105** (113), *114*
Bergmann, F. H., **180** (17), *199*, **185** (63), *201*
Bergquist, P. L., **287** (7), *299*, **315** (7, 12), *328*
Bernfield, H. R., **75** (64), *79*, **324**, *329*, **325** (65), *329*
Bernfield, M., **58** (16), *78*, **75**, *79*, **220** (25, 27), **221** (25), *231*, **234** (2), *249*, **234** (22), *250*, **245** (2), *249*, **245** (22), *250*, **247** (2), *249*, **298** (55), *301*
Bessman, M. J., **209** (3), *230*, **3** (6), *14*
Bianchi, P., **101** (92), *114*
Bibring, T., **101** (95), *114*
Bienfait, F., **101** (89), *113*
Biggs, D. R., **106** (137), *115*
Billen, D., **4** (11), *14*
Billeter, M. A., **23** (21), *37*, **110** (161), *116*, **123** (64), **124** (64), **129** (64, 78, 80, 83), **130** (86), **131** (64), **132** (64), **136** (80), **140** (83), **156** (64, 78, 80), *153*

Birnstiel, M. L., **99** (84), *113*, **316** (50), *329*
Bishop, J. M., **129** (84), **139**, *153*, **140**, *153*, **184** (57), *201*
Biswas, B. B., **316** (33), *329*
Björk, I., **53** (25), *54*
Black, L. M., **23** (23), *37*
Blackburn, G. M., **79**, **189** (83), **190** (83), **191** (83), **192** (87), *201*
Bladen, H. A., **149** (126), *154*
Boardman, N. K., **104** (127, 128), **105** (127, 128), *115*
Bock, R. M., **225** (30), **228** (30), *231*
Bode, V. C., **90** (53), **91** (53), *112*
Boedtker, H., **180** (24), *200*
Bogorad, L., **83** (29), **97** (29), *112*
Bollack, C., **327**, *329*
Bolle, A., **4** (9), *14*, **179** (1), *199*
Bollum, F. J., **3** (3, 4, 7), **4** (10, 13, 14, 15), **5** (16), **6** (13, 10), **12** (16), *14*, **15**, **102** (104), *114*, **312**
Bolton, E. T., **99** (85), *113*
Boman, H. G., **185** (62), *201*, **316** (36), *329*
Bonner, J., **98** (78), *113*
Bonner, Jr., W. D., **83** (22), **84**, **95** (22), **96** (22), **97** (22), *112*
Borek, E., **293** (40), *329*, **316** (31, 32, 40, 41, 42, 44, 48, 50, 51), **316**, *329*, **317** (60), *329*
Boriosova, D. F., **57** (14), *78*
Borsook, H., **185** (64), *201*
Borst, P., **23** (21), *37*, **81** (4, 5), *111*, **82** (24, 25), **83** (24, 25), *112*, **83** (4), *111*, **84** (24, 25), **85** (24, 25), *112*, **85** (4), **86** (4), **87** (4), **89** (4), **90** (4, 5), **91** (4, 5), **92** (4, 5), **95** (4), *111*, **95** (24, 25), **96** (24, 25), *112*, **100** (4), *111*, **100** (24), *112*, **101** (89), *113*, **105** (25), *112*, **110** (159, 161), *116*, **117** (9), *151*, **123** (64), **124** (64), *153*, **129** (64, 78, 80), **131** (64), **132** (64), **136** (80), *153*, **155**, **156**, (9), *151*, **156** (64, 78, 80), *153*, **177**
Bosch, L., **315** (11), *328*
Bourgeois, S., **105** (129), *115*
Boy de la Tour, E., **4** (9), *14*

Brachet, J., **101** (95), *114*
Bradley, D. F., **31** (28), *37*, **57** (7), *78*
Brahms, J., **52** (15,[1] 20), *54*, **57** (9), *78*, **66** (9), *79*, **295** (47), *300*
Brannberg, R. C., **255** (10), *260*
Brawerman, G., **82** (15), **83** (15), **97** (15), *111*, **315** (13), *328*
Bremer, H., **132** (96), **137** (96), **148**, *154*, **149** (127), *155*
Brenner, S., **3** (1), *14*, **36** (46), *38*, **158** (3, 4), **159** (3, 4), *176*, **179** (1, 3), **180** (11), *199*
Bretscher, M. S., **184** (59, 60), **189** (59, 60), *201*
Brimacombe, R., **58** (16), *78*, **220** (25, 27), **221** (25), *231*, **234** (2), *249*, **234** (22), *250*, **245** (2), *249*, **245** (22), *250*, **247** (2), *249*, **251** (2), *265*, **298** (55), *301*
Bronsert, U., **106** (139), *115*
Brow, R. S., **315** (15), *328*
Brown, D. M., **209** (1), *230*
Brown, F., **120** (35), **127** (35), **129** (35), *152*, **133** (100), *154*, **133** (35), *152*, **134** (100), *154*, **135** (35), *152*, **141** (35), *152*, **156**, *154*
Brown, G. M., **316** (38), *329*
Brown, G. L., **23**, *37*, **57** (5), *77*, **57** (6), **60** (45), *78*, **64**, *79*, *79*
Brownlee, G. G., *77*, **180** (26, 27), *200*, **303** (2), **308**, *312*
Buchanan, J. M., **13** (19, 20), *14*
Budowsky, E. I., **287** (21), **294** (43), *300*
Burdon, R. H., **23** (21), *37*, **110** (161), *116*, **123** (64), **124** (64), **129** (64, 78, 80), **131** (64), **132** (64), **136** (80), **156** (64, 78, 80), *153*
Burgi, E., **91** (52), *112*
Burma, D. P., **209** (8), *231*
Burns, R. O., **170** (16), *177*
Burton, K., **59**, *78*
Buchi, H., **210**, *231*, *232*
Byrd, C., **211** (21), *231*
Byrne, R., **149** (126), *154*

Cairns, J., **92** (56), *113*, **103** (105), *114*, **103** (56), *113*
Calendi, E., **34** (44), *38*

Callender, J., **118** (15), *151*
Calvori, C., **59** (36), *78*
Campagne, R. N., **109**, *115*
Campbell, P. N., **260**
Canevazzi, G., **34** (40), *38*
Cannon, M., **61** (50), *79*, **181** (39, 40), **187** (39), **191** (40), *200*
Cantarow, A., **95** (61), *113*
Cantoni, G. L., **295**, *301*, **295** (48, 49), *301*, **315** (14, 25), *328*
Capecchi, M. R., **60** (44), *78*, **109** (157), *115*
Carbon, J. A., **58** (25), *78*
Carbon, J., **230**
Carlton, B. C., **179** (2), *199*
Carlucci, A. F., **315** (5), *328*
Carrier, W. L., **102** (104), *114*
Cartwright, B., **120** (35), **127** (35), **129** (35), **133** (35), **135** (35), **141** (35), *152*
Cassani, G. R., **6** (17), *14*
Cassinelli, G., **34** (37, 38), *38*
Cecere, M. A., **149** (128), *155*, **179** (9), *199*, **252** (5), *260*, **264** (3), *265*, **297** (53), *301*
Cerna, J., **58** (21), *78*
Chamberlin, M., **132** (93, 95), *154*, **209** (7), *231*
Chan, M., **183** (53), *200*
Changeux, J. P., **173** (19), *177*
Chapeville, F., **180** (21), *200*
Chargaff, E., **18**, *37*, **315** (13), *328*
Cheng, T. Y., **101** (88), *113*, **101** (91), *114*
Cherayil, J. D., **225** (30), **228** (30), *231*
Chevalley, R., **4** (9), *14*
Chevremont, M., **101** (96), *114*
Chun, E. H. L., **82** (11), **83** (11), **97** (11), *111*, **101** (93), *114*, **104** (11), *111*
Clark, B. F. C., **60** (43), *78*, **221** (29), *231*, **232**, *231*, **246** (20), **247** (20), *250*, **264**, **265**, **266**
Clark, I. D., **116**
Clark, M. F., **105** (132), *115*
Clark-Wallace, G. D., **106** (137), *115*
Cline M. J., **118** (20), *152*
Cocito, C., **135** (105), *154*
Cohen, S. S., **117** (1), *151*

Coll, J. A., 70 (60), 72, 73, 79
Conway, T. W., 183 (52), 200, 189 (76), 201
Cooper, P. J., 19, 37
Cooper, S., 60, 78, 70 (61), 79
Corneo, G., 83 (16), 85 (16), 92 (16), 98 (16), 111, 101 (92), 114
Cornuet, P., 118 (25), 152
Cousin, D., 158 (10), 159 (10, 11), 176
Cousin, M., 295, 300
Cowan, C. A., 83 (31), 112
Cox, E. C., 180 (31), 200
Cramer, F., 59, 78, 240 (3), 249
Crawford, E. M., 85 (43), 90 (43), 112
Crawford, L. V., 85 (43), 90 (43, 51), 112
Crick, F. H. C., 18, 37, 36 (46), 38, 50, 54, 58 (15), 78, 131 (88), 153, 179 (10), 180 (10, 11, 19), 199, 199, 209 (2), 230, 223 (31), 231, 234, 264
Cuzin, F., 3 (1), 14, 158 (4), 159 (4), 176

Dales, S., 121 (49), 152
Dalgarno, L., 117 (6), 151, 118 (22), 152, 119 (6), 151, 119 (22), 120 (36), 152, 120 (6), 151, 122 (63), 123 (63), 153, 127 (6), 151, 127 (36), 152, 128 (6), 151, 141 (6, 36), 152, 143 (36), 152, 149 (6), 151
Daniel, T. M., 106 (136), 115
Darnell, J. E., 60 (41), 78, 110 (160), 116, 118 (19), 119 (30), 120 (34), 120 (30), 121 (19), 152, 121 (56), 153, 127 (34), 152, 135 (104), 140 (112), 154, 141 (34), 152, 147 (117), 154, 149 (30), 152, 149 (117), 154
Dasdia, T., 34 (39, 40), 38
Davidson, N., 101 (90), 114
Davies, D. R., 50, 54
Davies, J., 327 (67), 329
Davis, B. D., 186 (69), 201
Davis, F. F., 315 (5), 328
Davis, J. E., 135 (106), 154
Davison, P. F., 54, 54
Dawid, I. B., 101 (94), 114
DeDuve, C., 84 (37), 112

De Lamirande, G., 95 (61), 113
Delius, H., 120 (44), 152, 124 (65), 153, 129 (44), 152, 134 (44), 140 (44), 141 (44), 152
Denhardt, D. T., 133 (99), 141 (99), 154
Denhardt, D. H., 4 (9), 14
Deniset, J., 60 (45), 78
Desalle, L., 82 (28), 83 (28), 84 (28), 112
DeVoe, H., 19, 32, 37
DeWaard, A., 13 (21), 14
Diacumakos, E. G., 81 (3), 102, 111
Dickson, R. C., 83 (19), 111
Dieckman, M., 180 (17), 199, 185 (63), 201
DiMarco, A., 34 (39, 40, 44), 38
Dintzis, H. M., 184 (58), 201
Dirksen, M. L., 13 (19, 20), 14
Ditmars, W. E., 31, 32 (31), 37
Dobrov, E., 52, 54
Doi, R. H., 126 (71), 153, 135, 153, 141 (71), 153
Donahue, J., 57 (11), 78
Dorigotti, L., 34 (40), 38
Doty, P., 52, 54, 58 (19), 78, 65 (59), 69 (59), 79, 82 (35), 85 (44), 86 (44), 112, 91 (54), 113, 101 (35), 112, 149 (128), 155, 179 (9), 199, 252 (5), 253, 260, 260, 264 (3), 265, 288 (28), 300, 297 (53), 301
Duglas, H. C., 101 (101), 106 (101), 114
Dover, S. D., 25, 26, 29
Dubert, J. M., 105 (129), 115
Dubnau, D., 176 (23), 177
Dubuy, H. G., 98 (73), 113
Dunn, D. B., 315 (3, 4, 6, 8, 17, 18, 19, 20), 328
Düttirg, D., 57 (3), 77, 234 (28), 250, 271 (4), 284, 277 (9), 278 (11), 284 (12), 285, 287 (6), 293 (6), 297 (6), 299, 315 (28), 328, 315 (30), 329

Eason, R., 118 (20), 119 (32), 152
Ebel, J. P., 58 (23, 26, 30), 78, 327, 329
Eckert, K., 61 (51), 79, 181 (35), 187 (35), 203 (35), 200 234 (12), 237 (12), 242 (12), 243 (12), 245 (12), 246 (12), 250

AUTHOR INDEX

Edelman, M., **83** (20), *111*, **83** (31), *112,* **97** (20), **99** (20), *111*
Edgar, R. S., **4** (9), *14*
Edmonds, M., **316** (33), *329*
Egami, F., **289** (32), *300*
Eggers, H. J., **119** (29), **121** (49), *152,* **121** (57), *153,* **149** (29), *152*
Eidlic, L., **158** (6), **159** (6), **170** (6), **175** (6), *176,* **316** (52), *329*
Eigner, J., **91** (54), *113*
Eisenstadt, J. M., **82** (15), **83** (15), **97** (15), *111*
Eliasson, R., **53** (25), *54*
Elliott, W. H., **32** (35), *38*
Elper, J. L., **175** (22), *177*
Emerson, T. R., 57 (8), *78*
Emrich, J., **179** (6), *199,* **252** (6), *260*
Engelhardt, D. L., **109** (158), *115*
Engelhardt, V. A., **234** (1), *249*
Ennis, H. L., **149** (132), *155*
Eoyang, L., **118** (28), **122** (28), **125** (28), *152*
Ephrussi, B., **101** (97), *114*
Epstein, R. H., **4** (9), **13** (20), *14,* **157** (1), *176*
Epstein, H. T., **82** (12), **83** (20), *111,* **83** (31), *112,* **97** (12, 20), *111,* **99** (12, 20), *111,* **102** (103), *114*
Erikson, R. L., **82** (34), *112,* **120** (41), *152,* **126** (70), *153,* **126** (41), *152,* **129** (41), *152,* **129** (81), *153,* **131** (81), *153,* **132** (41), *152,* **133**, *153,* **136** (70), *153,* **141** (70), *153,* **156** (81), *153*
Eriksson, K. G., **316** (36), *329*
Estabrook, R. W., **95** (62), *113*
Evans, A., **209** (6), *231*
Evans, M. J., **116**
Everett, G. A., **57** (2), *77,* **234** (6), *250,* **271** (1), *284,* **274** (7), *285,* **287** (1, 2, 4, 5), **293** (5), **297** (2), *299,* **303** (1), *312,* **315** (23, 24), *328*

Fan, D., **15**, *330*
Fangman, W. L., **159**, *176,* *177*
Fasman, G. D., **52** (18), **53** (18, 27), *54*

Faulkner, R. D., **223** (30), **225** (30), **228** (30), *231*
Favelukes, S., **189** (77), *201*
Fawaz-Estrup, F., **58** (27), *78*
Fefferman, R., **316** (37), *329*
Feldmann, H., **57** (3), *77,* **188** (79), *201,* **234** (28), *250,* **271** (4), *284,* **278** (11), *285,* **315** (30), *329*
Fel'dman, M. Ya., **59** (31), *78*
Felix, **156**
Felsenfeld, G., **138** (111), *154*
Fenwick, M. L., **120** (41), *152,* **126** (70), *153,* **126** (41), **128** (41), **129** (41), *152,* **129** (81), *153,* **129**, *152,* **131** (81), *153,* **132** (41), *152,* **133** (81), *153,* **136** (70), **141** (70), **156** (81), *153*
Feughelman, M., **20** (12), *37*
Fiers, W., **135** (103), *154*
Filippovich, J. J., **104** (115), **105** (115), *114*
Fioretti, A., **34** (40), *38*
Flaks, J. G., **180** (31), *200*
Fleissner, E., **316** (42, 44, 50), *329*
Flesher, J. W., **183** (55), *200*
Flessel, C. P., **52** (21), *54*
Fletcher, M. S., **103** (111), *114*
Folk, J. E., **255** (10, 11), *260*
Fox, C. F., **203** (93), *201*
Fraenkel-Conrat, H., **108** (154, 155), *115*
Francesche, G., **34** (37, 38), *38*
Francki, R. J. B., **104** (127, 128), **105** (127, 128), *115*
Franklin, R. E., **18** (2), **19** (2), *37*
Franklin, R. M., **117** (2, 4, 8), **118** (14, 15, 17), **118** (8), *151,* **120** (41, 46), *152,* **123** (8), *151,* **126** (70), *153,* **126** (41), **128** (41), **129** (41), *152,* **129** (81), **131** (81), *153,* **132** (41), *152,* **133** (81), **136** (70), **141** (70), *153,* **147** (121), **148**, *154,* **156** (81), *152*
Frédéric, J., **97** (64), *113*
Fredericq, E., **203**
Freifelder, D., **52**, *54*
Fresco, J. R., **52**, *54,* **57** (12), *78,* **138** (110), *154*
Freundlich, M., **170** (16), *177*

Friedman, R. M., **120** (36), **127** (36), 141 (36), **143** (36), *152*
Frolova, L. Y., **57** (14), **58** (18), *78*
Frontali, L., **59** (36), *78*
Fujimura, R. K., **203** (98), *202*
Fuller, W., **19** (7), **20** (13), **24**, **25** (25), **26** (25), **26**, **32**, *37*, **33**, **34**, **36**, *38*, **38**, **39**, **155**
Furth, J. J., **209** (6), *231*

Gaeta, F. S., **101** (95), *144*
Gaetani, M., **34** (39, 40), *38*
Gaines, K., **149** (127), *155*
Galenter, Y., **285**
Galor, V, F., **203** (100), *202*
Gardner, R. S., **73** (62), *79*, **246** (26), **247** (26), *250*
Garnjobst, L., **81** (2, 3), *111*
Gavrilova, L. P., **203** (100), *202*
Gee, S., **315** (27), *328*
Geiduschek, E. P., **131** (89), *153*, **132** (94), *154*
Geirer, A., **134** (101), *154*
Gelboin, H. V., **295** (49), *301*, **315** (14), *328*
Gersch, N. F., **32**, *37*
Ghosh, H., **229**, *232*
Gibor, A., **81** (10), **97** (10), *111*, **97** (70), *113*, **99**, **101**, **102**, **109**, *111*
Gifford, Jr., E. M., **97** (66), *113*
Gilbert, W., **61** (50), *79*, **179** (5), *199*, **180** (33), **181** (38, 39), **184** (38), **187** (39),**188** (38), **191** (38), **192** (38), **193** (38), *200*, **316** (54), **317** (54), **324** (54), **327** (67), *329*
Gilham, P. T., **62** (54), **62**, **69**, *79*
Gillespie, D., **99** (83), *113*
Ginelli, E., **101** (92), *114*
Ginoza, W., **134** (102), *154*
Giorn, G., **34** (40), *38*
Gladner, J. A., **255** (10, 11), *260*
Glas, U., **95** (60), *113*
Glassmann, E., **185** (61), *201*
Godson, G. N., **141** (113), **147**, *154*
Goetz, O., **147** (118), *154*
Gold, M., **316** (43, 45, 46, 47, 49), *329*

Goldstein, J., **278** (10), *285*, **287**, *300*
Gomatos, P. J., **23**, **25**, *37*, **57** (10), *78*, **129** (79), *153*
Gonazo, F., **180** (14), *199*
Gorini, L., **158** (2), *176*, **171** (20), *177*, **171**, *176*, **327** (67), *329*
Gorshkova, V. I., **289** (31), *300*
Gosling, R. G., **18** (2), **19** (2), *37*
Gould, J. L., **61** (49), *79*, **128** (76), **129** (76), **141** (76), **142** (76), **147** (76), *153*
Grachev, M. A., **287** (21), *300*
Graham, A. F., **126**, **127**, **135**, **141** (73), *153*
Graig, L. G., **287** (22), *300*
Grandchamp, S., **101** (100), *114*
Granick, S., **81** (10), **97** (10), **99**, **101**, **102**, **109**, *111*
Gratzer, W. B., **70** (60), **72**, **73**, *79*
Graziosi, F., **101** (95), *114*, **131** (89), *153*
Grece, A., **83** (26), **95** (26), *112*
Green, D. E., **108**, *115*
Grein, A., **34** (40), *38*
Groeniger, E., 4 (15), *14*
Gros, F., **105** (129), *115*, **158** (9), **159** (9), **160** (9), **165** (9), **174** (9), *176*, **179** (5), *199*, **209** (10), *231*
Grossman, L. I., **52** (18), **53** (18), **54**, **83** (16), **85** (16), **92** (16), **98** (16), **103**, *111*
Gruber, M., **81** (5), **87** (5), **90** (5), **91** (5), **92** (5), *111*, **109**, *115*
Guest, J. R., **179** (2), *199*
Gupta, N., **229**, *232*
Guschlbauer, W., **138** (109, 110), *154*
Guthrie, S., **133** (98), *154*
Gutman, A. B., **293** (1), **295** (1), **315** (1), *328*
Guttman, H. N., **98** (72), *113*

Hagopian, H., **287** (11), **293** (11), *299*
Hall, R. H., **315** (21, 22), *328*
Hamilton, L. D., **19** (5, 6, 7), **20** (10, 12), *37*, **34**, **36**, *38*
Hamilton, M. G., **121** (55, 60), *153*
Hammarston, E., **53** (25), *54*
Hampel, A., **225** (30), **228** (30), *231*

AUTHOR INDEX

Hanawalt, P. C., **82** (14), *111,* **83** (21), *112,* **97** (14), *111,* **97, 99** (21), *112*
Harris, J. I., **180** (28), *200*
Harris, R. G., **119** (31), *152*
Hartmann, G., **185** (62), *201*
Haruna, I., **118** (27), **122** (27), *152,* **122** (61), **125** (61, 69), **130** (61, 69, 87), *153,* **156**
Haselkorn, R., **82** (28), **83** (28, 30), **84** (28), **97** (30), **100** (30), *112,* **120** (38), **129** (38), *152*
Haslbrunner, E., **96** (58), *113*
Hatfield, **79**
Hausen, P., **117** (11), **118** (11), *151,* **118** (18), *152,* **120** (37), **127** (37), **129** (37), *152,* **130,** *152,* **131** (37), **140** (37), **141** (37), *152*
Hawthorne, D. C., **101** (101), **106** (101), *114*
Hayatsu, H., **58** (17), **59,** *78,* **209** (14), **210** (14), **220** (26), **222** (26), **223** (30), **225** (30), **228** (30), *231,* **234** (24), **245** (24), **247** (24), *250,* **298** (56), *301*
Hayashi, M., **60,** *78,* **131** (90, 91), *154*
Hayashi, M. N., **131** (90, 91), *154*
Haywood, A. M., **60** (37), *78*
Hearst, J. E., **82,** *112*
Hecht, L. I., **180** (18), *199*
Hechter, C., **108,** *115*
Heeter, M., **140** (112), *154*
Heinrich, J., **58** (30), *78*
Heinz, R., **186** (68), *201*
Helge, H., **81** (8), *111,* **103** (110), **104,** **114, 104** (8), *111,* **104** (110, 119), *114,* **105** (8), *111*
Helinski, D. R., **179** (2), *199*
Heller, G., **181** (35), **187** (35), **203** (35), *200,* **234** (12, 14, 16), **237** (5, 12, 16), **240** (16), **242** (5, 12, 16), **243** (5, 12, 16), **245** (12, 16, 17), **246** (12, 16, 17), **247** (14, 16), **248** (16), **250** (17), *250,* **297** (52), *301*
Heller, G. Z., **61** (51), *79*
Henning, U., **179** (2), *199*
Hennix, U., **95** (57), **96** (57), **103** (57), *113*
Heppel, L., **240**
Herbener, G. H., **95** (60), *113*

Herbert, E., **287** (16, 17, 18, 19), *300,*
Herriott, S. T., **3** (6), *14*
Hershey, A. D., **91** (52), *112*
Hershey, J. W. B., **183** (54), *200*
Herskovits, T. T., **53** (26), *54*
Hiatt, H. H., **104** (120), *114,* **179** (51), *195*
Hickler, S., **98** (77), *113*
Higuchi, S., **23** (22), **25** (22), **26** (22), *37*
Hinds, H. A., **315** (15), *328*
Hinuma, Y., **121** (58), *153*
Hirth, L., **43** (5), **44** (5), **45** (5), **52** (5), *54*
Hitchbarn, J. H., **315** (20), *328*
Ho, N. W. Y., **62, 69** (55), *79*
Ho, P. P. K., **118** (24), *152*
Hoagland, M. B., **180** (20), *200,* **209,** *231*
Hofer, H., **83** (36), *112*
Hoffman, A., **106** (136), *115*
Hoffmann-Berling, H., **25** (26), *37,* **120** (43), **140** (43), **141** (43), *152*
Hofschneider, P. H., **120** (44), *152,* **124** (65), *153,* **129** (44), *152,* **134** (44), **140** (44), **141** (44), *152*
Hogness, D. S., **87** (48), *112*
Holcomb, D. N., **52** (54), *54*
Holland, I. B., **125** (69), **130** (69), *153*
Holland, J. J., **117** (5), **118** (16), *151*
Holley, R. W., **57** (2), *77,* **234** (6), *250,* **263,** **264, 269, 271,** *284,* **274** (7), *285,* **287,** *299,* **287** (1, 2, 3, 4, 5, 10), *299,* **287** (15), *300,* **291,** *299,* **292, 293,** *299,* **300,** **294, 295, 296, 297,** *299,* **299,** **303** (1), **309** (4), *312,* **315** (23, 24), *328,* **315** (29), *329*
Hollingworth, B. R., **180** (23), *200*
Holowinsky, A., **95** (62), *113*
Homma, M., **126, 127, 135, 141** (73), *153*
Honda, S. J., **97** (65), *113*
Hongladarom, T., **97** (65), *113*
Honig, G. R., **104** (121), *114*
Hooper, C. W., **19** (5, 6), **20** (12), *37*
Hooper, J. L., **147** (117), **149** (117), *154*
Horne, R. W., **121** (48), *152*
Horton, E., **117** (12), *151,* **118** (22), **119** (22), *152,* **123** (12), *151*
Hoshino, M., **127** (75), **136** (75), *153*
Hosokawa, K., **203** (98), *202*
Hotta, Y., **103** (106, 107), *114*

Hoyer, B. H., **99** (85), *113*
Hsu, R. Y., **203** (95), *202*
Huang, M., **106** (137), *115*
Hufnagel, D. A., **315** (13), *328*
Humm, D. G., **100** (86), *113*
Humm, J. H., **100** (86), *113*
Hummeler, K., **147** (118), *154*
Hurwitz, J., **209** (6), *231*, **316** (43), **316**, *329*, **316** (47, 49), *329*
Hutchinson, F., **23, 25** (25), **26**, *37*
Hutson, J. C., **13** (19), *14*
Huxley, H. E., **193** (88), *201*
Hülsman, W. C., **185** (12), *201*

Iitaka, Y., **23** (22), **25** (22), **26** (22), *37*,
Ikeda, Y., **128** (77), **140** (77), *153*
Ingraham, L., **91** (52), *112*
Ingram, V. M., **188** (81), *201*, **287** (8, 11) **291, 293** (11), *299*, **315** (26), *328*
Inman, R. B., **13** (18), *14*
Inouye, M., **179** (6), *199*, **252** (6), *260*
Isawa, M., **97** (70), *113*
Ishida, M. R., **82** (13), **83** (13), **97** (13), *111*
Ishida, T., **287** (20), *300*
Ishikura, H., **315** (25), *328*
Iwamura, T., **103** (108, 109), *114*

Jacherts, D., **89** (55), *112*
Jacob, F., **3** (1), *14*, **158** (3, 4, 9, 10), **159** (3, 4, 9, 10, 11), **160** (9), **165** (9), **174** (9), *176*, **179** (3, 4), *199*
Jacob, J., **316** (34), *329*
Jacob, T. M., **209** (13), **210** (13, 16, 18), **210, 220** (13), **222** (13), *231*, *232*
Jacobson, **312**
Jagendorf, A. T., **104** (123), *114*
Jansz, H. S., **90** (50), **91** (50), *112*
Jaouni, T., **220** (27), *231*, **234** (2), **245** (2), **247** (2), *249*, **298** (55), *301*
Jayaraman, J., **82** (17), **83** (17), **84** (17), **85** (17), **92** (17), **93** (17), **96** (17), **98** (17), **105** (17), *111*
Jehle, H., **132** (97), **137** (97), *154*
Jensen, R., **101** (90), *114*

Jesensky, C., **316** (56), **317** (56), *329*
Jinks, J. L., **101** (99), *114*
Joklik, W. K., **135** (104), *154*
Jones, D. S., **58** (17), *78*, **215** (23), **220** (23, 26) **222** (26), **223** (30), **225** (30), **228** (30), *231*, **234** (24), **245** (24), **247** (24), *250* **298** (56), *301*
Jones, O. W., **191** (85), *201*, **234** (10), **246** (20), **247** (20), *250*
Jordan, D. O., **32**, *37*
Josse, J., **15**
Jukes, T., **60** (46), *79*
Julien, J., **57**, *77*, **180** (25), *200*, **308** (3), *312*

Kadoya, M., **4** (12), *14*
Kaerner, H. C., **25** (26), *37*, **120** (43), **140** (43), **141** (43), *152*
Kaiser, A. D., **90** (53), **91** (53), *112*
Kaji, A., **177, 178, 181** (35, 36), **187** (35), *200*, **320** (63), *329*
Kaji, H., **181** (35), **187** (35), **203** (35), *200*, **203** (101), *202*, **320** (63), *329*
Kalf, G. F., **83** (26), **95** (26), *112*, **104** (117), *114*, **96** (59), *113*
Karau, W., **287** (6), **293** (6), **297** (6), *299*
Kataja, E., **158** (2), *176*
Katch, T., **101** (102), **103** (106), *114*
Katchalsky, E., **285**
Keith, G., **327**, *329*
Kellenberger, E., **4** (9), *14*
Keller, E. B., **183** (49), *200*
Kelly, R. B., **61** (49), *79*, **120** (42), **126** (42), *152*, **128** (76), **129** (76), **141** (76), **142, 145, 147**, *153*
Kemp, J. W., **315** (2), *328*
Kennedy, E. P., **203** (93), *201*
Kerr, I. M., **121** (55, 60), *153*
Khokhlov, A. S., **287** (14), *299*
Khorana, H. G., **6** (22), *14*, **58** (17), *78*, **180** (11, 15), *199*, **207, 209** (12, 13, 14), **210** (12, 13, 14, 15, 16, 17, 18, 19), **211** (20, 21, 22), **215** (20, 23), **220** (13, 20, 23, 26), **222** (13, 26), **223** (30), **225** (30), **228** (30), *231*, **231, 232, 234, 250, 245**

(24), 246 (7), 247, *250*, 251 (4), 259, *260*, 263, 264, 298 (56), *301*
Kieras, F. J., 83 (30), 97 (30), 100 (30), *112*
Kirk, J. T. O., 104 (112), 105 (112), *114*
Kiselev, L. L., 57 (14), 58 (18), *78*
Kislev, N., 83 (29), 97 (29), *112*
Kjellin, K., 316 (36), *329*
Kleinhauf, H., 234 (14, 16), 237 (16), 240 (16), 242 (16), 243 (16), 245 (16), 246 (16), 247 (14, 16), 248 (16), *250*, 297 (52), *301*
Kleinschmidt, A. K., 52, *54*, 89 (55), *113*
Kleinwachter, V., 32, *37*
Klotz, L., 138 (110), *154*
Knight, C. A. 129 (83), 140 (83), *153*
Knorre, D. G., 285, 287 (14), *299, 288, 300*
Kochetkov, N. K., 294 (43), *300*
Kohiyama, M., 158 (3), 159 (3, 11), *176*
Kolakofsky, D., 266
Konrad, M. W., 132 (96), 137 (96), *154, 148, 154,* 149 (127), *155*
Korenyako, A. T., 289 (33), *300*
Korn, M., 315 (16), *328*
Kornberg, A., 3 (2, 5, 8), 13 (18), *14*, 15, 185 (65), *201*, 209 (3), *230*
Kornberg, S. R., 15
Korner, A., 105 (130), *115*
Koudelka, J., 32, *37*
Kossel, H., 210, *231*, 221, *231*, 232
Krakow, J. S., 149 (124), *154*
Kroger, H., 209 (8), *231*
Krone, W., 58 (24), *78*
Kroon, A. M., 81 (5, 9), *111*, 82 (25), *112*, 83 (9), *111*, 83 (23, 25), 84 (25), 85 (25), *112*, 87 (5), 90 (5), 91 (5), 92 (5), 93 (9), *111*, 95 (25), *112*, 95 (9), *111*, 96 (25), 100 (23), *112*, 100 (87), *113*, 104 (116), *114*, 104, *111*, 105 (25), *112*, 105, (131), *115*, 106 (9), *111*
Krug, R. M., 61 (50), *79*, 120 (46), *152, 181* (39), 187 (39), *200*
Krutilina A. I., 234 (1), *249*, 287 (12, 14), *299*, 287 (21), 288 (25), 289 (31, 36, 37), *300*, 291 (12), *299*

Kuff, E. L., 83 (27), 85 (27), 96 (27), 97 (27), 103 (27), *112*
Kukhanova, M. K., 57 (14), 18 (18), *78*
Kurland, C. G., 179 (5), *199*
Kuwashima, S., 103 (108), *114*
Kübler, H., 245 (17), 246 (17), 250 (17), *250*
Kuntzel, H., 234 (14), *250,* 240 (3), *249,* 247 (14), *250,* 297 (52), *301*
Kyogoku, Y., 23 (22), 25 (22), 26 (22), *37*

Laduron, P., 135, *154*
Laipis, P., 90 (49), 91 (49), *112*
Lamfrom, H., 158 (3), 159 (3), *176*
Lane, B. G., 315 (10), *328*
Lang, D., 89 (55), *113*
Langridge, R., 19 (5, 6), 20 (12), 23, 24, 25, 26, *37*, 57 (10), *78*, 129 (78, 79), 156 (78), *153*
Laurent, T. C., 53 (25), *54*
Lawrence, M., 132 (92), *154*
Lawson, W. B., 288 (29), *300*
Lazar, D., 174 (21), *177*
Lazzari, M., 58 (26), *78*
Lazzarini, R. A., 320, *329*
Leahy, J., 184 (57), *201*
Lebowitz, J., 90 (49), 91 (49), *112*
Leboy, P. S., 180 (31), *200*
Leder, P., 58 (16), *78*, 75 (63), *79*, 202, 207, 220, *231*, 220 (25, 27), 221 (25), *231*, 231, 234 (2), *249*, 234 (22), 235 (21), 242 (21), *250*, 245 (2), *249*, 245 (22), 246 (20), 247 (20), *250*, 247 (2), *249*, 263, 264 (2), 265, 265, 266, 298 (54, 55), *301*, 312 (5), *312*, 320 (64), *329*
Lee, S., 57 (5), *77*, 57 (6), 60 (45), *78*
Leff, J., 82 (12), 97 (12), 99 (12), *111*
Legault, L., 105 (129), *115*
Lehman, I. R., 6 (23), 13 (20), *14*, 209 (3), *230*
Lehninger, A. L., 107 (140), 110 (140), *115*
Lengyel, P., 73 (62), *79*, 246 (26), 247 (26), *250*
Lerman, L. S., 31, *37*, 32, 34, *38*
Lermann, M. F., 203 (100), *202*

Levene, P. A., **209**
Levin, J. G., **149** (126), *154*
Levintow, L., **117**, *151*, **120** (45), **121** (52, 53, 54), *152*, **121** (59), **129** (84), **139** (84), **140** (84), *153*, **147** (117, 119), **148** (119), **149**, *154*
Levisohn, S., **98** (74), *113*
Li, L., **234** (1), *249*, **287** (12), *299*, **289** (37), *300*, **291** (12), *299*
Libonati, M., **130** (86), *153*
Lielausis, A., **4** (9), *14*
Lin, F. M., **98** (81), *113*
Lindblow, C., **52** (18), **53** (18, 27), *54*
Lindegren, C. C., **98** (77), *113*
Lindegren, G., **98** (77), *113*
Lindhal, T., **53** (25), *54*
Linnane, A. W., **106** (137), **107** (146), *115*
Lipmann, F., **180** (16), *199*, **180** (21), **182** (45, 46, 47, 48), **183** (45, 47, 51, 52), *200*, **185** (62), **188** (80), **189** (76), *201*, **195** (47), **196** (47), *200*, **209**
Lipsett, M. N., **52** (19), *54*
Littauer, U. Z., **38**, **266**, **285**, **316** (54, 55, 57), **317** (54, 55, 57, 59), *329*, **317**, **318** (57, 61), **319** (61), **320**, **322** (61), **323** (61), **324** (54, 59), **325** (59), **326** (59), **327** (59), *329*, **330**
Littlefield, J. W., **101** (93), *114*, **315** (3), *328*
Liu, S-L., **117** (12), *151*, **118** (22), **119** (22), *152*, **122** (63), **123** (63), *153*, **123** (12), *151*
Lodish, H. F., **60**, *78*, **125** (67), **130**, *153*
Lohrmann, R., **58** (17), *78*, **209** (14), **210** (14), **220** (26), **222** (26), **223** (30), **225** (30), **228** (30), *231*, **234** (24), **245** (24), **247** (24), **250**, **298** (56), *301*
Löw, H., **104** (124), *115*
Lowry, O. H., **240**
Lubin, M., **149** (132), *155*
Luborsky, S. W., **295** (48, 49), *301*, **315** (14), *328*
Lucas-Lenard, J., **182** (48), *200*
Luck, D. J. L., **81** (6), **83** (6), **92**, **104** (6), **105** (6), *111*, **106** (142), **107**, *115*
Luria, S. E., **13** (20), *14*

Luzzati, V., **31**, *37*, **41** (1, 2), **42** (1, 9), **43** (2, 3, 5, 6, 8), **44** (5, 8, 9), **45** (5, 9), **46** (8, 11), **47** (8), **48** (8), **49** (2, 12), **50** (2), **52** (5, 8, 11), **53** (8, 28), *54*
Lyman, H., **102** (103), *114*
Lynen, F., **203** (94), *202*

Mackler, B., **101** (100, 101), **106** (101), *114*
Maden, B. E. H., **192** (86, 89), **193** (86), **195** (86), *201*
Madison, J. T., **57** (2), *77*, **234** (6), *250*, **271** (1), *284*, **274** (7), *285*, **287** (1, 2, 5), **293** (5), *299*, **293** (40, 41), *300*, **293** (5), **297** (2), *299*, **303** (1), *312*, **315** (23, 24), *328*
Magasanik, B., **170**, *177*
Mahler, H. R., **82** (17), **83** (17), **84** (17), **85** (17), **92** (17), **93** (17), **96** (17), **98** (17), *111*, **101** (100, 101), *114*, **105** (17), *111*, **106** (101), *114*
Maizel, J. V., **60** (41), *78*
Maizel, Jr., J. V., **121** (59), *151*
Malec, J., **118** (13), *151*
Mandel, H. G., **120** (39, 40), *152*, **140** (40), **141** (40), *152*
Mandel, L. R., **293** (40), **316** (31, 32, 40, 41), *329*
Mandel, M., **82** (12), **97** (12), *111*, **98** (74), *113*, **99** (12), *111*
Marciello, R., **57** (13), *78*
Marcker, K. A., **60** (42, 43), *78*, **221** (29), *231*
Marinozzi, V., **105** (133), *115*
Markham, R., **23** (23), *37*
Marmur, J., **64** (57), *79*, **82** (35), *112*, **83** (16), **85** (16), *111*, **85** (44), **86** (44, 45), *112*, **92** (16), *111*, **98** (74), *113*, **98** (16), *111*, **100**, *111*, **101** (35), *112*, **103**, *111*, **176** (23), *177*
Marquisee, M., **57** (2), *77*, **234** (6), *250*, **271** (1), *284*, **274** (7), *285*, **287** (2), **297** (2), *299*, **303** (1), **309** (4), *312*, **315** (29), *329*

AUTHOR INDEX

Martin, E. M., **117** (6, 12), **118** (13), *151*, **118** (22), *152*, **119** (6), *151*, **119** (22), **120** (36, 47), *152*, **120** (6), *151*, **121** (55, 60), **122** (63), **123** (63), *153*, **123** (12), *151*, **125** (66), *153*, **127** (6), *151*, **127** (36), *152*, **128** (6), **141** (6), *151*, **141** (36), *152*, **142** (114, 115), *154*, **143** (36), *152*, **143** (115), *154*, **144** (116), **145** (116), *154*, **147** (66), **148** (66), *153*, **149** (6), *151*, **155**, **156**
Martin, R. G., **234** (10), *250*
Martin, S. J., **133** (100), **134** (100), **156** (100), *154*
Marvin, D. A. **19** (6), **20**, *37*
Masson, F., **31** (29), *37*, **41** (2), **43** (2, 7, 8), **44** (8), **46** (8), **47** (8), **48** (8), **49** (2, 7), **50** (2, 7), **52** (8), **53** (8), *54*
Massoulié, J., **138** (110), *154*
Mathieu, R., **95** (61), *113*
Mathis, A., **43** (8), **44** (8), **46** (8), **47** (8), **48** (8), **52** (8), **53**, *54*
Mattern, C. F. T., **98** (73), *113*
Matthaei, J. H., **61** (51), *79*, **181** (35), **187** (35), *200*, **191** (85), *201*, **203** (35), *200*, **207**, **209** (11), *231*, **232**, **233** (15), **234** (8, 9, 10, 12, 13, 14, 16, 19), **235** (13), **237** (5, 12, 16), **238** (9), **239** (19), *250*, **240** (3), *249*, **240** (16), **242** (5, 12, 16), **243** (5, 12, 16), **245** (12, 16, 17), **246** (11, 12, 16, 17, 27), **247** (11, 14, 16), **248** (16), *250*, **264**, **297** (52), *301*
Matthaei, H., **234** (14, 16), **237** (16), **240** (16), **242** (16), **245** (16), **246** (16), **247** (14, 16), **248** (16), **250** (17), *250*, **297** (52), *301*
Matthews, R. E. F., **120** (39, 40), **140** (40), **141** (40), *152*, **315** (7, 12), *328*
Matus, A., **120** (39, 40), **140** (40), **141** (40), *152*
McCandless, R. G., **107** (151), *115*
McCarthy, B. J., **99** (85), *113*
McCorquodale, D. J., **183** (53), *200*
McElvain, N. F., **121** (54), *152*, **147** (119), **148** (119), *154*
McLaughlin, C. S., **188** (81), *201*
McCullen, D. W., **57** (5), *77*, **57** (6), *78*

Mécs, E., **142**, *154*, **144** (116), **145** (116), *154*
Mednieks, M., **183** (56), *200*
Meek, J. A., **107** (147), *115*
Mehler, A., **315** (27), *328*, **316** (56), **317** (57), *329*
Melchers, F., **277** (9), **278** (11), **285**, **287** (6), **293** (6), **297** (6), *299*
Merker, H. J., **81** (8), **104** (8), **105** (8), *111*
Merrill, C. R., **57** (7), *78*
Merrill, S. H., **57** (2), *77*, **234** (6), *250*, **271** (1), *284*, **274** (7), *285*, **287** (1, 2, 10), *299*, **287** (15), *300*, **297** (2), *299*, **303** (1), *312*, **315** (24), *328*
Meselson, M., *179* (4), *199*, **203** (99), *202*
Michelson, A. M., **57** (8, 9), *78*, **66** (9), *79*, **105** (129), *115*, **295** (47), *300*
Milbauer, R., **316** (57), **317** (57), **318** (57), *329*
Miller, R. S., **73** (62), *79*
Mills, R. K., **82** (17), **84** (17), *111*
Mills, D., **125** (69), **130** (69), *153*
Mirzabekov, A. D., **234** (1), *249*, **287** (12, 13, 14), *299*, **287** (21), **288** (25), **289** (31, 33, 37), *300*, **291** (12), *299*
Misina, S. D., **288** (30), *300*
Mitsiu, H., **4** (12), *14*
Mitsui, Y., **23** (22), **25** (22), **26** (22), *37*
Miura, K., **58** (20), **59**, *78*, **287** (20), *300*
Moldave, K., **149** (125), *154*
Moller, W., **180** (24), *200*
Mondelli, R., **34** (37, 38), *38*
Monier, R., **57**, *77*, **180** (25), *200*, **308** (3), *312*, **315** (9), *328*
Monod, J., *179* (3), *199*
Monro, R. E., **181** (37), **182** (46), **183** (50, 51, 54), *200*, **186** (66), **187** (66), **189** (83), **190** (83), **191** (83, 84), **192** (66, 86, 87, 89), **193** (66, 86, 88), **195** (66, 86), **196** (84, 66), **197** (66), *201*, **202**
Montagnier, L., **117** (7), **120**, **123**, **126** (7), **129** (7), **133** (7), *151*
Moon, M. W., **209** (13), **210** (13, 15), **211** (21), **220** (13), **222** (13), *231*
Moore, C., **83** (16), **85** (16), **92** (16), **98** (16), *111*

Moore, P. B., **203** (96), *202*
Moore, P., **62**, *79*
Morgan, A. R., **210** (19), **221**, *231*, **232**
Morris, A., **188** (73), **189** (77), **191** (73), *201*
Morrison, R. A., **295** (48), *301*
Morton, R. K., **98**, *113*
Muench, K., **316** (54), **317** (54), **324** (54), *329*
Munn, R. Y., **107** (151), *115*
Muto, N., **103** (109), *114*
Mühlethaler, K., **107** (150), *115*
Myers, T. C., **183**, *200*
Mysina, S. D., **285**

Nagington, J., **121** (48), *152*
Nagler, C., **133** (98), *154*
Nakamoto, T., **132** (94), *154*, **266**
Nakamura, K., **183** (55), *200*
Nandi, U. S., **101** (90), *114*
Narang, S. A., **209** (13), **210** (13, 17, 18), **210**, *231*, **210** (19), **220** (13), **222** (13), *231*, **232**
Nass, G., **159** (14), *177*
Nass, M. M. K., **85** (39, 40, 41), **95** (39, 40, 41), *112*, **95** (57), **96** (57), **103** (57), *113*
Nass, S., **85** (39, 40, 41), **95** (39, 40, 41), *112*, **95** (57), **96** (57), **103** (57), *113*
Nathans, D., **60** (38), *78*, **149** (130), *155*, **182** (45), **183** (45, 51), *200*, **188** (78), **189** (75, 76), **196** (78), *201*
Naylor, R., **62**, **69**, *79*
Neidardt, F. C., **158** (6, 7), *176*, **159** (6), **159**, *176*, *177*, **164**, *176*, **170** (6), **175** (6), *176*, **316** (52), *329*
Neidle, A., **188** (78), **196** (78), *201*
Neth, R., **245** (17), **246** (17), **250** (17), *250*
Neu, H., **240**
Neubert, D., **81** (8), *111*, **103** (110), **104**, *114*, **104** (8), *111*, **104** (110, 119), *114*, **105** (8), *111*
Neupert, W., **106** (139), *115*
Newman, J. F. E., **119** (33), **127** (33), **141** (33), **146** (33), **149** (33), *152*

Nicolaieff, A., **41** (2), **43** (2), **49** (2), **50** (2), *54*
Nirenberg, M. W., **58** (16), *78*, **75** (63, 64), *79*, **79**, **149** (126), *154*, **191** (85), *201*, **207**, **209** (11), **220**, *231*, **221** (25), *231*, **234** (8, 9, 10, 19, 21), *250*, **234** (2), *249*, **234** (22), **235** (21), **238** (9), **239** (19), **242** (21), *250*, **245** (2), *249*, **245** (22), **246** (20), **247** (20), *250*, **247** (2), *249*, **247**, *249*, **259**, **263**, *265*, **298** (54, 55), *301*, **312**, *312*, **320** (64), **324**, *329*, **325** (65), *329*
Nishimura, S., **58** (17, 28), *78*, **211** (22), **215** (23), **220** (23, 26), **222** (26), *231*, **234** (24), **245** (24), **247** (24), *250*, **295**, *301*, **298** (56), *301*
Nishizuka, Y., **182** (47), **183** (47), **195** (47), **196** (47), *200*
Noll, H., **181** (43), **186** (43), **188**, *200*
Nomura, M., **203** (98), *202*
Nonoyama, M., **128** (77), **140** (77), *153*
Novelli, G. D., **295**, *301*
Novikoff, A. B., **106** (141), **107** (141), *115*
Novogrodsky, A., **285**
Nozu, K., **118** (27), **122** (27), *152*
Nussbaum, A. L., **6** (23), *14*

O'Brien, B., **240** (23), *250*
Ochoa, S., **110** (161), *116*, **118** (26), *152*, **123** (64), **124** (64), **129** (64, 80, 83), **131** (64), **132** (64), **136** (80), **140** (83), *153*, **149** (129), *155*, **155**, **156**, *152*, **156** (64, 80), *153*, **179** (7, 8), *199*, **207**, **209** (8), *231*, **246** (25, 26), **247** (26), *250*, **251** (1, 2, 3), **252** (1, 3), **254** (2), **255** (2), **257** (2), **258** (2), *260*, **260**, **261**, **264**, **265**, **266**, **269**, **297** (51), *301*, **327** (66), *329*
Ofengand, E. J., **180** (17), *199*, **185** (63), *201*
Ogur, M., **98** (77), *113*
Ohtaka, Y., **118** (27), **122** (27), *152*
Ohtsuka, E., **58** (17), *78*, **232**, **234** (24), **245** (24), **247** (24), *250*, **298** (56), *301*
Okada, Y., **179** (6), *199*, **221**, *231*, **252** (6), *260*

AUTHOR INDEX

Okamoto, T., **61** (48), *79*, **181** (34), *200*
Olivera, B. M., **101** (90), *114*
O'Neal, C., **58** (16), *78*, **220** (25), **221** (25), *231*, **234** (22), **245** (22), *250*
Orezzi, P., **34** (37, 38, 39), *38*
Orgel, A., **36** (46), *38*
Orr, C. W. M., **3** (6), *14*
Osawa, S., **4** (12), *14*
O'Sullivan, A., **97** (63), *113*
Otaka, E., **4** (12), *14*
Othtsuka E., **209** (13, 14), **210**, *231*, **210** (19), **211** (20, 21), **215** (20), **220** (13, 20, 26), **222** (13, 26), **223** (30), **225** (30), **228** (30), *231*

Palade, G. E., **121** (49), *152*
Pardon, J., **20**, *37*
Park, R. B., **108** (152), *115*
Parmeggiani, A., **245** (17), **246** (17), *250* (17), *250*
Parsons, J. A., **83** (19), *111*, **85** (42), **103**, *112*
Paul, A. V., **13** (21), *14*
Peacocke, A. R., **36**, *38*
Penman, S., **110** (160), *116*, **118** (19), **119** (30), **120** (30), **121** (19, 50), **149** (30), *152*
Penniston, J. T., **58** (19), *78*, **65** (59), **69** (59), *79*
Penswick, J. R., **57** (2), *77*, **234** (6), *250*, **271** (1), *285*, **287** (3), **297**, *299*, **303** (1), *312*
Pepper, C. R., **98** (80), *113*
Perevertajlo, G. A., **52**, *54*
Pestka, S., **246** (20), **247** (20), *250*
Peterkofsky, A., **315** (27), *328*, **316** (56, 58), **317** (56), **320**, *329*
Petermann, M. L., **180** (22), *200*
Pigram, W., **34**, *38*
Plagemann, P. G. W., **127**, **136** (75), *153*, **156**
Plaut, W., **97** (67), *113*
Polakis, E. S., **107** (147), *115*
Pollard, C. J., **97** (71), *113*
Polli, E., **101** (92), *114*

Poole, F., **23** (18), *37*
Pons, M., **129** (85), **133** (85), *153*
Portatius, V. H., **288** (28), *300*
Porter, J. W., **203** (95), *202*
Potter, V. R., **3** (4), *14*
Pouwels, P. H., **90** (50), **91** (50), *112*
Preer, J. R., Jr., **83**, *111*
Preiss, J., **180** (17), *199*
Price, T. D., **315** (15), *328*

Rabinowitz, M., **82** (28), **83** (28), **84** (28), *112*, **104** (121), *114*
Radloff, R., **90** (49), **91** (49), *112*
Raison, J. K., **98**, *113*
Raj Bhandery, U. L., **271** (3), *284*
Ralph, R. K., **120** (39, 40), **140** (40), **141** (40), *152*
Ramakrishnan, T., **170** (17), *177*
Rammler, D. H., **188** (80), *201*
Rancourt, M. W., **98** (80, 81), *113*
Ray, D. S., **82** (14), *111*, **83** (21), *112*, **97** (14), *111*, **97**, **99** (21), *112*, **116**
Ray, W. J., **180** (21), *200*
Reggiani, M., **34** (44), *38*
Reich, E., **34**, **36**, *38*, **81** (6), **83** (6), **92**, **104** (6), **105** (6), *111*, **117** (2), *151*
Reshetov, P. D., **287** (14), *299*
Rether, B., **58** (23, 30), *78*
Revel, J. P., **104** (120), *114*
Revel, M., **104** (120), *114*, **317** (59), **318** (61), **319** (61), **322** (61), **323** (61), **324** (59), **325** (59), **326** (59), **327** (59), *329*
Rich, A., **20**, **23**, **25**, *37*, **50**, *54*, **82** (11), **83** (11), **97** (11), **104** (11), *111*, **181** (41), *200*
Richards, B. M., **20**, *37*
Richards, E. G., **52** (21), *54*
Richards, E. G., **70** (60), **72**, **73**, *79*
Richards, H. H., **295** (49), *301*, **315** (14, 25), *328*
Richardson, C. C., **3** (8), **13** (18), *14*, **252**, *260*
Richardson, J. P., **85** (43), **90** (43), *112*
Ridge, W. M., **107** (148), *115*
Riley, F. L., **98** (73), *113*
Riley, W. J., **59**, *78*

Ris, H., **85** (38), **95** (38), *112*, **97** (67), *113*, **110** (38), *112*
Risebrough, R. W., **179** (5), *199*
Roberts, W. K., **119** (33), **127** (33), **141** (33), **146**, **149** (33), *152*
Robertson, J. M., **287** (7), *299*
Rodeh, R., **317**
Roodyn, D. B., **104** (125), *115*
Rörsch, A., **158** (5), **159** (5), *176*
Rosner, J., **117** (4), *151*
Rosset, R., **57**, **77**, **180** (25), *200*, **308** (3), *312*
Rottman, F., **58** (16), *78*, **220** (25), **221** (25), *231*, **234** (22), **245** (22), *250*
Roubein, J. F., **315** (5), *328*
Rownd, R., **64** (57), *79*, **86** (45, 46), *112*
Rozijn, T. H., **98** (76), *113*
Rueckert, R. R., **119** (33), **127** (33), **141** (33), **146** (33), **149** (33), *152*
Ruttenberg, G. J. C. M., **81** (4, 5), *111*, **82** (24, 25), **83** (24, 25), *112*, **83** (4), *111*, **84** (24, 25), **85** (25), *112*, **85** (4), **86** (4), **87** (4, 5), **89** (4), **90** (4, 5) **91** (4, 5), **92** (4, 5), **95** (4), *111*, **95** (24, 25), **96** (24, 25), *112*, **100** (4), *111*, **100** (24), *112*, **101** (89), *113*, **105** (25), *112*
Rychlik, I., **58** (21), *78*, **197** (90, 91), **197**, *201*, **202**
Ryter, A., **159** (11), *177*

Sager, R., **82** (13), **83** (13), **97** (13), *111*, **101** (98), *114*
Salas, M., **149** (129), *155*, **179** (7, 8), *199*, **246** (25), *250*, **251** (2, 3), **252** (3), **254** (2), **255** (2), **257** (2), **258** (2), *260*, **261**
Saludjian, P., **49** (12), *54*
Salzman, N. P., **117** (3), *151*
Sampson, M., **103** (106), *114*
Sanadi, D. R., **83** (16), **85** (16), **92** (16), **98** (16), *111*, **103** (111), *114*
Sandakhchiev, L. S., **285**, **287** (14), *299*, **287** (21), **288** (30), *300*
Sandell, S., **104** (124), *115*
Sanders, F. K., **117** (7), *151*, **119** (31), *152*, **120**, **123**, *151*, **126** (7), **129** (7), **133** (7), *151*
Sanger, F., **77**, **60** (42), *78*, **180** (26, 27), *200*, **202**, **207**, **269**, **303** (2), *312*, **308**, **313**
Sanuida, S., **101** (102), *114*
Sarabhai, A. S., **179** (1), *199*
Sarkar, P. K., **52** (16), *54*
Sarnat, M., **131** (89), *153*
Sato, T., **23**, **25**, **26**, *37*
Scarpinato, B. M., **34** (39, 44), *38*
Schactschnabel, D., **58** (24), *78*
Schaeffer, J., **181** (42), **186** (42), *200*, **186** (68), *201*, **187** (42), *200*
Scharff, M. D., **120** (45, 53), *152*, **147** (119), **148**, *154* **153** (59), *152*
Schatz, G., **96** (58), *113*, **107** (149), *115*
Schejde, D. A., **107** (151), *115*
Scherrer, K., **118** (19), **121** (19), *152*
Schiff, J. A., **82** (12), **83** (20), *111*, **83** (31), *112*, **97** (12, 20), **99** (12, 20), *111*, **102** (103), *114*
Schildkraut, C. L., **3** (8), *14*, **64** (57), *79*, **82** (35), **86** (45), *112*, **98** (74), *113*, **101** (35), *112*
Schleich, T., **278** (10), *285*
Schlesinger, S., **170**, *177*, **180** (23), *200*, **209** (10), *231*
Schneider, M. C., **129** (83), **140** (83), *153*
Schneider, W. C., **83** (27), **85** (27), **95** (27), **96** (27), **103** (27), *112*
Schöch, G., **245** (17), **246** (17), **250** (17), *250*
Schweet, R., **149** (131), *155*, **181** (42), *200*, **184** (57), **185** (61), *201*, **186** (42), *200*, **186** (68), *201*, **187** (42), *200*, **188** (73), **189** (77), **191** (73), *201*
Schweiger, H. G., **104** (113), **105** (113), *114*
Schweiger, M., **287** (9), *299*, **288** (29), **293** (42), *300*
Scott, J. F., **288** (26), *300*
Seaman, E., **53** (27), *54*
Seeds, W. E., **19** (6), **20** (12), *37*
Seidel, H., **59**, *78*
Setlow, R. B., **58** (27), *78*, **102** (104), *114*, **312**

AUTHOR INDEX

Shapiro, L., **118** (28), **122** (28), **125** (28), *152*, **125** (68), *153*
Shapiro, M. B., **57** (7), *78*
Shatkin, A. J., **117** (2), *151*, **121** (53), *152*
Shephard, D. C., **97** (69), *113*
Sheridan, W. F., **97** (68), *113*
Shin, D. H., **149** (125), *154*
Shipp, W. S., **83** (30), **97** (30), **100** (30), *112*, **120** (38), **129** (38), *152*
Shugar, D., **VII, 116, 327,** *329*
Silverstrini, R., **34** (39), *38*
Simmons, J. R., **87** (48), *112*
Simms, E. S., **209** (3), *230*
Simon, L., **118** (26), **156** (26), *152*, **183** (56), *200*
Sinclair, J., **82** (28), **83** (28), **84** (28), *112*
Singer, B., **108** (155), *115*
Singer, M. F., **240** (23), *250*, **251** (2), *265*, *295* (48, 49), *301*, **315** (14), *328*
Sinsheimer, R. L., **52, 54, 60** (37), *78*, **61** (49), *79*, **120** (42), **126** (42), *152*, **128** (76), **129** (76), *153*, **132** (92), **133** (98, 99), **135** (103, 106), **141** (99), *154*, **141** (76), *153*, **141, 142** (76), **147,** *154*
Sirlin, J. L., **316** (34), *329*
Sissakian, N. M., **104** (115), **105** (115), *114*
Sjöquist, I., **287** (11), **293** (11), *299*
Sjöquist, J. A., **287** (8, 11), **291,** *299,* **293** (11), *299,* **315** (26), *328*
Skerrett, J. N. H., **36,** *38*
Slater, J. P., **260**
Slayter, H. S., **85** (43), **90** (43), *112*
Slonimski, P. P., **101** (100), *114*
Sluyser, M., **116, 315** (11), *328*
Sly, W. S., **246** (20), **247** (20), *250*
Smellie, R. M. S., **118** (20), **119** (32), *152*
Smith, C. J., **287** (16, 17, 18, 19), *300*
Smith I., **176** (23), *177*
Smith J., **121** (51), *152*
Smith J. D., **184** (67), **189** (67, 83), **190** (83), **191** (83), *201*, **315** (4, 8), *328*
Smith, M. A., **149** (129), *155*, **179** (7, 8), *199,* **189** (82), *201*, **246** (25), *250,* **251** (2, 3), **252** (3), **254** (2), **255** (2), **257** (2), **258** (2), *260*

Smith-Sonneborn, J. E., **98** (74), *113*
Soffer, R., **105** (129), *115*
Sokol, F., **39**
Soldati, M., **34** (39, 40), *38*
Söll, D., **58** (17), *78,* **209** (14), **210** (14), **220,** *231,* **222** (26), **223,** *231,* **225** (30), **228** (30), *231,* **234** (24), **245** (24), **247** (24), *250,* **298** (56), *301*
Sonnabend, J. A., **120** (36), **127** (36), **141** (36), *152,* **142,** *154,* **143** (36), *152,* **143** (115), **144** (116), **145,** *154*
Sorm, P., **58** (21), *78*
Spahr, P. F., **315** (8), *328,* **316** (54), **317** (54), **324** (54), *329*
Spalla, C., **34** (40), *38*
Spencer, D., **104** (126), *115*
Spencer, M., **20** (10), **23, 24, 25** (25), **26,** *37*
Speyer, J. F., **73** (62), *79,* **246** (26), **247** (26), *250*
Spiegelman, S., **60,** *78,* **99** (83), *113,* **118** (27), **122** (27), *152,* **122** (61), **125,** *153,* **125** (69), **126** (71), **130,** *153,* **130** (69, 87), **131** (90, 91), *154,* **135, 141** (71), *153,* **156**
Spirin, A. S., **203** (100), *202*
Sponar, J., **116**
Spyrides, G. J., **189** (76), *201*
Sred, S., **118** (13), *151*
Srinivasan, P. R., **316** (48, 51), **317** (60), *329*
Staehelin, M., **202** (99), *202,* **252** (7), *260,* **287** (9), **293,** *299*
Stahman, M. A., **189** (82), *201*
Stanley, W. M., **179** (7, 8), *199*
Stanley, W. M., Jr., **149** (129), *155,* **246** (25), *250,* **251** (2, 3), **252** (3), **254** (2), **255** (2), **257** (2), **258** (2), *260,* **261.**
Starman, B., **133** (98), *154*
Starr, J. L., **316** (35, 37, 53), *329*
Steffensen, O. M., **97** (68), *113*
Steinberg, C. M., **4** (9), *14*
Steiner, R. F., **45** (10), *54*
Stent, G. S., **138** (107), **148,** *154,* **149** (127), *155,* **155**
Stephenson, M. L., **180** (18), *199,* **288** (26, 27, 28), *300,* **315** (9), *328*

Stern H., **103** (106, 107), *114*
Stern, R., **320**
Stevens, A., **209** (5), *230*
Stevens, C. L., **138** (111), *154*
Stocking, C. R., **97** (66), *113*
Stodolsky, M., **131** (89), *153*
Stokes, A. R., **18** (1), **20** (12), *37*, **57** (5), *77*, **57** (6), *78*
Stone, A. B., **155**
Storck, R., **106** (134), *115*
Streisinger, G., **179** (6), *199*, **221**, *231*, **252**, *260*
Stretton, A. O. W., **179** (1), *199*
Stuart, A., **6** (14), *14*, **271** (3), *284*
Subirana, J. A., **86** (47), *112*
Sueoka, N., **82**, *112*, **97** (63), **101** (88), *113*, **101** (91), *114*
Summers, D. F., **60** (41), *78*, **121** (52, 54), *152*, **129** (84), **139** (84), **140** (84), *153*
Sundararajan, T. A., **149** (128), *155*, **179** (9), *199*, **252** (5), *260*, **264** (3), *265*, **297** (53), *301*
Susman, M., **4** (9), *14*
Suyama, J., **83** (22), **84**, **95** (22), **96** (22), **97** (22), *112*
Suyama, U., **83**, *111*
Suzuka, I., **181** (36), *200*
Suzuki, H., **4** (12), *14*
Svensson, I., **316** (36), *329*
Svetailo, E. N., **104** (115), **105** (115), *114*
Swan, R. J., **57** (8), *78*
Swenson, P. A., **58** (28), *78*
Swift, H. H., **82** (28), **83** (28, 29), **84** (28), **97** (29), *112*
Swim, H. E., **127** (75), **136** (75), *153*, **156**
Szafranski, P., **261**
Szer, W., **327** (66, 68), *329*, **330**
Szybalski, W., **82** (34), *112*

Tada, M., **288** (29), *300*
Takagi, Y., **4** (12), *14*
Takahashi, K., **289** (32), *300*
Takanami, M., **58** (20), *78*, **60**, **61** (47, 48), *79*, **181** (34), *200*

Tamm, I., **23** (24), *37*, **119** (29), **121** (49), *152*, **121** (57), *153*, **149** (29), *152*
Tanaka, K., **315** (25), *328*
Tandler, C. J., **316** (34), *329*
Tatarskaya, R. I., **289** (33, 34), **294**, **295** (34), *300*
Tatum, E. L., **81** (2, 3), *111*, **117** (2), *151*
Taylor, M. M., **106** (134), *115*
Tecce, G., **59** (36), *78*
Tener, G. M., **289** (38), *300*
Terzaghi, E., **179** (6), *199*, **252** (6), *260*
Tewari, K., **82** (17), **83** (17), **84** (17), **85** (17), **92** (17), **93**, **96** (17), **98** (17), **105** (17), *111*
Thach, R. E., **149** (128), *155*, **179** (9), *199*, **252** (5), **253**, *260*, **260**, **264** (3), *265*, **297** (53), *301*
Theil, E. C., **316** (39), *329*
Thenius, E., **83** (36), *112*
Thiebe, R., **278** (11), *285*, *287*, *300*
Thorén, M. M., **121** (54), *152*, **147** (117, 119), **148** (119), **149** (117), *154*
Tikchonenko, T. I., **52**, *54*
Timasheff, S. N., **42** (9), **44** (9), **45** (9), *54*
Tinoco, I., **19**, **32**, *37*, **52** (17), *54*
Tissières, A., **66** (58), *79*, **105** (129), *115*, **180** (23), *200*, *209*, *231*
Tobey, R. A., **126**, **127**, **129** (72), *153*
Tocchini-Valentini, **131** (89), *153*
Todd, A. R., **209** (1), *230*
Tomita, K. I., **23**, **25**, *37*
Tomlinson, R. V., **289** (38), *300*
Tonino, G. J. M., **98** (76), *113*
Traut, R. R., **180** (32), **182** (32), *200*, **184** (66), **186** (66), **187** (66), **189** (83), **190** (83), **191** (83), **192** (66, 86, 87), **193** (66, 86), **195** (66, 86), **196** (66), **197** (66), **198** (92), **203** (92), *201*
Trim, A. R., **315** (20), *328*
Truman, D. E. S., **105** (130), **106** (138), *115*
Trupin, J., **58** (16), *78*, **220** (25, 27), **221** (25), *231*, **234** (2), *249*, **234** (22), **250**, **245** (2), *249*, **245** (22), *250*, **247** (2), *249*, **298** (55), *301*
Tsuboi, M., **23** (22), **25** (22), **26** (22), *37*

Tsugita, A., **179** (6), *199*, **180** (13), *199*, **252** (6), *260*
Tubbs, R. K., **31, 32,** *37*
Tuppy, H., **81** (1), *111*, **96** (57), *113*, **102**, *111*, **104** (114), **105** (114), *114*,

Uchida T., **289** (32), *300*
Ulbricht, J. L. V., **57** (8), *78*
Umbarger, H. E., **170** (16), *177*

Valentini, L., **34** (39, 44), *38*
Van Bruggen, E. F. J., **81** (5), **87** (5), **90** (5), **91** (5), **92** (5), *111*
Van Dillewijn, J., **158** (5), **159** (5), *176*
Van Holde, K. E., **57** (9), *78*, **66** (9), *79*, **295**, *300*
Van de Putte, P., **158** (5), **159** (5), *176*
Van Winkle, Q., **32,** *37*
Varney, N. F., **59,** *78*
Vasquez, D., **106** (135), *115*, **181** (37), *200*, **187** (71), *201*
Vaughan N. H., Jr., **82** (11), **83** (11), **97** (11), **104** (11), *111*
Venabbe, J., **28**
Vendrely, J. N., **98** (79), **99** (79), *113*
Venkstern, T. V., **234** (1), *249*, **287** (12, 13), *299*, **289** (33, 34, 35, 36, 37), **291** (39), *300*, **291** (12), *299*, **295** (34), *300*
Verwoerd, D. W., **117** (11), **118** (11), *151*, **118** (18), *152*
Vinograd, J., **82**, **90** (49), **91** (49), *112*
Vinuela, E., **130** (86), *153*
Vogt, M., **234** (14), **247** (14), *250*, **297** (52), *301*
Voigt, H.-P., **234** (14, 16), **237** (16); **240** (16), **242** (16), **243** (16), **245** (16, 17), **246** (16, 17, 27), **247** (14, 16), **248** (16), **250** (17), *250*, **297** (52), *301*
Von Ehrenstein, G., **180** (14), *199*, **180** (21), **183** (51), *200*
Von Der Decken, A., **104** (124), *115*

Wahba, A. J., **73** (62), *79*, **149** (129), *155*, **179** (7, 8), *199*, **246** (25, 26), **247** (26),

250, **251** (2, 3), **252** (3), **254** (2), **255** (2), **257** (2), **258** (2), *260*
Wallace, F. G., **98** (72), *113*
Wallace, J., **99** (84), *113*
Wallace, P. G., **107** (146), *115*
Waller, J. P., **180** (28, 29, 30), *200*
Walters, C. P., **118** (24), *152*
Wang, J. C., **101** (90), *114*
Waring, M. J., **32**, *37*, **32** (36), **33**, *38*
Warner, J. R., **181** (41), *200*
Warner, R. C., **209** (8), *231*
Wasson, G., **203** (95), *202*
Watanabe, K., **121** (58), *153*
Watanabe, Y., **121** (58), **129** (82), *153*
Watson, J. D., **18**, **20**, *37*, **50**, *54*, **61** (52), *79*, **131** (88), *153*, **179** (5), *199*, **180** (23), **182** (44), **186** (44), *200*, **209**, *231*
Watson, R., **90** (49), **91** (49), *112*
Watts-Tobin, R. J., **180** (11), *199*
Webster, R. E., **109** (158), *115*
Wecker, E., **147** (118, 120), **148** (120), *154*
Weil, J. H., **58**, *78*
Weill, J. D., **209** (8), *231*
Weisberger, A. S., **106** (136), *115*
Weisblum, B., **180** (14), *199*, **180** (21), *200*
Weiss, S. B., **131** (89), *153*, **132** (94), *154*, **209** (4), **222** (4), *230*
Weissmann, B., **293** (1), *328*, **295** (1), *328*, **315** (1), *328*
Weissman, C., **23** (21), *37*, **110** (161), *116*, **117** (9), *151*, **118** (26), *152*, **123** (62, 64), **124**, **129** (64, 80, 83), **130**, **131** (64), **132** (64), **136**, **140** (83), *153*, **155**, **156**, *151*, **156** (26), *152*, **156** (62, 64, 78, 80), *153*, **156**
Wells, R. D., **210** (19), **211**, **215** (20), **220** (20), **221**, *231*, **232**
Wettstein, F. O., **181** (43), **186** (43), **188**, *200*
Whitfeld, P. R., **295** (45), *300*
Wiberg, J. S., **13** (20), *14*
Widholm, J., **101** (90), *114*
Wilczok, T., **116**
Wildman, S. G., **97** (65), *113*, **104** (126, 127, 128), **105** (127, 128), *115*
Wildner, G., **81** (1), **102**, *111*

Wilkie, D., **98** (82), **99** (82), **101** (82), **102** (82), *113*
Wilkins, M. H. F., **18, 19** (5, 6, 7), **20, 23, 24, 25** (25), **26, 29,** *37*
Will, S., **101** (101), **106** (101), *114*
Williams, R. C., **108** (154), *115*
Williamson, A. R., **149** (131), *155*
Wilson, C. W., **287** (17, 18), *300*
Wilson, H. R., **18** (1), **19** (5, 6, 7), **20** (12), *37*
Wilson, J. F., **81** (2), *111*
Wilson, R. G., **118** (23), *152*
Wintersberger, E., **81** (7), **104** (7), *111*, **104** (114, 118), **105** (114, 118), *114*, **105**, *111*
Witz, J., **42** (9), **43**, **44** (5, 8, 9), **45** (5, 9), **46** (4, 8, 11), **47** (8), **48** (8), **49** (4), **52** (5, 8, 11), **53** (8), *54*
Witzel, H., **295** (44, 45), *300*
Wolf, M. K., **31** (28), *37*
Wolfenden, R., **188** (80), *201*
Wong, J. T., **287** (19), *300*
Work, T. S., **117** (12), **118** (13), *151*, **118** (22), **119** (22), **120** (47), *152*, **121** (55, 60), **122** (63), **123** (63), *153*, **123** (12), *151*

Yan, Y., **60** (46), *79*
Yang, J. T., **52** (16), *54*
Yaniv, M., **158** (9), **159** (9), **160** (9), **165** (9), *176*, **170** (18), *177*, **174** (9), *176*, *177*
Yanofsky, C., **179** (2), *199*, **229, 230**

Yanofsky, S., **285**
Yoneda, M., **4** (10, 15), **6** (10), *14*
Yoshikawa, H., **97** (63), *113*
Yotsuyanagi, Y., **107** (145), *115*
Yu, C. T., **58** (22), *78*

Zachau, H. G., **57** (3), *77*, **58** (29), **59** (29), *78*, **180** (16), *199*, **188** (79), *201*, **234** (28), *250*, **263, 271** (4), *284*, **274** (6), **277** (9), **278** (6, 11), *285*, **284** (12), *285*, **285**, **287** (6, 9), *299*, **287**, *300*, **288** (29), *300*, **293** (6), *299*, **293** (42), *300*, **296**, *299*, **297** (6), *299*, **301, 312, 315** (28), *328*, **315** (30), *329*
Zahn, R. K., **89** (55), *113*
Zamecnik, P. C., **58** (22), **59**, *78*, **180** (18), **183** (49), **189** (74), **203** (97), *202*, **209**, *231*, **285, 288, 300, 315** (9), *328*
Zamenhof, S., **316** (39), *329*
Zamir, A., **57** (2), *77*, **234** (6), *250*, **271** (1), **274** (7), **284**, *285*, **287** (1, 5), *299*, **293** (40), *300*, **293** (5), *299*, **303** (1), **309** (4), *312*, **315** (23, 24), *328*, **315** (29), *329*
Zillig, W., **58** (24), *78*
Zimmerman, E. F., **140** (112), *154*
Zimmerman, S. B., **15**
Zinder, N. D., **60**, *78*, **70** (61), *79*, **109** (158), *115*, **125** (67), **130**, *153*
Zubay, G., **57** (13), *78*, **61** (47), *79*, **138** (108), *154*, **155**, **187** (70), *201*
Zubkoff, P. L., **287** (10), *299*

Subject Index

N_6-acetylcytidine, 271
  isolation from sRNA, 276
  in serine tRNA, 271, 273
  structure of, 272
  UV spectrum of, 275
Acridines, interaction with DNA, 31-36
Acriflavine,
  cytoplasmic mutations and, 116
  fluorescence quenching by DNA, 32
  oxidative phosphorylation and, 105
  protein and RNA synthesis and, 104, 105
Acrylamide gel electrophoresis,
  of CMEC-oligonucleotides, 72, 73
    molecular weight and, 73
Actinomycin,
  chemical structure of, 35
  host-cell RNA synthesis and, 127, 128
  hydrogen bonding to DNA guanosine, 36
  mammalian RNA synthesis and, 117
  protein and RNA synthesis of organelles, 104, 105
  viral replication and, 143
  viral RNA synthesis and, 117
Adaptor hypothesis, 284
  protein synthesis and, 180
  scheme of, 283
Adenine,
  perphthalic acid oxidation, 59
  polynucleotide structure and, 59
S-adenosyl methionine, methyl donor for tRNA and, 316
Alanine transfer ribonucleic acid, 295, 301
Alanyl soluble ribonucleic acid synthetase,
  linear protein synthesis and, 165, 166
  mutations, genetic location of, 174, 175
Amino acid adapting system,
  assay method for, 237, 238
  chelating agents and, 242, 243

  ionic environment and, 242, 243
  polyribonucleotides and, 245
  triplet assignments and, 246, 247
  universality of, 248
Aminoacyl soluble ribonucleic acid, 184, 185
  binding of human ribosomes, 248
  codon-directed binding with ribosomal 30S subunit, 181
  hydrolysis of, 185
  hydrolysis by divalent cations, 243
  preparation of, 242
  purification of, 242
  receptor site, 187, 188
  substitution of puromycin for aminoacyl adenosine, 188
Aminoacyl soluble ribonucleic acid synthetases,
  different types of, 162
  mutational sites of, 175
  temperature sensitive mutants and, 159
  thermoresistant recombinats and, 166
Anticodon,
  adaptor hypothesis and, 283, 284
  alanine tRNA and, 284
  protein synthesis and, 180, 181, 184
  serine tRNA and, 284
  valine tRNA and, 284
Anticodon sequence,
  in alanine tRNA, 263
  in serine tRNA, 263
Arginine transfer ribonucleic acid, reaction with CMEC, 74

Bacteriophage,
  f2, 59
    temperature-sensitive mutants, 125
    viral RNA synthesis and, 125
  ØX174, conservative replication of, 132, 133

SUBJECT INDEX

M12, infective RNA, 133, 134
MS2, RNA polymerase of, 124
  interjacent RNA and, 141, 142
μ2, ^{32}P-labelled, preparation of, 70
  purification of, 70
  reaction with CMEC-iodide, 68
  ribonucleic acid of, 70
    dihydrouridylic acid in, 60
    ^{32}P-labelled, preparation of, 70
    reaction with CMEC-iodide, 68
    single-stranded regions of, 70 ff.
Qβ, and RNA polymerase, 125
R17, and RNA synthesis, 126, 128
Base-pairing schemes, for DNA,
  adenosine–uridine, 18, 272
  guanosine–cytidine, 18, 272
β-benzyl-N-carboxy-L-aspartate, 285
  formation of polypeptidyl tRNA, 285
Buoyant density,
  chloroplast DNA, 83
  E. coli reference DNA, 82
  mitochondrial DNA, 83
  mouse liver DNA,
    mitochondrial, 86
    nuclear, 86
  nuclear DNA, 83
  organelle DNA, 83

Carboxypeptidase,
  A and B, 255
  lysine-asparagine peptides and, 257
  lysine-threonine peptides and, 257
  peptide chain analysis and, 252, 255
Carboxypeptidases, and hydrolysis of peptides, 255, 257
Chain initiation, polypeptide, 229, 265
  methionyl sRNA and, 265, 266
  peptidyl sRNA and, 266
  ribosomal sites and, 265, 266
Chain termination, polypeptide, 265
  codewords, 228, 265
Chloramphenicol,
  binding to 50S ribosomal subunit, 187
  protein and RNA synthesis and, 104–106

RNA synthesis and, 164
viral RNA synthesis and, 125
Chloroplast deoxyribonucleic acid 82, 83, 97 ff.
  buoyant density of, 82, 83
  hybridization with total cell DNA, 100
  melting curves of, 97
  molecular weight of, 97
  separation from nuclear DNA, 82, 97
Chloroplasts, biosynthesis of, 108
Chromatography, paper, 237
  fast separations with, 237
CMEC, see also N-cyclohexyl, N'-β-(4-methylmorpholinium)ethyl carbodiimide
  adducts, resistance to nucleases, 69 ff.
CMEC-guanosine-5'-monophosphate, 63
CMEC-iodide, 62, 64
  reaction with DNA, 65
  reaction with specific bases, 64
  reaction with tRNA, 65
    specificity of, 66
  reaction with uridine, 62
CMEC p-toluene sulfonate, 62
CMEC-uridine-5'-monophosphate, 63
Codewords (or triplets, or codons)
  alanine tRNA and, 263
  chain initiation and, 229, 265
  chain termination and, 265
  N-formyl-methionyl-tRNA and, 264
  synonymous, 263, 264
  5'-terminal, 255, 258, 264, 265
Coding, mechanisms of,
  enzymes involved in, 233
  polynucleotides and, 233
Coding nucleotide sequences, see Triplet assignments
Coding triplets, and secondary structure, 297
Codon, protein synthesis and, 180, 184, 185
Codon-anticodon interaction, 184, 185
Codons, assignment of,
  from amino acid incorporation, 221, 222

SUBJECT INDEX

from aminoacyl-sRNA binding to ribosomes, 220, 222
Codons, base sequence assignment of, 259
Coliphages, ribonucleic acid-containing, 59
   coat protein and, 60
   RNA polymerase and, 60
   sequence of protein synthesis, 60
Column chromatography, of oligonucleotides from valine tRNA, 290
Conditional mutants,
   genetics of viral replication and, 157
   macromolecular synthesis and, 157, 158
   temperature sensitive, 158
      classes of, 159
Corepression,
   activation as prerequisite for, 170
   amino acid activating systems and, 174
   mechanisms of, 174
Counter-current distribution,
   E. coli tRNA, 287
   purification of tRNA, 269
   yeast valine tRNA, 287, 288
Cycloheximide,
   protein and RNA synthesis and, 104, 105
   viral RNA synthesis and, 147, 148
N-cyclohexyl,N'-β-(4-methylmorpholinium)ethyl carbodiimide, *see also* CMEC
   labelled, preparation of, 63
   reaction with iodine, 62
   reaction with uridine and guanosine, 62
Cytoplasmic deoxyribonucleic acid,
   genetic function of, 101, 103, 104, 109
   replication of, 103
   synthesis of, 103
Cytoplasmic mutations,
   methods of study, 102
   nature of, 102

Daunomycin,
   chemical structure of, 34
   interaction with DNA, 34
Density gradient analysis of,
   poly dA:dT and dAT:dAT, 8

polydeoxyadenylate replication product, 8, 9, 11
viral RNA, 124, 128, 138, 142, 144–146
viral RNA intermediates, 123–125, 127–130, 136 ff.
Deoxynucleotidyl transferase, terminal,
   addition of homopoly dA to DNA, 12
   homopolymer templates and, 4
Deoxyribonuclease, protein and RNA synthesis and, 105
Deoxyribonucleic acid (DNA), *see also* Chloroplast DNA, Cytoplasmic DNA, Mitochondrial DNA, etc.
   A conformation, 19–21, 23, 39
   A to *B* structural transition, 19
   alkali salts of, 19
   amount per chloroplast, 97–99
   amount per mitochondrion, 86, 87, 93, 95–98
   base-pairing in, 18, 23
   base-pair sequence and stacking interaction, 19
   base-pair spacing in, 30
   B conformation, 19, 20, 22, 35, 36
   C conformation, 20
   complex formation with actinomycin D, 35, 36
   conformational energy of, 20
   conformation and humidity, 19
   control mechanisms of replication of, 1 ff.
   dipole-dipole interactions in, 19
   heat stability of, 25
   hydrophobic interaction and structure of, 19, 20
   interaction with antimetabolites, 29 ff.
      intercalation mechanism of, 30–36
   intercalated molecule in,
      helix pitch and, 30, 31
      stacking forces of, 30
   ionic concentration and structure of, 19
   melting temperature and GC content, 19
   molecular models of, 21, 22
   organelle synthesis and, 109

of phage ØX174, 52, 53
  structure of, 53
protein synthesis and, 233
replication, defective mutants of, 13
RNA synthesis and, 233
shape, in solution, 43, 44
  effect of formaldehyde, 46, 48
  effect of temperature on, 44
  ethylene glycol and, 48, 49, 53
single-stranded, hydrogen bonds with water of, 19
stacking interactions and mutational "hot spots", 19
structure,
  Crick–Watson model for, 17, 18
  effect of relative humidity on, 19
  X-ray diffraction studies of, 18 ff.
sugar-phosphate chain in, 18, 23, 30
thermal denaturation of, 52
uncoiling by intercalation of daunomycin, 34
UV absorption and temperature, 44
Deoxyribonucleic acid polymerase,
  calf thymus,
    acetylated poly dT as template for, 6
    action on native and denatured templates, 5
    electron microscopy of product of, 2
    reaction catalyzed by, 2, 4
    templates and, 4
  complex formation with polyadenylate in replication process, 11
  initiation defective enzyme, 1, 2, 12, 13
  initiation positive enzyme, 1, 2, 13, 15
  kinetics of, 212, 213
  models of products of, 5
  templates for — see Templates
  types of reactions catalyzed by, 211–215
Deoxyribopolynucleotides,
  nearest neighbour analysis of, 213, 214
  repeating sequences, synthesis of, 210, 211, 215
4,5 (or 5,6)-dihydrouridine,
  serine tRNA and, 273, 277

structure of, 272
UV spectrum of, 275
Dihydrouridylic acid,
  chain-initiating codon and, 60
  μ2 bacteriophage RNA and, 60
  in serine tRNA, 273, 277
  in tRNA, 292, 293
  in valine tRNA, 296
N$_6$-γ,γ-dimethylallyladenosine (or isopentenyladenosine), 271
  alanine tRNA and, 284
  serine tRNA and, 271, 273, 277
  structure of, 272
  UV spectrum of, 275
Dimethylamino guanosine,
  serine tRNA and, 272, 277
  UV spectrum of, 275
Dinucleoside monophosphates,
  CMEC adducts of, 71
    enzymatic hydrolysis of, 71
  enzymatic hydrolysis of, 71
  primers in polyribonucleotide synthesis and, 241

Electron microscopy,
  chick liver mitochondrial DNA, 87, 89, 90
  chick liver nuclear DNA, 87, 89, 90
  mitochondrial DNA, 87, 89, 92–94
  phage ØX174 DNA, 52, 53
Encephalomyocarditis virus, 117
  effect of, on protein and RNA synthesis, 118
  replicative infectious RNA from, 133
  RNA polymerase of, 119, 124
Ethidium,
  chemical structure of, 32
  intercalation between DNA base-pairs, 33
Ethidium bromide, interaction with DNA, 32–34

Fingerprint analysis, of valine tRNA, 289, 291
Fingerprinting technique, two-dimensional,

SUBJECT INDEX

characterization of nucleic acids, 308, 313
separation of radioactive (oligo) nucleotides, 303 ff.

N (6)-(N-formyl-α-aminoacyl)-adenosine, 272
Flotation technique, for purification of tRNA, 289

p-fluorophenylalanine, and poliovirus synthesis, 147, 149
N-formyl methionyl sRNA, 60
   binding to ribosomes, 266
   chain-initiating codon, and, 60, 264
   as chain initiator, 265, 266
β-galactosidase activity, in thermosensitive mutants, 163, 177
Genetic code, nature of, 179, 180, 184
Genetic message, direction of reading of, 252
Genetic suppression, mechanism of, 229, 230
Genetic translation process, requirements for, 179
Guanosine-5'-monophosphate, CMEC adduct, 63
Guanosine-5'-triphosphate,
   hydrolysis of, 183, 185
   protein synthesis and, 182, 183, 186, 187, 192, 196, 198
Guanylyl methylenediphosphate,
   protein synthesis inhibition and, 183, 197
   structure of, 183

Hybridization,
   chloroplast and total cell DNA, 100
   mitochondrial and nuclear RNA, 100

Infectious ribonucleic acid,
   inactivation by strand breakage, 134
   from picornaviruses, 117, 133, 134
   replicative forms and, 133, 135

Inheritance, cytoplasmic, 81, 101–103, 109
Initiation reaction, in initiation defective DNA polymerase system, 13
Initiator region, of DNA polymerase templates, 12
Initiators, of polynucleotide synthesis, hexathymidylate as, 5, 6
Inosine,
   alanine tRNA and, 284
   anticodon of serine tRNA and, 284
   structure of, 272
   UV spectrum of, 275
   valine tRNA and, 284
Ionophoresis, of nucleotides and oligonucleotides, 304 ff.

Kinetoplast deoxyribonucleic acid, 97, 98
   buoyant density of, 98

Lipoprotein complexes, and viral RNA replication, 149
Lysine transfer ribonucleic acid,
   anticodon of, 75
   reaction with CMEC, 74

Macromolecular synthesis, in thermo-sensitive mutants, 158, 175
Melting point,
   chick liver mitochondrial DNA, 116
   DNA–acridine complex, 32
   poly dAT : dAT and dA : dT, 11
   polyribothymidylic acid, 327
Mengovirus, 118
   L-cell system, 126
   viral RNA polymerase and, 118
Messenger ribonucleic acid (mRNA),
   binding of phage μ2 RNA and, 75, 76
   minor bases and, 60
   protein synthesis and, 179–182, 184, 186, 233
   role in adaptor hypothesis, 283
   specific association with 30S ribosomal subunit, 181, 182
Methionyl soluble ribonucleic acid,
   binding to ribosomes, 265

chain initiator, 229, 265
formylated, 265, 266
Methylases, of RNA, 316
  specificity of, 316
Methylated bases, *see also* Minor bases,
  S-adenosyl-methionine and, 316
  coding properties and, 327, 328
  in RNA, 315
  in tRNA, 272, 315
  tRNA methylases and, 316
    species specificity and, 316
    specificity of, 316
5-methylcytidine,
  in serine tRNA, 273
  UV spectrum of, 275
5-methylcytidylic acid, in valine tRNA, 291, 292
2'-0-methylguanosine, 271
  in serine tRNA, 273, 277
1-methylinosine, in alanine tRNA, 284
2'-0-methyluridine, 271
  in serine tRNA, 273, 277
Minor (or odd) bases,
  sequences in tRNA and, 274
  structures of, 272
  in transfer RNA, 271
  UV spectra of, 275
Mitochondria, information content of, 93 ff., 106, 109
Mitochondrial deoxyribonucleic acid, 81 ff.
  analytical ultracentrifugation of, 87, 88, 92
  base composition of, 82, 84
  base sequence of, 98–100
  buoyant density of, 82–85, 95, 98, 116
  circular form of, 89, 92–94
  CsCl gradient analysis of, 85, 86
  cytochemical evidence for, 85
  denaturation of, 85–88, 90, 116
  dual nature of, 87 ff.
  electron microscopy of, 87, 89, 92–94
  half-life of, 103
  homogeneity of, 85, 86, 95, 116
  hyperchromicity of, 116
  hypertwisted, 87, 89–92

identification, 81 ff.
identification criteria, 84, 85
length frequency distribution, of circular, 90, 92
melting curves of, 82, 85, 87, 95, 116
molecular weight of, 85, 87, 90–93, 95, 97, 116
of *Neurospora crassa*, 92, 93, 102
physicochemical properties of, 85 ff.
redundancy in, 99
relative to nuclear DNA, 98 ff.
renaturation of, 85–88, 92, 95, 116
  and homogeneity, 85, 116
replication of, 103, 104
resistance to DNase, 84, 95
sedimentation coefficients of, 91–93
Mitochondrial ribonucleic acid, 81, 100, 104–106
Mutants,
  aplastidic, 82, 89
  cytoplasmic, 101, 102
  temperature sensitive,
    cessation of RNA synthesis, 159
    energy yielding processes, 159, 160
    inhibition of cell bipartition, 159
    inhibition of DNA synthesis, 159
    non-conditional, 159
      and drug resistance, 159
    suppression of protein synthesis, 159, 160
    valyl sRNA synthetase, 167

Nearest neighbour analysis,
  in deoxyribopolynucleotides, 213, 214
  in ribopolynucleotides, 216–219
Nuclear deoxyribonucleic acid,
  buoyant density of, 82, 83
  electron microscopy of, 89
  relation to mitochondrial DNA, 98 ff.
Nucleic acids, *see also* Deoxyribonucleic acid and Ribonucleic acid
  major developments in chemistry and biochemistry of, 209
  partial specific volume, 49
  role of hydrogen bonding in, 53

Nucleohistone, chromosomal, X-ray diffraction studies of, 20
Nucleoprotamine, sperm heads, X-ray diffraction studies of, 20

Odd bases, *see* Minor bases
Oligodeoxynucleotide initiation, chain length requirements for, 7
Oligonucleotides,
 amino acid incorporation and, 254–256, 258
 common to various tRNA's, 293, 297
 enzymatic preparation of, 253
 for determination of codon base sequence, 258, 259
 ionophoresis of, 304 ff.
 isomers, separation of, 308
 as messengers, 251
 as primers for polynucleotide phosphorylase, 253
 spontaneous breakdown, 293
  and dihydrouridylic acid, 293
  from valine tRNA, 291, 292, 294, 296 ff.
Oligoribonucleotides, synthetic messenger, direction of reading of, 251
Organelle deoxyribonucleic acid, 81 ff.
Organelle ribonucleic acid, template function of, 105, 110
Osmium tetroxide, and anti-codon in tRNA, 59

Peptide bond formation, 179 ff.
 enzyme–(poly)peptidyl transferase-responsible for, 179, 185, 187, 192, 193, 197, 198, 202
 localization, 188, 193
 mechanism, 184, 185
 steps involved in, 184
Peptide chain growth, 179, 183, 184
 model of, 183, 184
Peptide synthesis, direction of, 182, 252
Peptidyl soluble ribonucleic acid, 184–188
Peptidyl (polypetidyl) transferase, 179 ff.

Phenylalanine transfer ribonucleic acid, *see* Transfer ribonucleic acid
Phosphodiesterase,
 snake venom (PDE), 271
  hydrolysis of tRNA, 274, 276, 277, 282, 284
 of *Vipera lebetina*, 289
Phospholipids, in virus infected cells, 121
 possible function of, 121
Picornaviruses,
 actinomycin D and, 117
 cellular localization, 119
 classes of, 117
 inhibition of cellular RNA synthesis, 118
 initial effects following infection by, 118
 input RNA, 126, 127
 parental RNA, 126, 127
Poly adenylic acid,
 conformation, and circular dichroism, 52
  and optical rotatory dispersion, 52
  and pH, 45
 dual forms of, 45, 46
 effect of chain length on lysine incorporation, 260, 261
 small-angle X-ray scattering, 46
Poly dA:dT, strand separation of, 12
Polydeoxyadenylate,
 replication of, 10
  density gradient analysis, 9
  self initiation catalyzed by calf thymus DNA polymerase, 7
Polymers, DNA-like,
 DNA polymerase and, 212
 nearest neighbour analysis of, 213, 214
 with repeating sequences, 211, 215
 synthesis of ribopolynucleotides, 215 ff.
 templates for DNA polymerase, 214, 215
Polynucleotide(s),
 chain, loop formation in, 28
 dipole-dipole interactions in, 28
 hydrophobic interactions in, 28
 interbase hydrogen bonds in, 28
 single-stranded, 28, 29
  X-ray diffraction pattern of, 29

SUBJECT INDEX

two-stranded, X-ray diffraction pattern of, 28, 29
Polynucleotide phosphorylase, 236
  electrophoresis of, 240, 250
  M. lysodeikticus, 253
  nuclease activity, 240
  polyribonucleotide synthesis and, 253
  primers and, 241
  purification of, 240
Polypeptides,
  chain, termination of, by triplets, 228
  enzymatic reaction sequence for synthesis of, 209, 210
  synthesis, and adaptor hypothesis, 283
Polyribonucleotides, *see also* Ribopolynucleotides
  characterization of, 253, 254
  dinucleoside monophosphate primed, 240
  enzymatic synthesis of, 236
  preparation and purification, 241, 242
Polyribothymidylic acid, secondary structure of, 327
Polythymidylate, self initiation catalyzed by calf thymus DNA polymerase, 7
Poly uridylic acid,
  binding to ribosomes, 323
  directed synthesis of polypeptides, 317
  secondary structure of, 327
Proflavine, chemical structure of, 32
Proteins,
  ribosomal, 180–182
    dual nature of, 182
    heterogeneity of, 180
    supernatant, as component of protein synthesis complex, 181, 182, 195, 196
    fractions of, 182, 196
Protein synthesis, 179 ff.
  E. coli system, 179, 189 ff.
  inhibitors, 197
  in isolated chloroplasts, 104, 105
  in isolated mitochondria, 81, 104–106
  properties in subcellular systems, 105

30S and 50S ribosome subunits and, 182
  terminology, 187
Protein, virus capsid, 120, 121
  soluble antigens and, 119, 120
  synthesis of, 120
    localization of, 120, 121
  viral RNA synthesis and, 119, 120
Pseudouridine, *see* 5-ribosyl uracil
Puromycin,
  inhibition of viral RNA polymerase, 125
  inhibition of viral RNA synthesis, 147, 148
  as part of amino acyl sRNA, 188
  mechanism of action, 188, 189
  peptide bond formation and, 179, 188, 189, 191, 192, 194, 196
  protein and RNA synthesis and, 105
  reaction with peptidyl sRNA-ribosome complex, 184
  stimulation of release of peptides by, 188, 189
  structure, 188
  viral parental RNA and, 147
Puromycin 5'-β-cyanoethyl phosphate, 188
  ^{32}P-labelled, 189
    attachment to lysine peptides, 189–191, 197
    attachment to polyphenylalanine, 191, 192, 197
    biological action of, 189
    structure, 188

Renaturation, of mitochondrial DNA, 85 ff.
  alkali denatured and, 88, 116
  homogeneity and, 85
  thermal denatured and, 86, 87
"Repair enzymes", 102
Ribopolynucleotides,
  with alternating sequence, polypeptide synthesis directed by, 220
  methylation and secondary structure of, 327, 330
  with repeating sequences, 215–220

SUBJECT INDEX

nearest neighbour analysis of, 216–219
preparation of, 220
Ribonuclease,
  degradation of RNA by, 39
  protein and RNA synthesis and, 105
Ribonuclease, pancreatic,
  activity and secondary structure, 295
  degradation of serine tRNA by, 273, 274, 277, 282
  digestion of valine tRNA, 292, 293
  hydrogen bonding and, 297
  initial point of attack in tRNA, 298
  interjacent RNA and, 144, 145
  Mg^{++} ions and, 295
  parental viral RNA and, 126
  resistant RNA, 127–129, 142
  sequence of tRNA and, 277 ff.
  viral RNA intermediates and, 133 ff.
Ribonuclease, T_1,
  from *Actinomycetes*, 289
    digest of valine tRNA, 292, 294, 296
    isolation of, 289
    purification of, 289
    specificity of, 289
  degradation of serine tRNA, 273, 274, 277, 282
  digestion of phenylalanine tRNA, 311
  digestion of ribosomal RNA, 305, 307
  digestion of tRNA, 305
  hydrolysis of CMEC adducts by, 71
  hydrolysis of dinucleoside monophosphates by, 71
  sequence of tRNA and, 277 ff.
Ribonucleic acid (RNA), *see also* Aminoacyl RNA, Messenger RNA, Mitochondrial RNA, Viral RNA, etc.
  active sites of, 57
  base-pairing in, 23, 39
  base-pair stacking in, 23
  characterization by fingerprinting, 308
  conformations of, 57
  double-stranded, as template, 123, 124
  heat stability of, 25
  helical content and loops, 52
  helical models for structure of, 24, 25, 27, 28
  intermolecular bonds in, 25, 28
  intramolecular bonds in, 25, 28
  meridional reflections of, 25, 26
  10- and 11-fold models of, 23–28
  molecular packing in, 27
  off-meridional reflections of, 26
  reaction with CMEC, 65
  reovirus, X-ray diffraction patterns of, 23, 24, 26, 39
  replicative intermediate (MS2), X-ray diffraction patterns of, 23, 39
  ribosomal,
    5S component, sequence of, 308
    composed nature of, 180
    X-ray diffraction studies of, 38, 39
  rice dwarf virus, X-ray diffraction patterns of, 23, 25
  shape of, in solution, 44, 45
    effect of temperature on, 44, 45
  small-angle X-ray scattering, 44, 45
  sugar-phosphate chain in, 24–26, 39
  synthesis,
    in isolated chloroplasts, 104
    in isolated mitochondria, 81, 104, 105
  transfer-ribosomal interaction, role of intermolecular hydrogen bonds in, 28
  various types of, 57
  viral, two stranded, α and β forms, 25–28
    cylindrical Patterson synthesis of, 25
    infrared dichroism of, 25
  virus-specific, properties of, 119, 120
  wound tumour virus, X-ray diffraction patterns of, 23
  X-ray diffraction studies of, 20 ff.
  yeast, X-ray diffraction patterns of, 23, 24, 26
Ribonucleic acid polymerase,
  polynucleotide synthesis and, 236
  viral, activity in infected ascites cells, 119
    dual character of, 123–125, 143

E. coli, synthesis of DNA–RNA hybrids, 132
  functions of, 123, 124
  inhibition by drugs, 124, 125
  lability of, 148
  in mengovirus-infected L cells, 118
  of picornaviruses,
    localization of, 119
    properties of, 118, 119
    products of, 122, 123
    requirements for template RNA, 125
  of Semliki forest virus, 143–145
Ribonucleic acid, soluble (sRNA or tRNA), *see also* Soluble RNA and Transfer RNA
  codon recognition and, 223–228
  N-formylmethionine-, as chain initiator, 228
  redundancy, 228
  stimulation of binding to ribosomes by trinucleotides, 223–227
Ribose thymine,
  in serine tRNA, 273, 277
  UV absorption spectrum of, 275
Ribosomal organization of peptide assembly, models of, 186, 187
Ribosomal ribonucleic acid,
  formaldehyde inactivation, 62
  Mg^{++} and dissociation, 61, 62
  methylated bases in, 315
  single-stranded regions of, 76
    and messenger binding, 75, 76
  structure of, 38, 39
  as template for viral RNA polymerase, 125
Ribosomes,
  active sites of, 61
  chemical composition, 180
  chloroplast and bacterial, 105, 106
  chloroplast and mitochondrial, 105, 106
  E. coli, 238
  effect of formaldehyde treatment, 62
  human, 239, 240
    binding of aminoacyl RNA, 248

  interactions with components of protein synthesis system, 180
  measurement of activity of, 237, 238
  messenger binding site, 75, 76
    inactivation with CMEC, 75, 76
    protection by poly A, 76
  mitochondrial and bacterial, 106
  nature of, 180
  phage f2 RNA and, 61
  proteins of, 180
    and functions, 182
  purification of, 238, 239
  role in adaptor hypothesis, 283
  role in coding system, 233
  sites for chain initiation, 265, 266
  30S subunits, 180, 181, 187, 188, 193, 198
  30S and 50S subunits, 180
    conditions for combination of, 180
    functions of, 181
    and protein synthesis, 182
  50S subunits, 179–181, 187, 188, 193, 195, 198
    ^{14}C-polyphenylalanine, charged, preparation of, 193
    role in peptide bond formation, 179, 180
  70S, 180, 193, 195
5-ribosyl uracil,
  CMEC adducts, 62, 63
    and hydrolysis of, 63
  in serine tRNA, 273, 276, 277
  structure of, 272
  in tRNA's, 293
  UV spectrum of, 275
  in valine tRNA, 292

Satellite deoxyribonucleic acid, 83, 97, 100, 101, 116
Semliki forest virus, 142, 143
  replication of, 134
  and actinomycin, 134
Serine transfer ribonucleic acid, 171, 301
  anticodon of, 284
  column types for preparation of fragments of, 278–282

isolation of, 274
purification of,
 by counter-current distribution, 271
 by column chromatography, 274
sequence of, 273
sequential degradation of, 272, 277 ff.
Seryl transfer ribonucleic acid,
 reaction with CMEC, 74
Snake venom phosphodiesterase,
 hydrolysis of CMEC adducts, 71
 hydrolysis of dinucleoside monophosphates, 71
Soluble ribonucleic acid (sRNA, tRNA),
 see also Ribonucleic acid, soluble,
 and Transfer ribonucleic acid
 binding to 50S ribosomal subunit, 181, 182
 conformational changes of, 180
 peptide attachment to, 184
 protein synthesis and, 180–182, 184–187
 release from ribosome by puromycin, 191
 translocase, 186
Synthetases, leucyl RNA, 316, 317
 and aminoacylation of tRNA, 316

Templates, for deoxyribonucleic acid polymerase,
 acetylated poly dT as, 6, 7
 homopolymers as, 4
 models of sites for replication initiation, 13
 modified, 11, 12
 native and denatured DNA as, 5
 poly dA as, 6, 7
Templates (mRNA), for protein synthesis, 184, 185
 relative movement of, 185–187
Thermal transition profiles,
 of chick liver mitochondrial DNA, 87
 of chick liver nuclear DNA, 87
 of chloroplast DNA, 97
2-thiouridine,
 structure of, 272
 UV spectrum of, 275

Threonine deaminase,
 derepression, 170
 in thermosensitive mutants, 170, 171, 173
Thymidine, reaction with CMEC, 65, 66
Transfer ribonucleic acid (tRNA, sRNA),
 see also Serine tRNA, Valine tRNA, etc.
 acceptor recognition site, 58
 arrangement of anticodons, 283, 284
 coding site, 58
 of $E.\ coli$, 73, 74
  CMEC inactivation of, 73–75
  N-formyl methionine-, as chain initiator, 229
 fractionation of,
  by chemical methods, 285
  by column chromatography, 274
  by counter-current distribution, 271
 inactivating reagents and, 58
 "methionine-starved", 316, 317
  as amino acid acceptor, 316
  chromatographic fractionation, 317, 318
 methylases, 316
  specificity of, 316
 methylated bases in, 315
  function of, 316, 317
 methyl-deficient, 316 ff.
  isolation of, 317–320
 minor bases in, 271 ff.
  structure of, 272
  UV spectra of, 275
 osmium trioxide and, 59
 phenylalanine,
  coding properties of, 324
   magnesium ions and, 327
  methyl-deficient,
   binding to ribosomes of, 325, 326
   chromatography of, 319, 320
   coding properties of, 324 ff.
   magnesium ions and, 327
   identification of, 320 ff.
   purification of, 312, 319
   transfer activity of, 317

role in adaptor hypothesis, 283
sequences of bases in, 273, 296, 312
as template for viral RNA polymerase, 125
triplex model of, 138, 139
X-ray diffraction pattern of, 29
Triplet assignments, methods for, 234
Triplets, chain initiating, sequence of, 228, 232
Tyrosine tRNA, 301
Tyrosyl transfer RNA, reaction with CMEC, 74

Ultraviolet absorption, of thin films, 43
Uridine,
CMEC-adduct, hydrolysis of, 63
reaction with CMEC-iodide, 62
Uridine-5'-monophosphate, CMEC-adduct, 63

Valine,
bacteriostatic effect in *E. coli* K_{12}, 169
resistant revertants, 169
thermosensitive mutants and, 169
Valine transfer ribonucleic acid, 285 ff.
from *E. coli*, 287
from yeast,
fragments of, 294
nucleotide composition of, 292
nucleotide sequence of, 296, 301
preparation of, 287
purification of, 287, 288
two fractions of, 288
various species of, 287
Valyl sRNA synthetase,
mutations, genetic locations of, 174, 175
thermal inactivation *in vivo*, 167, 168
kinetics of, 168
in thermosensitive mutants, 167
Viral deoxyribonucleic acid, 85, 87, 90–92
Viral replication, 120, 121
asymmetry of, 131
interjacent RNA and, 141 ff., 146, 147

localization of, 119
mechanism of, 131 ff.
conservative, 132–136
viral RNA synthesis and, 118 ff.
Viral ribonucleic acid,
associated proteins and, 59
coat protein and, 60
double-stranded, 117, 123 ff.
base composition of, 140
foot-and-mouth disease, replication of, 133
infectious, from picornaviruses, 117, 133
interjacent, 141 ff.
function of, 146, 147
as viral RNA messenger in phage MS2, 147
mechanism of replication of, 131 ff.
as messenger RNA, 59
picornavirus,
functions of, 121
properties of, 119, 120, 123
synthesis of, 118–120
synthesis of capsid protein by, 121
replication, inhibition by drugs, 147, 148
in vivo, 126 ff.
protein synthesis and, 147–150
three-stranded model, 138, 139, 155, 156
ribonuclease-resistant, nature of, 129
secondary structure, 61
synthesis,
inhibitors of protein synthesis and, 147 ff.
thymidine analogues and, 117

"Wobble" hypothesis,
alanine anticodons and, 264
base pairing and, 264

X-ray diffraction,
calculated, for single- and double-stranded RNA, 29

DNA conformation and, 18 ff.
nucleohistone and, 20
nucleoprotamine and, 20
ribosomal RNA, 38, 39
RNA conformation and, 18, 23 ff.
virus RNA and, 23 ff.

X-ray scattering,
of DNA solutions, 43, 44
small-angle,
discussion of techniques, 49, 50
of DNA solutions, 43, 44
and temperature, 43, 44
experimental, 41, 42
of polynucleotides, 50
theory, 41–43

Yeast deoxyribonucleic acid, 92, 93, 95, 96
Yeast, petite mutants of, 84, 85, 98, 99, 102, 106, 116